Interpreting Nature

gROUNDWORKS |
ECOLOGICAL ISSUES IN PHILOSOPHY AND THEOLOGY

Forrest Clingerman and Brian Treanor, *Series Editors*

Series Board:

Harvey Jacobs   Catherine Keller   Norman Wirzba

Richard Kearney   Mark Wallace   David Wood

Interpreting Nature

The Emerging Field of
Environmental Hermeneutics

EDITED BY

Forrest Clingerman, Brian Treanor,
Martin Drenthen, and David Utsler

Fordham University Press | *New York 2014*

Fordham University Press has no responsibility for the persistence or accuracy of URLs for external or third-party Internet websites referred to in this publication and does not guarantee that any content on such websites is, or will remain, accurate or appropriate.

Fordham University Press also publishes its books in a variety of electronic formats. Some content that appears in print may not be available in electronic books.

Library of Congress Cataloging-in-Publication Data
Interpreting nature : the emerging field of environmental hermeneutics / edited by Forrest Clingerman, Brian Treanor, Martin Drenthen, and David Utsler. — First edition.
     pages cm. — (Groundworks: ecological issues in philosophy and theology)
     Includes bibliographical references and index.
     ISBN 978-0-8232-5425-5 (cloth : alk. paper) — ISBN 978-0-8232-5426-2 (pbk. : alk. paper) 1. Human ecology—Philosophy. 2. Hermeneutics. I. Clingerman, Forrest, editor of compilation.
     GF21.I58 2014
     304.201—dc23

                                                                    2013004643

Printed in the United States of America
16 15 14 5 4 3 2 1
First edition

Contents

Acknowledgments   *xi*

Introduction: Environmental Hermeneutics   1
   David Utsler, Forrest Clingerman,
   Martin Drenthen, and Brian Treanor

Part I: Interpretation and the Task of
Thinking Environmentally

1   Environmental Hermeneutics Deep in the Forest   17
    John van Buren

2   Morrow's Ants: E. O. Wilson and Gadamer's Critique
    of (Natural) Historicism   36
    Mick Smith

3   Layering: Body, Building, Biography   65
    Robert Mugerauer

4   Might Nature Be Interpreted as a "Saturated
    Phenomenon"?   82
    Christina M. Gschwandtner

5   Must Environmental Philosophy Relinquish the
    Concept of Nature? A Hermeneutic Reply to
    Steven Vogel   102
    W. S. K. Cameron

Contents

Part II: Situating the Self

6   Environmental Hermeneutics and Environmental/
Eco-Psychology: Explorations in Environmental
Identity   123
    David Utsler

7   Environmental Hermeneutics with and for Others:
Ricoeur's Ethics and the Ecological Self   141
    Nathan M. Bell

8   Bodily Moods and Unhomely Environments: The
Hermeneutics of Agoraphobia and the Spirit of Place   160
    Dylan Trigg

Part III: Narrativity and Image

9   Narrative and Nature: Appreciating and
Understanding the Nonhuman World   181
    Brian Treanor

10 The Question Concerning Nature   201
    Sean McGrath

11 New Nature Narratives: Landscape Hermeneutics
and Environmental Ethics   225
    Martin Drenthen

**Part IV: Environments, Place, and the
Experience of Time**

12 Memory, Imagination, and the Hermeneutics of Place   245
    Forrest Clingerman

13 The Betweenness of Monuments   264
    Janet Donohoe

14 My Place in the Sun   281
    David Wood

15 How Hermeneutics Might Save the Life of
(Environmental) Ethics   297
    Paul van Tongeren and Paulien Snellen

Notes 313

A Bibliographic Overview of Research in
Environmental Hermeneutics 365

List of Contributors 373

Index 377

Acknowledgments

The editors would like to thank the following people and organizations for generously allowing the use of material that originally appeared elsewhere.

John van Buren's "Environmental Hermeneutics Deep in the Forest" (Chapter 1) is an edited revision of "Critical Environmental Hermeneutics," which first appeared in *Environmental Ethics* vol. 17, no. 3 (Fall 1995): 259–275. Reprinted with permission from *Environmental Ethics* 17:3, pp. 259–275.

The image of the monument to Amy Biehl in Gugulethu Township, Cape Town, South Africa, in Chapter 13 was taken by Dr. Steven Gish.

The editors would also like to express their heartfelt thanks to Helen Tartar and Thomas Lay of Fordham University Press, whose unflagging support and encouragement kept our enthusiasm for this project very high during the long process of collecting and editing the papers.

Interpreting Nature

Introduction

Environmental Hermeneutics

David Utsler, Forrest Clingerman, Martin Drenthen, and Brian Treanor

Friedrich Nietzsche famously stated: "There are no facts, only interpretations."[1] Perhaps this could be slightly rephrased: no facts go uninterpreted. There are simply no bare facts, at least if a fact is to be meaningful. Every fact has meaning only in relation to other facts, to context, and to the human understanding itself. In other words, at the heart of every confrontation of concept and perception is the issue of *hermeneutics*: the art and science of interpretation.

The present volume uncovers some of the ways that interpretation takes place in the human relationship to the environment. This collection brings together essays on the questions that hermeneutics raises for environmental philosophy. In the public sphere, much of the focus on "the environment" is concerned with discovering scientific facts and then reporting how policy can act on these facts. On its face, philosophical hermeneutics might appear to be an unrelated enterprise. But this volume follows Nietzsche in arguing that even the facts of the sciences are given meaning by how humans interpret them. Of course this does not mean that there are no facts, or that all facts must come from scientific discourse. Rather, one point of agreement among the essays presented here is the need for mediation—the mediation that grounds the interpretive task of connecting fact and meaning through a number of different structures and forms. This has practical implications, not simply intellectual ones. Ostensibly bare facts are contextualized by a variety of individual and social relations, and responsive actions emerge as a matter of consequence. For example, the science of the human body may seem to be only a collection of factual data, but what someone does with that data (or the ignorance of the data) in terms of habits, behaviors, and practices all reflect interpretations that involve value and meaning. Already then we can see how philosophical hermeneutics recommends itself to the topic of the environment: *philosophical hermeneutics offers a unique reflection on the human mediation of the meaning of environments*, and, equally, *hermeneutics assists in understanding the practical implications of our encounters with the world.*

Defining the Place of Environmental Hermeneutics

Throughout this volume, the term "hermeneutics" balances between a broad and a narrow meaning—both meanings are often operative in the individual essays collected here. To explain, it is helpful to be reminded of the place of philosophical hermeneutics as a tradition within philosophy. In a narrow sense, the present volume is interested in the specific tradition of discourse called "philosophical hermeneutics"—a modern dialogue over the nature of interpretation that begins with Schleiermacher and is carried into contemporary philosophy through figures such as Dilthey, Heidegger, Gadamer, Habermas, and Ricoeur.[2] This school of thought emerged from more general concerns with how to understand texts—before the modern era, most theorizing about the task of interpretation was concerned with the proper understanding of the Biblical text (Augustine's *On Christian Doctrine* might thus be considered the ancient ancestor of philosophical hermeneutics). From this background, the historical trajectory of modern philosophical hermeneutics can be explained simply: it is an investigation that began with a narrow concern around understanding the authorial intent of written texts, and gradually moved toward the recognition of the inevitable interpretation of our historical, factical existence itself.

But hermeneutics has a broader sense as well, which is sometimes also employed here: Hermeneutics is commonly defined as the reflection on the "art and science of interpretation," not simply of written texts, but as a form of thinking itself. Most broadly, the question of interpretation is not merely asking about a technique for discerning a single meaning or finding one interpretation that is the right one. Nor is it concerned simply with the imposition of meaning on an object by a subject, making any interpretation possible and, therefore, acceptable. As Robert Mugerauer noted in his 1995 book,

Interpreting Environments, hermeneutics is a matter of "finding the valid criteria for polysemy within the fluid variety of possibilities."[3] This statement also reveals an inherent critical element in hermeneutics. Not all interpretations are valid; but there is more than just one valid interpretation possible, and interpretation is a structurally open project that never comes to final closure. Hermeneutics aims at opening all possible worlds, for our encounter with the world is always already rendered through our interpretations of it.

In light of both narrow and broad meanings, what exactly is *environmental* hermeneutics? The answer to that question is as multivalent as the notion of philosophical hermeneutics itself. "Environmental hermeneutics" includes a number of different, often-overlapping possibilities and approaches, among them the following.

1. Environmental hermeneutics is *the extension of principles of interpretation to environments of any kind (natural, built, cultural, etc.)*. This definition is both abstract and wide-ranging. As a result, hermeneutics is a rationale and framework for interpretive activity in general, whether that interpretation is done by visitor, inhabitant, botanist, artist, farmer, architect, construction worker, or someone merely looking out a window.

2. Environmental hermeneutics is *the interpretation of actual encounters of or within environments*. Most often this type of interpretation is meant to deepen our understanding of places with which we have direct interaction—hence the near omnipresence of case studies and concrete examples in essays engaging environmental hermeneutics. Examples of such interpretation include informational signs at nature preserves or historical markers, both of which convey the interpretation of "experts" for the benefit of visitors. But indirectly, this would also include activities such as the construction and development of walking trails, which assume a certain relation to the landscape.

3. Environmental hermeneutics refers to *a form of nature writing*. Archetypal examples include Aldo Leopold, Henry David Thoreau, John Muir, and Annie Dillard, among others. This is perhaps a more personalized account of the previous category. Both in the respect that nature writing is an interpretation of nature by the author (referring back to the encounter in the second definition) as well as the interpretive action of the reader of the text concerning nature. This also can include the notion of the ways in which nature can be grasped or experienced in the text.

4. Environmental hermeneutics provides accounts of *the approach of various disciplines to environments*. Environmental hermeneutics, therefore, can be genuinely interdisciplinary in scope. Different disciplines interpret the natural environment in different ways according to their own internal logic. Thus, there are geological interpretations, economic interpretations, technological interpretations, agricultural interpretations, and on and on it goes. Environmental hermeneutics can critically mediate between different disciplinary interpretations so as to suggest a fuller and more robust understanding of environments.

5. Perhaps in its most robust sense, environmental hermeneutics is *a philosophical stance which understands how the inevitability of what Gadamer called our "hermeneutical consciousness" informs our relationship with environments*. This final sense of environmental hermeneutics is concerned not simply with techniques to interpret landscapes but with the ontological framework that necessitates such interpretation. This presumption of environmental hermeneutics, it should be noted, is implicit throughout the present collection.

These five possibilities are not mutually exclusive. Furthermore, there are certainly other possibilities of connecting interpretation and environment. As environmental hermeneutics is further developed and explored, many other aspects are sure to emerge. For example, the sciences, theology and religious studies, leisure studies, and other fields provide additional perspectives for environmental hermeneutics.

In the present volume, each of these five approaches are at least in passing touched upon; however, it is the fifth approach that suggests the cohesion of this collection. For the editors and authors, this offers a working definition of environmental hermeneutics found at the intersection between philosophical hermeneutics and environmental thought. This area of study has variously been called "ecological hermeneutics," "ecohermeneutics," "environmental hermeneutics," "hermeneutics of place," "hermeneutics of landscape," and "biological hermeneutics."[4] What tie all of these conceptions together are the intersections where philosophical hermeneutics (in both the narrow and broad meanings described previously) comes into contact with environmental thinking. So those working in environmental hermeneutics may address any number of a wide variety of topics involving natural entities and ecosystems; land- and seascapes; wild, rural, or urban environments; and indeed any other conceptions or meanings of "environment" where interpretation is involved.

Interpretation as the Ground of Environmental Philosophy

Environmental hermeneutics offers a fresh way of looking at traditional problems of environmental philosophy and environmental ethics—areas of discourse that hitherto have not been influenced by philosophical hermeneutics to any great degree. Rather than simply debating the nature of nature or whether it exists, for example, hermeneutics offers the possibility of pondering and reflecting upon the experience in environments as a form of interpretation. Environmental hermeneutics offers an implicit critique of many forms of environmental philosophy, in other words, based on the idea that ". . . there is no unmediated encounter with nature."[5] Because of this, environmental hermeneutics advocates mediation as the appropriate stance for any environmental philosophy. On one hand, many early works of environmental philosophy advanced arguments that rely on essentialist notions of "nature," "wilderness," or similar terms in order to advocate specific ethical positions (what might be considered a first, ethically oriented wave of environmental philosophy); human responsibility in relation to such idealist notions seems a shadow of lived experience, however. On the other hand, a more recent, second wave of environmental philosophy advances a social constructionist or phenomenological notions of nature. While this overcomes the difficulties of a reified and essentialist notion of "nature," it is in danger of ignoring the reality of a world outside human determinations of meaning. It also underdetermines the complexity of the reflexive nature of encountering the meaning of nature itself. In contrast, environmental hermeneutics concentrates on the "conflict of interpretations" that exists in our intersubjective encounters with the material, emotional, and intellectual world. For hermeneutics, the issue is not a binary of a pure-versus-constructed encounter with the environment. Rather, hermeneutics is interested in understanding the mediated experience, the in-between place characterized by detours that result from our historically situated place of human finitude.

Not only does environmental hermeneutics offer a mediated perspective, it thereby also expands the concerns of environmental philosophy. Philosophical hermeneutics encounters the world in a variety of ways, and thus environmental hermeneutics applies to myriad contexts. The title *Interpreting Nature* might seem to indicate hermeneutics aims at what is commonly referred to as the "natural environment." Of course, this is already problematic. As many debates within environmental philosophy and related disciplines reveal, "pure" nature, free from human intervention, does not exist. Further, even the definition of what "nature" is in the first place is not understood univocally. This hinders debate: For example, without a consensus on the meaning of nature, environmental thought becomes mired in discussions of dualisms such as the nature/culture divide and the acceptable place of anthropocentrism versus non-anthropocentrism. As the editors and authors of this book see it, environmental hermeneutics encompasses a much broader understanding of environment. An environment may refer to a physical environment, sociocultural environment, a built or architectural environment, or virtually any other way an "environs" can be construed. From this understanding, it is possible to take a stance that acknowledges the difficulties of defining nature and the natural environment, and thereby open up space for a productive dialogue in response to these aporias.

But isn't hermeneutics about the methods of textual interpretation? What does hermeneutics have to do with interpreting nature? The present collection is not an attempt to argue that past hermeneutic philosophers are in truth environmental activists; but only an inattentive or fragmented reading of hermeneutical philosophers, especially Gadamer and Ricoeur, would lead one to conclude that contemporary philosophical hermeneutics is limited to actual texts. Contemporary hermeneutics rests on Heidegger's "hermeneutics of facticity," and there has long been a strong interest in questions of existence and meaning. In the foreword to the second edition of his monumental *Truth and Method*, Hans-Georg Gadamer employs Heidegger's "temporal analytic of Dasein" to indicate that "understanding" (or interpretation) is not one way of being but is "the mode of being of Dasein itself. . . ."[6] Therefore, for Gadamer, hermeneutics ". . . denotes the basic being-in-motion of Dasein that constitutes its finitude and historicity, and hence embraces the *whole of its experience in the world*."[7] Without question, our experience in the world includes that of environments. One might even argue that "environmental hermeneutics" is redundant insofar as all hermeneutics is concerned with experience in the world, which is already always environmental.

While we wish to preserve a broad inclusive understanding of environmental hermeneutics (in order to be able to address specific environments among the plurality of environments), the intersection of hermeneutical thinking with environmental thought and contemporary environmental concerns requires that we maintain the distinction. Otherwise, we risk reducing decades of hermeneutics to the contemporary insights and meditations that have led to a relatively new field. The primary point is that hermeneutics is not reducible to the interpretation of texts. Rather, that "the text" can be seen as a model for what hermeneutics is aiming at, which is the interpretation of experience in the world. What happens when we encounter a text reflects the realities of what happens in experience in general.

This is one reason that both Gadamer and Ricoeur noted repeatedly that philosophical hermeneutics is not primarily concerned with the intention of an author. In the case of a text, the work takes on an autonomy that permits it to possess many valid meanings apart from the author. The reader, or interpreter, brings her own "horizons" to the text and, in the encounter, may have those horizons expanded. Or, with regard to the inherent prejudices of the horizons, may have her horizons entirely obliterated as a new understanding of the world emerges, and as a consequence change, one hopes, for the better. What we begin to see is that the experience one has with a text bears many of the features that one has in almost any encounter in the world. Of course, our experience of the world is not identical with our experience of written texts. For instance, we may have "direct" and powerful bodily experiences of the world we find ourselves in. What environmental hermeneutics will stress, though, is that as soon as we ask what these experiences *mean* to us, we are confronted with the same issues as when we ask what a particular text has to say to us. Given this stance, environmental hermeneutics proceeds from an understanding of hermeneutics which draws on the work of both Gadamer and Ricoeur.

While the present collection acknowledges the distinction between philosophical hermeneutics in general and environmental hermeneutics in particular, both do share an important trait: hermeneutics is dialogical. In his editor's introduction to Gadamer's *Philosophical Hermeneutics*, David E. Linge writes, "Hermeneutics has its origin in the breaches of intersubjectivity. Its field of application is comprised of all those situations in which we encounter meanings that are not immediately apparent but require interpretive effort."[8] These "breaches of intersubjectivity" begin in what Gadamer refers to as experiences of alienation between an I and thou that are joined together by common experience or a "deep common accord," which makes communication possible and out of which arises understanding.[9] While there are limits to the notion of "the breaches of intersubjectivity," our interest is in hermeneutics as it is located in the "in between" of "strangeness and familiarity," that place between the distanced, alienated object and the interpreter. Hermeneutics operates in this "intermediate position."[10]

This dialogical structure might seem to pose a problem, however, for environmental hermeneutics. After all, since we can't have verbal communication with natural entities that do not possess language, isn't the possibility of a (hermeneutic) relationship ruled out? Our wager is that it is not, because the stance of hermeneutics remains one of mediation. On one hand, humans provide the "language of nature," through descriptions and interpretations that are shared within the human community. Thus what started as an environmental hermeneutics in truth becomes a dialogue within culture *about* nature. But on the other hand, what becomes clear later in *Truth and Method* is that, for hermeneutics, language does not always refer to only human language but rather to "any language that things have."[11] While it is true that philosophical hermeneutics has primarily been concerned with human language and therefore human discourse, language can also refer to the presentation of a being from itself to others. For environmental hermeneutics, the intermediate location of hermeneutics is the place where meaning is discovered beyond the binaries and dichotomies between humans and "nature" that have long obfuscated much environmental thinking and from which environmental philosophy has not been able to completely free itself.

The nearly generation-long debates over anthropocentrism and the various non-anthropocentrisms have often failed to define these terms and almost completely ignored that multiple understandings are possible. For example, is all anthropocentrism about the value of humans? Can anthropocentrism be understood as epistemic? Are there ecocentric and biocentric aspects to being human? Are human beings so "other" than nature that there is nothing natural about us at all? Or are we perhaps a kind of nature? With certain exceptions, of course, much environmental thinking falls into a paradigm of a human/nature divide, which in its more extreme forms is simply another dualism. When they fail to recognize the complexity and complementarity of their respective positions, both anthropocentrists and non-anthropocentrists fall into patterns of thinking that both assume and reinforce this dualism. Acknowledging the conflict of these different interpretations, however, leads us to see that there is a complex interweaving of all beings that really isn't well explained or understood when reduced to simple binaries.

In addition, environmental hermeneutics may also offer something else to contemporary environmental discourse—a way to apply theoretical understanding in a manner that makes a difference. In other words, if the purpose of environmental philosophy in general and a concern of environmental hermeneutics in particular is to address itself the real-world environmental crises that we face, then it cannot be merely abstract or theoretical. We would argue, to borrow from J. Baird Callicott, that environmental hermeneutics is no doubt a form of environmental activism.[12] Gadamer makes a forceful argument that hermeneutics is fully about the "real world" and not an abstraction. "The principle of hermeneutics simply means that we should try to understand everything that can be understood."[13] And the "what" of understanding is not found in abstract concepts but in actual encounters in the world. The very universality of hermeneutics is present in the experience of the world from where the meanings of those experiences obtain.[14] Hermeneutics actually works from and within concrete, historical realities and is thus intended to speak to those same realities and "real-world" situations. Thus, Gadamer can say that "we consider application to be just as integral a part of the hermeneutical process as are understanding and interpretation."[15] Understanding, interpretation, and application—taken together as a "unified process"— are important features of environmental hermeneutics. We think environmental hermeneutics, if it is to mean anything at all, should matter! This is where environmental hermeneutics has a close affinity with eco-phenomenology. Environmental hermeneuticists and eco-phenomenologists contend that philosophy can and must motivate for concrete change, in defiance to certain aspects of modern day "green speak" that suggest we can have our cake and eat it, too, and leaving the future of the environment to "green" consumption.[16]

And concrete change can only occur when philosophy examines the world concretely. Rather than remain at the level of abstract discourse, environmental hermeneutics is concerned with both interpretation per se and lived particularities—something seen in this volume by the use of and reference to numerous and varied case studies. Because hermeneutics, as noted previously, characterizes the being-in-motion of Dasein and is about the entirety of experience in the world, it only makes sense that case studies provide a prime locus for hermeneutical reflection. Moreover, what we are calling "environmental hermeneutics" focuses upon the dialogical relationship between humans and environments, which is likewise well expressed by the attention to case studies. Holding to the claim at the beginning of this introduction that no facts are left uninterpreted, hermeneutics takes the data of case studies in order to uncover meaning and unveil understanding.

Environmental understanding is contextual understanding. It does not find itself in abstract space but is situated in concrete places or locations, and always within the particular cultural setting belonging to that place. Moral meanings do not exist *in abstractum*, but only as part of moral language and within the constant flow of interpretation and reinterpretation that we call traditions. Without active debates and disputes about the meaning of environments, that is, without a vivid culture and a vivid moral tradition (conversation or dispute about transmitted interpretations), moral meanings cannot exist, and moral culture becomes numb. Hermeneutics in general seeks to articulate and reflect upon the (moral) understandings of a historically situated moral community; an *environmental* hermeneutics focuses on the "emplaced" situatedness of a given understanding of the environment. By explicating and critically reflecting on an "emplaced" understanding of the environment, environmental hermeneutics is not primarily a theoretical endeavor but truly practical philosophy: critically reflecting on our practices and our understanding of the environment, articulating dormant meanings that have remained hidden from view, opening new avenues of interpretation. Doing so means that hermeneutics matters: It helps to deepen and broaden our moral environmental understanding of the place-based context of a given moral community, and in a sense helps *create* a culture of place. Hermeneutics, in other words, helps to create moral communities by breathing life into our moral language.

One example of such concrete prescriptions is the field of environmental justice, to which environmental hermeneutics is more than congenial. If environmental justice teaches us anything, it is that environments mean nearly everything to us as human beings—our health, our culture and way of life, individual and collective identities, traditional knowledge and practice, and so forth are all woven into our emplacement (we would even say our embodiment)

in environments. Secondly, environmental hermeneutics pertains
to the way environmental justice issues play out in terms of activism,
law and policy making, and communicative reason amongst all
players. To the former, environmental hermeneutics addresses itself
on the theoretical plane to understanding the human relationship
to environments in terms of the concerns of environmental justice.
This should be evident by what has already been said in this introduction, and additional insights can be found throughout this
volume. In terms of activism and the "realities on the ground," environmental hermeneutics provides resources for deliberation and creative
means for thinking through problems. Thus the potential links
between environmental hermeneutics and environmental justice
offer broad horizons for future research and scholarship.

The Literature of Environmental Hermeneutics

The remarkable diversity and variety of essays in this volume reflect
this wider understanding of environmental hermeneutics as an
"emerging field" within environmental philosophy in particular and
environmental thought in general. In fact, *environmental hermeneutics might be thought of as the "third stage" of environmental
philosophy*. Emerging out of classic texts by luminaries such as
Henry David Thoreau, John Muir, and Aldo Leopold, the first works in
the field were almost exclusively oriented toward applied ethics.
In other words, the first stage of environmental philosophy saw the
field as a topic within applied ethics, and the issues discussed
reflected this narrow concern. While these were foundational, a
second stage of thinking about the environment emerged: one
that recognized the fecundity of intellectual questions beyond
ethics. Scholars working on aesthetics, ontology, theology, and other
disciplines brought these questions to bear on environmental
issues. A greater diversity of philosophical methods was used, and, in
particular, the continental philosophical tradition was engaged in
the dialogue. Among other things, this second stage both globalized
the dialogue as well as opened up the fundamental question of how
humans are "in the world." This provides entry to what we believe
is a third stage of environmental philosophy. Increasingly philosophers are confronted with the "conflict of interpretations" on issues
of environment. Acknowledging the need to better understand
how we understand our environments, environmental philosophy
has recently begun to address questions of how environments are

mediated in our intellectual, moral, and perceptual experience. The essays on environmental hermeneutics that make up this volume are, we hope, harbingers for many fruitful discussions that are beginning to arise.

By characterizing environmental hermeneutics as a new stage of environmental philosophy, it is important to recognize the work already done by philosophers on this issue. Don Ihde has been credited by many for applying the term "hermeneutics" to environmental concerns, and his works are among the earliest in the field. Likewise, the work of Robert Mugerauer and John Van Buren has influenced several of the essays collected here. While it cannot claim comprehensiveness, the bibliography appended to this volume is meant to be a starting point to explore relevant literature already in print. What these earlier foundational works coupled with the appearance of the essays in this volume suggest is that the role of hermeneutical thinking in environmental philosophy and other environmental discourses is coming into its own.

Prior, of course, to earlier works that set the stage for the field of environmental hermeneutics is the foundation of philosophical hermeneutics itself. The editors and authors of this volume would not limit or restrict the field of environmental hermeneutics to any particular articulation of hermeneutical thinking or specific thinkers. Our hope is that environmental hermeneutics as an emerging field will continue to be elucidated from a wide spectrum of hermeneutical thinking, past and contemporary. But the collective exposition of environmental hermeneutics in this volume owes a significant debt to Paul Ricoeur and Hans-Georg Gadamer (and in a somewhat lesser way, Martin Heidegger). Ricoeur, in particular, with his work on such themes as narrative, identity and selfhood, the conflict of interpretations, and memory, just to name a few, has provided the authors of this volume with a rich field of hermeneutical tools with which to construct an interpretive matrix for environmental philosophizing. While it is true that *Interpreting Nature* is not a book about Ricoeur and environmental philosophy, it is certainly the case that Ricoeur, along with Gadamer, has greatly influenced philosophers—including the editors of this volume—who set out to explore the intersection and interfacing of philosophical hermeneutics and environmental philosophy. We would contend that environmental hermeneutics offers environmental philosophy completely new worlds of thought and expression not previously available to it, in order to address the increasing complexity of environmental challenges facing the world today (see, for example, the final essay in this volume by Paul van Tongeren and Paulien Snellen). Hence, we have proposed environmental hermeneutics as a third stage of environmental philosophy.

The Genesis of *Interpreting Nature*

One of the unique features of this volume is the means through which the collection came into being. The four editors became acquainted through the discovery of a shared interest in the intersection of philosophical hermeneutics and environmental thought. This included what might very well be considered the beginnings of a tradition for environmental hermeneutics: many common philosophical texts, concern with similar questions, and shared vocabulary. What followed was a long period of communication and dialogue that encouraged the notion of environmental hermeneutics as a distinct field within environmental philosophy and complementary to other philosophical approaches to environmental philosophy.

After a number of meetings at conferences and email communications, it became apparent that the progress of thought in environmental hermeneutics called for a collection introducing it as an "emerging field." The question quickly became "how to proceed?" One option was to invite potential contributors to simply write essays and collect these into a volume. Such an approach is cost-effective, but the disconnectedness of the resulting contributions was deemed by the editors to be contrary to the idea of environmental hermeneutics itself: We must be attuned to the dialogical nature of hermeneutics and the "fusion of horizons" to which Gadamer refers. What would be much preferred is a colloquium where potential contributors could present their work and interact with others so that all could be mutually enriched and challenged in our distinct but related philosophical inquiries into environmental hermeneutics. The prospect of such a gathering seemed next to impossible, especially in an environmentally responsible manner. In the first place, how to get all the contributors together in one place at the same time? Multiple schedules and additional commitments, not to mention the question of funding, presented serious obstacles to bringing such a group together. Moreover, we were confronted with the aporia that constantly haunts academics with environmental commitments: that countless miles are traveled and resources are consumed, vastly multiplying one's carbon footprint, in order to give a twenty-minute presentation about saving the environment!

The deliberation of the editors concerning these issues brought about the idea of "virtual" seminars. This fulfilled the desire for a shared virtual space/place at the same time, without the difficulties of funding, scheduling, or environmental costs. Thanks to the graciousness of Ohio Northern University (home institution of one of the coeditors, Forrest Clingerman), which provided the use of their technological capabilities, an ongoing virtual seminar took place over the 2010–2011 academic year. Throughout the fall and spring, one or two seminar sessions were held each month, attended by a core group of fellow contributors. Each seminar was devoted to one paper, sent out beforehand. At the virtual seminar, the author was given a few minutes to offer some introductory remarks before a brief response from another of the contributors, who acted as a designated respondent for the purpose of focusing the discussion. This was followed by a lively discussion by the rest of the group, usually lasting over an hour. The virtual seminars allowed for a live video and audio feeds of upward of sixteen persons. This format permitted open dialogue and debate within this little virtual/cyber community that aided all participants in crafting a more polished version of their contribution. It also established an ongoing intellectual discussion, which we hope will extend beyond the publication of this volume.

In the end, the virtual seminars did not replace all of the dynamics of typical conference. After all, our spirited discussion of actual places was occurring in virtual space, which meant we could not continue this discussion over a shared meal or coffee, as might have been the case otherwise. However, the seminars did provide their own unique dynamics, something that grew more apparent each month as the seminars progressed. This virtual space led to a significant intellectual engagement with each other; essays were read with more depth and seriousness, and later papers were influenced by earlier ones. That is to say, the community and the liveliness of the intellectual connections made through this process were substantive and meaningful in ways that recommend the process. It is the hope of the editors and contributors that what started as an experiment in the construction of this volume can promote a model that others will use in the future, especially those working in environmental philosophy and other such related disciplines.

PART

I

Interpretation and the Task of Thinking Environmentally

1

Environmental Hermeneutics
Deep in the Forest

John van Buren

Paul Ricoeur said that the main task of hermeneutics is to clarify and mediate "the conflict of interpretations" in the world.[1] If this is true, hermeneutics should be well suited for dealing with heated environmental conflicts, such as local, national, and international conflicts over the use of forests. For their part, these frequently stalemated conflicts between logging companies, government, environmentalists, native peoples, local residents, recreationalists, and others—for example, the old controversy over the spotted owl in the old-growth forests of the Northwestern United States or the ongoing conflict about rain forests in South America—have shown the need for philosophical analysis to help clarify the basic issues and points of contention involved. Joining other scholars who address the issue of forests, my particular approach takes the form of outlining a *critical environmental hermeneutics* by applying hermeneutics, narrative theory, and critical theory to environmental philosophy and ethics, and then using this hermeneutical theory as a method to illuminate the "deep" underlying issues relating to the perception and use of forests. In applying this method, I first take up the analytical problem of identifying, clarifying, and ordering the different interpretive narratives about forests in terms of the underlying epistemological, ethical, and political issues involved. I then address the critical problem of mediating and resolving conflicts between these different interpretations of forests by working out a set of legitimation criteria which all of the stakeholders would ideally be able to subscribe to and use to work out a resolution to their "conflict of interpretation."

Philosophical hermeneutics was developed in the twentieth century by Martin Heidegger, Hans-Georg Gadamer, and Paul Ricoeur, and was carried forward in still different directions within the critical theory of Jürgen Habermas and the American pragmatism of Richard Rorty.[2] Generally, hermeneutics can be defined as the philosophical study of the most common aspects of interpretation, of what people do when they interpret something. These aspects include such things as intentionality, being-in-the-world, language, sociality, time, and narrative. As noted, the more practical task that hermeneutics addresses is that of finding ways to deal with "the conflict of interpretations" in the world. Accordingly, hermeneutics has already been applied to such special philosophical fields as ethical theory, aesthetics, and political philosophy, as well as to particular disciplines in the natural sciences and social sciences and humanities, including geography.[3] What I will be arguing is that hermeneutics can be applied just as fruitfully to environmental philosophy and ethics. The result of this application, which can be called *critical environmental hermeneutics* or *hermeneutical environmental ethics*, addresses the general features of interpretation, specifically of the environment, and attempts to clarify and help us cope with the epistemological, ethical, and political conflicts which arise here.[4]

Environmental Interpretations, Narratives, and the Hermeneutic Arc

Let me describe very quickly the most general features of this environmental hermeneutics. It studies primarily the *sense* or *meaning* of the environment for perceivers, and is thus unlike the natural sciences which are focused primarily on the biophysical aspects of the environment. This sense of the environment is both interpretive and narrative. There are many different interpretations of the meaning of, for example, a forest. I may see a forest as a religious "sanctuary," that is, interpret it in terms of the presuppositions of my religious tradition. Someone else might see it as "timber" and "wood fibre." Moreover, this interpretive sense of environment usually also has the form of a *narrative* or story, since it usually entails views of past, present, and future, and these function like the beginning, middle, and end in the unified plot of a narrative, as defined classically by Aristotle.[5] For example, consider Amerindian experiences of forest as "land" or "dwelling place." Here forest might be the setting for a narrative with its beginning (the creation of earth as home and sanctuary, the role of people as stewards), its middle (what people are now doing with forests), and its end (future salvation or catastrophe). Likewise, the lumber company's view of woodland as "lumber" and "resource" might be bound up with a frontier narrative of conquering an unruly wilderness and using it for the benefit of human "progress." The perspective on woodland as leisure or recreation (e.g., as a site for one's summer cottage) can take place within a narrative of original innocence (original unity with nature), fall (artificiality of modern technological society), and periodic release from big city life (weekends at the cottage). A biologist or a environmentalist might see woodland within a narrative about a biotic community and its physical basis; for example, the story of a valley or other bioregion told from the standpoint of the evolution of landscape, climatic changes, and species' populations, including the human species, for instance, James Lovelock's "Gaia hypothesis" which sees the entire global ecosystem within the plot of an original generation of life-forms (beginning), present global change and crisis (middle), and future homeostasis in which the human species may or may not disappear (end). It is these environmental narratives that provide human actors with a great deal of their self-understanding, identity, and roles. So, in the interpretation of the forest as "lumber," I might understand myself as "logger," "forester," or "lumberman." In other interpretations, I might take up the role of "resident," "conqueror of the wilderness," "hunter," "hiker," "advocate" of nonhuman life, and so on. As Holmes Rolston puts this, the human self is a "storied-residence" on earth.[6]

In studying the interpretive and narrative meaning of the environment, environmental hermeneutics also studies the "texts" that express this meaning, and here "texts" (ensembles of signs or signifiers) is taken in a very broad analogical sense. Originally, hermeneutics in figures such as Dilthey was primarily the study of the expression of meaning specifically in *written* texts, but with Paul Ricoeur[7] it becomes the study of any and all ensembles of signs understood as texts or quasi-texts: oral discourse, electronic media, art, human action, and so forth. To this list of texts, we can add cultural and built environments as well as "pristine nature," since all these things mean or signify something to perceivers and users and are thus signifiers, signs, physical carriers of meaning. Landscapes, buildings, and actions can be as it were "read" and interpreted. Thus, for environmental hermeneutics, there are three inseparable elements in the environment it studies: (1) the biophysical environment (referent), (2) its meaning or sense, and (3) text, the carrier of meaningful references.

As Ricoeur emphasized, the study of interpretive, narrative meaning expressed by texts should follow a three-step "hermeneutic arc." This begins with the study of the concrete referent in the life-world of immediate experience (e.g., this particular biophysical forest or an environmental problem such as deforestation). It then arcs back to the study of the interpretive sense or meaning (e.g., of a forest), analyzing this level of sense descriptively, critically, and in an interdisciplinary fashion with the methods of disciplines in philosophy, the other humanities, and the social sciences, both those that employ "understanding" (empathetic understanding of a particular person's or society's world of meaning) and those that employ "explanation" (causal explanation in terms of general laws and structures). To this list can be added (though Ricoeur himself did not do so) the explanatory methods of the natural sciences, since evolutionary biology, for example, can make important contributions to understanding human behavior and worlds of meaning.[8] More specifically, to the study of the interpretive meaning of the environment in environmental hermeneutics would be brought the full array of environmental disciplines employed today in the interdisciplinary field of environmental studies: the environmental sciences of geology, evolutionary biology, and ecology; environmental history, anthropology, economics, and politics; environmental geography, aesthetics and literature, theology, and philosophy. According to Ricoeur, critical analysis should not remain at the level of meaning but rather should arc all the way back to the level of the referent, to the interpretive application and appropriation of this meaning back in the concrete lifeworld. It should ultimately issue in the projection of possibilities or possible worlds that can be lived in real life by real people. In this application, there is a dynamic back-and-forth movement in which the meaning gleaned from texts (e.g., forest as "sanctuary" or "sacred grove" in ancient times) is creatively interpreted so as to be fitting in the context of a particular lifeworld (e.g., twenty-first-century technological society), and this context is in turn interpreted in the light of the meaning in question (e.g., making room for the sacred in modern life through nature preserves, ecological restoration, or city gardens). Especially relevant to this application in the case of environmental hermeneutics would be the applied arts and sciences employed in environmental studies (e.g., conservation biology, land use and urban planning, architecture, communications and media, education, the arts, sustainable business), as well as the policy-oriented disciplines of environmental politics and law.

Interpretation begins on the surface (the lifeworld referent), proceeds to the deep background level of sense or meaning, and after a long journey returns back to the surface. This suggests an analogy between the "deep ecology"/"shallow ecology" distinction and a distinction between a deep hermeneutics and a shallow hermeneutics. Deep ecology claims to be concerned primarily with changing our fundamental paradigms of understanding the environment, whereas "shallow" or "reform ecology" is supposedly concerned mainly with changing legislation, practices, and lifestyles, but without understanding fully that these are the consequence of our underlying traditional paradigms of understanding.[9] Similarly, a distinction can be made between a deep hermeneutics, which analyzes the underlying epistemological, ethical, and political sense of practices and interpretations of the environment, and a surface hermeneutics, which would be focused mostly on what really derives from the application of the former, in other words, on specific knowledge, legislation, policy, and practices regarding such things as silviculture, establishing wildlife sanctuaries, hunting regulations, pollution laws, and so forth. Surface hermeneutics is the downward way on the hermeneutic arc from the unconscious and unexamined level of basic underlying assumptions to specific knowledge and practices. Deep hermeneutics is the upward way from the latter to a fundamental reflection on the former. It has at least three main tasks: (1) environmental epistemology (describing and critically evaluating the different views of what the environment is), (2) environmental ethics in a narrow sense (describing and evaluating views of the value of environment and our obligations regarding it), and (3) environmental politics (describing and evaluating who has or should have political power in the environment).

In applying environmental hermeneutics to the specific problem of forests, my discussion will be focused on a deep hermeneutics of the underlying epistemological, ethical, and political issues, and leaves for another occasion the necessary work of returning to surface hermeneutics on the hermeneutic arc.

Environmental Epistemology, Ethics, and Politics

As it relates to forests, *environmental epistemology* means not only the analysis of how such things as intentionality, being-in-the-world, language, mood, history, and so forth[10] function in our interpretations of woodland but also the descriptive typology of the different interpretive views or knowledges ("epistemes")[11] of woodland. The basic perspectives on forest can be broken down into "land," "life," "lumber," and "leisure."[12] Each of these will, of course, contain a plethora of subtypes. Under forest as "land" in the sense of dwelling place or home can be found different interpretations of forest by such groups as year-round residents in small communities, Amerindians, settlers and farmers, and those using the forest as sanctuary, sacred land, burial ground, sacred grove, hermitage (e.g., the monastery, the back-to-nature movement begun in the sixties, the solitary trapper).[13] The view of forest as "life" encompasses different interpretations by natural scientists and biocentric ecologists, who both see the environment as a biotic community or as a single organism. Under forest as "lumber" can be found interpretations of forest by such groups as local logging and sawmill operators, national and multinational lumber companies, the daily logger and lumber worker, alternative or holistic forestry, government forestry departments, and Amerindian small-scale harvesting operations. The view of forest as "leisure" encompasses such different perspectives on forest as recreational hunting and fishing area, park and game reserve, hiking and skiing area, the forest of boy and girl scouts' movements, and the forest of the summer cottage.

For any of these specific types of forest interpretation, one can investigate such characteristics as its historical aspect (the constellations of cultural presuppositions at play in it), its narrative aspect (the narratives about woodland which it expresses), and its existential aspect (the types of narrative "roles" for human and nonhuman actors, that is, the *genres de vie*, which it provides). In addition to these rather broad analytic categories, we also need, of course, concepts which are more specific and conducive to empirical research, and thus more in touch with the environmental disciplines that have been underway for years in branches of geography, biology, psychology, political science, law, sociology, anthropology, economics, history, literary criticism, and theology.

Interpretations and narratives express not only different and conflicting knowledges of what a forest is but also, of course, conflicting values or ethics. The beginning-middle-end structure in the plot of narratives usually involves an "end" in the sense of a value, or a future goal, which tells us where the story is going or should be going, if it is to have a "happy ending."[14] As it applies to the specific study of forests, one of the tasks of environmental ethics is thus the analysis of forest narratives for the different values they express.

An initial classification of such values can be found by seeing how each of the four basic perspectives on woodland analyzed previously—land, life, lumber, and leisure—expresses a cluster of basic values or goods in forests, which can in turn be broken down into subtypes.[15] The land perspective includes the values of home or dwelling, including all associated social, community, and religious values. The values of the integrity ("intrinsic value") of biotic forest life, biotic home and community, and general life-support are expressed in the ecological life perspective. The biological life perspective includes natural-historic values (forests as natural museums providing a sense of antiquity, duration, and continuity), scientific values (discovery of new knowledge about animal and plant species, climate, waters, whole ecosystems, etc.), and medical values (indigenous folk medicine, discovery of new species with medical benefits). The lumber perspective mainly expresses the economic values of forests as resource. And, finally, the leisure perspective encompasses the values of recreation and outdoor life, including aesthetic values and character-building values.

One can see these values at work even in the concern with "sustainable development" and, as this relates specifically to forests, sustainable forest management. The Bruntland Commission defined sustainable development as "development that meets the needs of the present without compromising the ability of future generations to meet their own needs."[16] The concepts of "development" and "needs" in this definition are, of course, not purely descriptive terms but rather also value terms. "Needs" refer to things needed and valued, and "development" refers to cultivation of the means to realize these needs, values, or goals. Sustainability means that the present realization of needs or values should not threaten future realization. As many have pointed out, there is a real problem in determining exactly what "needs" and "development" mean in the Bruntland Commission's definition.[17] One cannot simply assume, as many do, that these terms primarily have an economic meaning. These controversial terms have more than one meaning and involve real valuative decisions. In the specific case of sustainable forest development, it is thus easy to see that each of the above four clusters of values corresponding to the land, life, lumber, and leisure narratives about woodland expresses a different and potentially conflicting concept of sustainable development. The latter can, therefore, mean the development or realization of either human dwelling (land perspective); biotic integrity, community, and

life-support (ecological life perspective); natural history, scientific
study, and medical values (biological life perspective); economic
resources (lumber perspective); or, finally, recreation and outdoor
life (leisure perspective). Value conflicts regarding forests thus cash
out as conflicts about the definition and practice of sustainable
development in this case.

Particular values are always affirmed within a general moral
orientation, paradigm, or framework of justifying principles, in other
words, a morality that tells us why a particular value is important.
The standard primary division in environmental ethics these days is
between anthropocentric moralities that affirm the hierarchical
centrality of human species and non-anthropocentric egalitarian
moralities that affirm the whole biological community on the
planet, of which the human species is only one member. In most
Greek ethical theories, nonhuman nature and thus also forests
would be valuable only as a means to realizing the excellence (*arete*)
of human life, which is the highest purpose in the hierarchical-
teleological order of nature. Likewise, most traditional forms of
Christian ethics take nature and forests—in particular, "creation"—
to be valuable only as a means for the soul's journey to God. Modern
utilitarianism would see a forest as valuable in relation to its utility
as a dwelling place, life-support, resource, and recreation for realizing
the greatest good of the greatest number of human beings. Modern
rights theories would see a forest as valuable insofar as it provides
the means of realizing the alienable rights of humans; for example,
a forest as life-support and resource guarantees the right to life and
to a decent standard of living, a forest as a home is the exercise of the
right to property, and a forest as recreation is the exercise of liberty.

One group of non-anthropocentric egalitarian moralities comprises
varieties of ecological utilitarianism, biocentrism, the land ethic,
and ecofeminism, most of which share the view that individual
biota in forests, including human beings, are valuable in relation to
their promoting the health, well-being, and community of the whole
forest ecosystem. In turn, entire forest ecosystems are valuable in
relation to their promoting the health and community of the entire
planetary ecosystem. Ecological rights theories, on the other hand,
would argue that individual forest biota have basic rights, the primary
one of which is the right to life, and that these rights cannot be
overridden except in special circumstances. Another group includes
postmodern anarchism, decentralism, and bioregionalism, which
argues that all individual life-forms or individual communities

should be autonomous or self-governing and should not be oppressed and marginalized by external authorities.

Interpretations of forests express not only different basic knowledges and values but also political narratives about which of these knowledges and values are or should be represented, valorised, and empowered in the public sphere and allowed to determine public discourse, policy, and management concerning the environment. A first task of environmental politics is therefore to analyze interpretations regarding which social groups are being politically represented and empowered in them. Each interpretation of forests involves an answer to the question: Whose interests does it represent? What social group does it empower? One needs to map out a typology not only of forest knowledges and forest moralities but also of forest politics, that is, of the different and conflicting interest groups involved.

We can divide human actors related to forests into at least three social categories, namely, "people," "officials," and "academics." By "people," I mean not only "the common people" but that whole diverse mass of those who do not fit into the categories of "officials" and "academics": for example, residents, loggers, native peoples, alternative forestry groups, grassroots environmental groups, artists, recreational users, and so on. Officials include those involved in all levels of policymaking and management in government, business, and forest-related companies. Under academics one finds, of course, radically different groups: natural scientists, social scientists, and those in the humanities. Obviously, there cannot be an exact mapping of the land-life-lumber-leisure perspectives and their corresponding home-integrity-resource-recreation values onto the academics-officials-people groups. Members of each of the three social groups could be found to fit into any of the different perspective and value categories. Defining these categories stipulatively would not have much value for the study of the real workings of human society. A better approach would be to focus on the perspective and value orientations held by the majority in each of the social groups.

In the world of academics, natural scientists tend to represent the biological life and lumber perspectives, as well as the corresponding natural-historic, scientific, medical, and life-support values. Social scientists tend to emphasize the land, leisure, and lumber perspectives and the dwelling, recreation, and resource values. Academics in the humanities tend to be attracted to the land, leisure, and ecological life perspectives, as well as the dwelling and recreation values. An important issue for consideration is which academic

group has traditionally been given greater power both within the
political structure of the university community itself and within
the university-government-business triangle (with respect to, for
example, funding for facilities, research, hiring, interdisciplinary
programs, and international network links). Experience shows that
it has usually been the life-lumber-resource narrative of natural
scientists and a sector of social scientists that has functioned as the
official and favored voice for the university in the public realm.

In the sphere of officials, one finds that the majority in business
represent the lumber perspective and the resource value of forests.
In government, it is again primarily the lumber perspective and resource
value that are affirmed, with an attempt to accommodate the other
land, life, and leisure perspectives and their corresponding values.

The social category of people contains a wide diversity of perspectives and values. Residents tend, of course, to represent the land
perspective and the dwelling value. In native peoples, one finds
predominantly the land and ecological life perspectives and the
dwelling, biotic community, integrity, and life-support values. The
majority of forest industry workers, along with all those who depend
indirectly on the forest industry for their livelihood, tend to affirm
the lumber perspective and resource value. Alternative forestry
groups tend to identify with both the lumber and ecological life
perspectives, as well as a mix of the resource, biotic community,
integrity, and life-support values. Environmental movements advocate
the ecological life perspective and the biotic community, integrity,
and life-support values. Recreational users naturally express the
leisure perspective and recreation values. If we ask which group or
groups in the people category have traditionally had their forest
narratives represented and politically empowered in the public
realm, it is easy to see that first come forest industry workers and
those who depend economically on this industry, then residents,
and finally recreational users. Native peoples, environmental groups,
alternative forestry groups, the women's movement, and other
groups are still struggling to have their stories heard. Regarding the
relations of the three main groups to each other, it seems that traditionally officials have taken priority over both academics and people.

Environmental Dialogue and Legitimation

Such identification, clarification, and ordering of the conflicting
knowledges, values, and politics regarding forests serves to illuminate

just *what* the underlying problems are and *which* different approaches have been and can be taken. This has the merit of helping to ensure that the problems and conflicts are addressed in a clear and intelligent manner without the different stakeholders involved talking, as they often do, at cross-purposes. But ultimately this descriptive analysis only highlights and makes more urgent the *critical* task that environmental epistemology, ethics, and politics have ultimately to address, namely, judging the "truth-value" of different environmental knowledges, values, and politics about forests and arbitrating "conflicts of interpretation" between them. How can a *critical* environmental hermeneutics provide a method for fulfilling this task of conflict mediation and resolution as it relates specifically to forests?

A problem often pointed out in controversies over the use of forests is the lack of willingness in the stakeholders involved to "talk to each other" and "listen to each other," as well as the lack of institutional arrangements for such dialogue. A first step in dealing with the critical problem of arbitrating between conflicting perspectives is thus the cultivation of communicative openness in institutional forums for dialogue in which all knowledges, values, and social groups can participate. Strategies for dialogue need to be explored in and between all three institutional sectors of academics, officials, and people. The university functions as a setting for seminars and conferences in which there can be wide involvement. In government, forums for dialogue can be cultivated through cooperation between business, labor, and community; environmental hearings and conferences; constitutional reform conferences including questions of an "environmental bill of rights"; native land claim negotiations; referendums on environmental issues; and international exchange, for instance, in the United Nations. Among people, community groups and popular movements can play a large role in promoting dialogue and exchange. The growth of such forums could provide more opportunities for epistemological, ethical, and political conflicts to be fought out and worked out at the table in rational debate—and not with guns on logging roads and proposed development sites.

But such calls for more dialogue, consultation, and participation only go halfway. What is also needed is the willingness of the stakeholders concerned to become conscious of and to argue for their underlying assumptions (e.g., the strictly economic definition of sustainable development that is usually bandied about). One cannot simply take these assumptions for granted as self-evident and use them to draw inferences or explore points within the parameters opened up by the assumptions (e.g., by arguing for a particular method of realizing economic sustainable development or merely supplementing this economic definition with other apparently less important definitions such as the sustainability of *genres de vie*). There is a fundamental social demand in our "rational" civilization going back to Socrates and the rhetorical tradition that opinions be backed up and legitimated with rational argumentation, as opposed to basing them merely on the status quo language of the public realm, on the authority of senior academics, or on folk tradition, allegiances, habit, feeling, and intuition. Though human understanding would never get off the ground without them, all these dimensions provide us with only starting points that can and should be raised to the level of reflection, analyzed, and critically defended.

This demand for rationality or, better said, reasonableness in our communicative dealings with others would, of course, be empty without acknowledged norms or criteria for legitimating our viewpoints. We need to be able to appeal communally to criteria that allow us to decide about the truth of interpretations and to settle disputes between conflicting interpretations. What prevents environmental interpretations from becoming arbitrary? Are we thrown into relativism and anti-science here? There are, in fact, a number of criteria available for assessing interpretations of the environment and of forests in particular. In the following, I sketch out biophysical, historical, technical, and ethical-political criteria. The biophysical criterion (does the interpretation fit the biophysical reality of the bioregion in question?) and the historical criterion (does it fit in and cohere with individual and shared traditions?) relate to environmental epistemology. The technical or pragmatic criterion (does it work, is it efficient, does it produce the desired end?) and the ethical-political criterion (does it satisfy fundamental ethical-political norms?) relate to environmental ethics and politics.

The Biophysical Criterion

To begin with, it is important to acknowledge that, in spite of the anxieties of positivistic science and philosophy, for which the world consists of naked "facts" determined by the natural sciences and all other views are to be dismissed as merely "subjective," it is simply a *fact* of life that the biophysical world lends itself to a number of interpretations as to its *sense for* human beings, and that these interpretations, to one degree or another, all correspond to reality and reveal some aspect of it. One and the same forest can, obviously, lend itself to and support all four types of interpretation: namely, land, life, lumber, and leisure. Unlike the positivist, a tree has no qualms about being interpretively related to as "shade" for picnickers, as "log" for the logger, as "ladder" for a would-be Romeo, as "home" for the squirrel, or as "matter" for the physicist.

The biophysical criterion stipulates only that interpretations must be "fitting" to the bioregion, that they must fit the biophysical world to which they refer. Truth here means interpretive fittingness and adequacy.[18] Even though a particular perspective on the environment is creatively interpretive and obviously goes beyond what is there from a purely biophysical standpoint, it must nonetheless be adequate to the biophysical reality. This definition of interpretive truth both harkens back to and makes an advance on the classical definition of truth as *adequatio intellectus et res*, the adequation (correspondence) of intellect and thing. It is neither the realist definition of truth (simple correspondence of mind to the biophysical) nor the idealist definition (correspondence of the biophysical to the mind), but rather a third middle way for the definition of truth as the interpretive fittingness or adequacy of mind to the biophysical. Truth means *creative* correspondence, *interpretive* adequacy, because, even though a viewpoint has to fit the biophysical world, it still mediates and interprets this physical world in terms of the realm of cultural sense or meaning.

Let me stress that truth as interpretive adequacy does *not* mean that "anything goes," that all interpretations are right. In the first place, even though there may be many right interpretations, there are some interpretations that are obviously wrong—they just don't fit. To take a silly example, the paranoid's interpretation of the forest as a hostile spirit disguised in the form of trees and plotting against the human race is obviously wrong from the standpoint of the biophysical criterion (though it may be right from the standpoint of other criteria, for instance, artistic ones in a poem or surrealist painting). Secondly, regarding settling disputes between conflicting interpretations, some are more right (fitting) than others. For example, in a forest bioregion with healthy harvestable trees, where the number of animal species and individual species populations are very low, the planner's view of the forest as potential game reserve might be less right/fitting than the forest industry's view of it as lumber. In another kind of bioregion where trees are few in number or the majority are of a noncommercial species, the resident's view of it as home may be more right/fitting than the logging company's view. In another forest bioregion, a multiple-perspective approach involving multiple-use of the forest might be the most fitting interpretation. Conflicting interpretations all reveal some aspect of the forest and are therefore true to some degree (i.e., fit to one degree or another), but some reveal more than others about the biophysical forest (i.e., they fit better).

It is, of course, experience, research, and expertise both in the environmental sciences and in nonacademic sectors (e.g., experienced foresters, native peoples, local residents) that can determine, through dialogue, the biophysical truth of interpretations of forest and of other environments. In academic disciplines, this means studies of the biosphere (human and nonhuman life), the atmosphere (air, climate in its local, regional, and global aspects), lithosphere (soils, minerals), and the hydrosphere (waters). Such studies can take the specific form of environmental assessment and impact studies that research the effects on ecosystems from acting out different views of forest, and thus help determine whether the views in question are fitting or unfitting to the biophysical world.

The Historical Criterion

As explained previously, the reality of "forest" means not only its biophysical aspects but also its sense within historical traditions. The reality of forests in relation to which interpretations are supposed to be fitting is, therefore, also this body of social (economic, moral, religious, aesthetic, and so on) traditions regarding forests, including the local *genres de vie* of forest bioregions. This historical criterion for deciding the fittingness of interpretations is a version of the "coherence theory of truth," since it looks to how an interpretation coheres or fits with historical traditions and to what extent it involves either creative growth or alienating disruption (culture shock). A real danger is that new and different views of the forest will lead to cultural displacement, alienation, and homelessness for traditional users. Here again truth means creative adequacy in the sense of an interpretation that, while creatively modifying traditions, nonetheless is fitting to these traditions.[19] Environmental assessment and impact studies using this historical criterion are obviously the domain not of biophysical disciplines but of environmental disciplines in the social sciences and humanities, which can research the effects on social traditions from acting out different views of forests and thus determine which of the views in question are fitting or unfitting. By itself, the historical criterion does not get us that far and needs to be guided by other ethical and political criteria. What if the tradition with which an interpretation fits is morally and politically questionable?

The Technical or Pragmatic Criterion

Another criterion for deciding the truth of interpretations is the technical or pragmatic criterion. Here one asks: Does the interpretation work? Is a certain perspective—for example, the lumber perspective—efficient in relation to the end in view that is to be produced or realized (material well-being)? Does it provide the right means (say, a program of tax incentives and penalties) for realizing the desired end (sustainable economic development of forests)? As William James put it, here truth means what works, what is pragmatic, what is fitting and adequate to pre-given ends. This technical criterion is the domain of instrumental reason, or what the Greeks called *techne*, technique, technical knowledge.[20] There are two basic characteristics of this type of instrumental rationality that point to its shortcomings. First, the ends to be realized, which are usually economic or organizational in nature, are simply assumed as pre-given, and the real concern is simply with finding efficient means for producing these ends. Instrumental rationality is concerned solely with efficiency of use and manipulation of the world to produce desired ends. In other words, it cannot itself supply us with the ends or goals of human life, but rather simply takes these from the status quo and serves them. The technical criterion can certainly help us decide which are the better means for a given end, but it cannot choose between alternative ends themselves and rationally legitimate this choice. Just because we have the technical ability to do something (e.g., in genetic biotechnology), does not mean *ipso facto* that we *should* do it. Second, this technical knowledge may become concentrated in an elite of "experts" or "technicians," so that the sphere of activity in question comes under the sway of "technocracy" or "expertocracy," that is, authoritarian rule. When the sphere of activity in question concerns people directly, as in a technical approach to social activity and organization, then the human sphere can become the object of manipulation and production in political technocracy. Persons are here treated merely as means and not as "ends in themselves" (Kant), that is, they are depersonalized, dehumanized. Like the historical criterion, the technical criterion needs to be guided by ethical and political criteria.

The Communicative Ethical-Political Criterion

Since instrumental reason is concerned solely with means and not ends, and by itself tends to reduce persons to the realm of manipulable things, it has to be "ruled" by practical or ethical reason (*phronesis*), as Aristotle put it in book VI of *Nicomachean Ethics*. Practical reason is concerned with the ends of human action that are worked out in rational discourse (*logos*) between free citizens in the public or political sphere. Aristotle's practical reason has therefore come to be called communicative reason in contemporary thought.[21] Instrumental reason, on the other hand, does not communicate with others through dialogue and debate about the ends of life—it simply tells you what to do or else uses manipulation and direct force "without wasting words."

Communicative rationality supplies us with a communicative ethical-political criterion to legitimate environmental interpretations and narratives and arbitrate between them when they conflict. Here one asks if the adoption in the public sphere of the views, values, and politics of an interpretation (e.g., forests as lumber) or a particular combination of interpretations (e.g., hierarchical multiple-use of forests with lumber leading the way) has been or can be arrived at in conformity with the fundamental procedural norms or ideals for making decisions democratically in society. Jürgen Habermas and others have argued that these norms are built right into the very nature of human being as *zoon logon echon* (animal rationale), a living being capable of rational social discourse. As procedural democratic norms, they concern not so much *what* is being talked about and decided upon in our communicative praxis with one another as *how* it is being talked about and decided. These norms include universal participation in decision-making versus monopoly by particular interest groups; free versus coerced discussion; equality of opportunity versus a priori privileges; respect for others; tolerance; and consensus which ideally should be universal, but realistically has to be in the form of a majority. In applying these communicative ethical-political norms, one asks: To what extent was or can the adoption and empowerment of a particular view or hierarchical combination of views of forests be arrived at through free and universal discussion involving equality, respect, and tolerance, and universal consensus?[22] Or, conversely, one can ask: To what extent does a view of forests—let us say the European-colonialist grand narrative of North American forests as untamed wilderness to be conquered for human progress—falsely universalize itself by claiming to be the whole truth, misrepresent its interests as the interests of all (ideology), and thus *ipso facto* marginalize and exclude other views, for example, those of native peoples?[23] Ethical-political truth here means fittingness or adequacy to communicative ethical-political norms that are meant to insure the "common good."

The application of these norms means the democratizing of forests and the environment in general. They allow for the participation in decision-making of all sectors of human society: people, officials, academics. And this should include groups that traditionally have been marginalized and politically silenced: women, non-whites, indigenous peoples, children, the unborn, the poor, the third world, and so forth. Participation becomes cross-cultural when these norms are applied in international political institutions (e.g., the United Nations). Participation becomes still wider and even planetary if nonhuman forms of life are seen as a legitimate group of perceivers and users of the environment and are no longer related to merely on the level of instrumental reason as "resource," but rather welcomed into the dialogue community of all living things on the planet, even though the practical reality of "communication" remains problematic here (as it is in the case of, say, "the unborn").[24]

Critical environmental hermeneutics stands "beyond objectivism and relativism"[25] in its treatment of environmental issues, insofar as it searches for a balance between the objective and the relative. Relative elements are acknowledged in the creative role of historical and cultural interpretation in the perception and use of forests. Objective elements are affirmed in the application of the four criteria outlined previously for judging the "truth-value" of interpretations of the environment. They were also affirmed earlier in my analysis of the general aspects of environmental interpretation and in my typologies of forest knowledges, values, and politics. Other approaches to environmental issues often suffer from being either one-sidedly objectivistic or one-sidedly relativistic. Environmental positivism, which reduces sense to physical facts, and early environmental phenomenology,[26] which reduces sense to intentional essences, fits the former category. Environmental postmodernism,[27] which affirms radical "difference" and relativity in the environment, often seems to fit the latter category by obscuring the possibility of appeal to such shared criteria as I have outlined here. Critical environmental hermeneutics, on the other hand, travels midway between identity and difference, objectivism and relativism.

Toward Communicative Environmental Reason

In looking at controversies over the perception and use of forests, I have tried to pinpoint the important issues of contention (epistemology, ethics, politics); the conflicting knowledges (land-life-lumber-leisure), values (home-integrity-resource-recreation), and stakeholder groups (people-officials-academics-nonhumans); and, finally, legitimation criteria for resolving such conflicts within dialogue. In exploring these criteria, my aim is not, to begin with at least, to become another party to the epistemological, ethical, and political conflicts between the different views. Rather, it is primarily descriptive and not prescriptive. The role I am pursuing here is more like that of an observing referee in the "conversation of mankind" [sic] who advocates fair play, tries to keep the conversation going by suggesting avenues of dialogue, and urges the participants to strive for consensus that is arrived at on the basis of criteria acknowledged by all parties concerned.

Another concluding point to consider is that all four criteria should be applied in concert with one another, since they need to act as correctives to one another. A view (e.g., the view of a forest as lumber) may be biophysically true in the scientific realm, that is, fitting to the biophysical realities, but communicatively invalid in the political realm, since it has not been adequately worked out in conformity with ethical-political norms and is in fact based on relations of manipulation, distortion, and oppression (e.g., in colonialism and cultural imperialism). Similarly, a view (e.g., forest as leisure organized by big business) may be technically true in the sense that it is efficient for the pre-given end of creating jobs in the economic technosphere, but at the same time it can be historically/culturally invalid insofar as it threatens to destroy traditional genres de vie in small communities. Or a democratically chosen approach in forest management might be based on a gross misunderstanding of the biophysical world. And so on. None of the criteria on its own gives us the whole truth. The ideal here would be to satisfy all four criteria or find a creative compromise between them (one might also have to consider if some of the criteria are more important than others). Similarly, with regard to the substantive content of decisions made, the ideal would be to realize genuine multiple-use of forests, that is, a creative compromise between the different views, values, and interest groups.

But communicative ethical-political norms have a kind of meta-function in relation to the other criteria. They function as a standard not only for political decisions in the public realm but also for applying each of the other three criteria. These latter criteria can work only if there is free and open discussion striving for rational agreement among and between academics, people, and officials about which environmental perspectives are most fitting for the bioregion in question (biophysical criterion), for the traditional cultural realities and local *genres de vie* (historical criterion), and for the realization of chosen goals (technical criterion). The ethical-political norms of communicative environmental reason are really the conditions of the possibility of all realms of rational environmental discourse.

The workings of communicative environmental reason in effect provide us with a metanarrative about particular environmental narratives, that is, about what we should do in the face of the conflict of narratives about forests and the environment in general. But this story of communicative discourse, of "getting together and talking things out," is a metanarrative not in the sense of a particular, substantive, and homogeneous perspective that ideologically marginalizes all other perspectives but rather in the sense of a nonsubstantive, procedural narrative that, without falling into subjectivism, makes room for radical heterogeneity and localism in environmental narratives. It espouses coexistence, communication, compromise, cooperation, and consensus. Heidegger spoke of the historical sense or meaning of the world as "the house of being" in which we "dwell."[28] Thinking, he suggested, is a kind of "original ethics" in the literal sense of "the art of home," or even "ecology" in the literal sense of "discourse concerning the household." In terms of this metaphor, communicative environmental reason amounts to a metanarrative of ecumenism and hospitality; of opening doors to rational debate; mediation and resolution between different environmental knowledges, values, and stakeholder groups in that inevitable "conflict of interpretations" which Paul Ricoeur placed at the center of hermeneutics.

2

Morrow's Ants: E. O. Wilson and Gadamer's Critique of (Natural) Historicism

Mick Smith

What shall it tell me if a timeless insect
Says the world wears away?
　　　　　　　—*Dylan Thomas, "Here in this Spring"*

In 1975 Edward Hyams, novelist, gardener, broadcaster, anarchist, and a long-time advocate of the need for agriculturally sustainable societies, wrote a political novel, *Morrow's Ants*.[1] It tells the story of billionaire businessman Graham Morrow's attempt to build a futuristic city modeled on his intensive study of ant colonies.[2] The Hive, a massive, largely underground, complex powered by tidal and nuclear energy, will house and feed two hundred thousand people in an entirely self-sustaining manner. But this development, as his opponent (the embittered revolutionary Evans), suggests, comes with a price—the loss of individual liberty, freedom, and creativity. It is "designed to establish, permanently, the power of the caste of industrial managers by . . . creating an entirely new and massively stable form of society" (48). It will offer a "safe and comfortable slavery which generation after generation . . . will be increasingly educated to accept as the only order of things" (52). The Hive effectively completes and supersedes both capitalism and communism. In Evans's words, "It is Morrow, not Marx, who will set up the Hegelian State; the State as God; the perfect Synthesis" (51).

The ethical and political results of this "end of history" hardly matter to Morrow since he considers ideas like freedom to be "hampering anachronisms," "the nature of which prevents us from getting clear of the clutter and rubbish of the past" (76). Their loss is, in any case, Morrow argues, an inevitable consequence of historical progress. This progress is driven by increasing population pressures that will compel us to "make ourselves elements of a community with a new thoroughness." "The truth is we have no choice; and once we recognize that truth, we could move forward into the final phase of man's socialization with less agony" (77). Morrow's Law holds that "freedom destroys itself by destroying the conditions in which it can exist" (20).

> The concept of individual freedom of action remained viable with a population of up to one per square kilometre; of freedom to 'opt out' at up to 5 per square kilometre [. . .] of freedom of speech and printing at up to 25 per square kilometre; of freedom of thought at, perhaps, up to 500 per square kilometre, the more active freedoms having, by then, been lost of course. The price of liberty is not eternal vigilance: it is restraint of the philo-progenitive instinct. The love of children is the death of liberty and at population densities of more than 1000 per square kilometre while the life of the Hive remains possible, the life of Socrates and Thoreau is not. Men cease to exist: Mankind takes their place. (77)

For Morrow, increasingly dense forms of sociality necessarily require ant-like behaviors, but they are also the source of a new and greater form of collective intelligence, for his observations convince him that "[t]hose ants which had solved the most, and the most difficult, scientific and economic problems, were the species that formed very large colonies. . . ." "[D]rawing analogues from cybernetics," Morrow explains how such ants overcame their neurophysiological limits by effectively "contributing all their minds to a supermind. . . . In our brains the individual elements are neurons; in an ant's nest, *which is amongst other things a brain*, the elements are individual ants" (78). The inhabitants of Morrow's Hive will similarly be so closely integrated that their very individuality will be subsumed into the larger functional whole.

Morrow's technocratic survivalism elicits a particular (quasi-Hegelian) reading of "Natural History," that is, of evolution *and* history combined in an overarching synthesis.[3] For example, the apparently inescapable progenitive *instinct* provides the biological basis and motor for his account of *social* evolution and then, of course, there is the cybernetic interpretation of ant colonies which

Hyams almost certainly borrowed from the work of biologist Rémy Chauvin, who in turn drew on Von Neumann's cybernetic work (both of whom Hyams's has Morrow refer to). Indeed, Chauvin's *The World of Ants: A Science-Fiction Universe* was translated by Hyams's occasional coauthor George Ordish[4] and its book-jacket blurb states:

> The ants' society may resemble man's [*sic*] in many ways, but the tiny brain of the individual ant does not allow it the independence of the individual man. Professor Chauvin shows how the ants' apparent social organization and intelligence is really a combination of chance and cybernetics, resulting from the enormous numbers involved and the variety of their adaptive mutations.

"Natural History" as Eschatology

In discussing what he terms the "four pinnacles of social evolution" (namely, "the colonial invertebrates, the social insects, the nonhuman mammals, and man [*sic*]"), Edward O. Wilson's influential text *Sociobiology: The New Synthesis* (also first published in 1975) raises the very question that Morrow's social experiment seeks to answer.[5] "At what point does a society become so nearly perfect that it is no longer a society" (383) but can, instead, be considered a "superorganism"?[6] What is more, Wilson, who like the fictional Morrow is a myrmecologist (a specialist in ants),[7] presents a different, but hardly less biologically deterministic and reductive account of human social evolution based on his own reading of "Natural History." For, having defined sociobiology as the "study of the biological basis of all social behavior" (4), including human social behavior, he suggests that the "principal aim of a general theory of sociobiology should be an ability to predict features of social organization from a knowledge of . . . population parameters combined with information on the behavioral constraints imposed by the genetic constitution of the species" (5). And since this theory is explicitly intended to "reformulate the foundations of the social sciences" (4), this sounds suspiciously like an attempt to develop Morrow-like laws. Indeed, Wilson almost seems to echo Morrow when he writes:

> To maintain the species indefinitely we [humans] are compelled to drive toward total knowledge, right down to the levels of

the neuron and the gene. When we have progressed enough to explain ourselves in these mechanistic terms, and the social sciences can come to full flower, the results might be hard to accept. (575)

It seems then that Wilson, like Morrow, regards his own work as furthering a form of inevitable intellectual progress grounded *in* and *on* a biological truth that, once fully revealed, we will have no choice but to accept, whatever our ethical or political qualms. (Wilson even provides a time frame suggesting that we "still have another hundred years" [575] before sociobiology will have completely divested us of our comforting unscientific illusions and values.) Each, in their own way, regards themselves as being on the side of (a progressive) history intimately interconnected with the unfolding and revealing of underlying natural laws. To be sure, Wilson's universalizing theoretical ambitions envisage nothing as specific as Morrow's totalizing social ambitions, but Hyams's novel does, perhaps, suggest why many of *Sociobiology*'s contemporary critics regarded its reading of "Natural History" as posing potential threats to individual freedoms, to ethics, and to politics.[8]

Much of this criticism had (and still has) to do with the biological "determinism" and genetic "reductivism" underlying Wilson's sociobiological arguments.[9] Here the issue is one of how far a reading of "Nature" can and should be taken to have sociohistorical, ethical, or political implications. Take, for example, Wilson's claim, from the very beginning of *Sociobiology*, that evolutionarily speaking, the "individual organism counts for almost nothing" (3) because they are just temporary carriers of genes. Now place this statement next to his statement toward the book's end that a code of ethics accurately based on genetics would be "completely fair" (575). An obvious question to ask is, *if* evolutionary genetics regards individuals only as instrumental *means* serving genes' *ends*, should (or could) ethical values reflect this and if so how?

Wilson argues that our increasing genetic and evolutionary knowledge means we will "elect a system of values on a more objective basis."[10] Luckily for environmentalists, this "objective basis" apparently includes an innate developmental tendency to find "personal fulfillment" and "freedom" through interactions with diverse living things—a tendency he terms *Biophilia*.[11] But while talk of electing values, of personal fulfillment, and freedom might suggest that we can *choose* to adopt such values, claims about objectivity and genetic predispositions seem to suggest that such values have some kind of "natural" authority over us. To be sure, Wilson elsewhere recognizes that our contemporary ethical values are based on "strong innate feeling *and historical experience,*"[12] but he still has no doubt that "[s]cience for its part will . . . in time uncover the bedrock of the moral and religious sentiments."[13] The "human spirit" (and that is indeed the term he uses) requires that "[w]e must know, we will know."[14] And while *Sociobiology* only mentions politics occasionally, this reading of "Natural History" clearly has profound political implications. For example, Hyams would no doubt have been interested in the fact that Wilson *explicitly* rules out anarchism as a "biologically possible" form of society.[15]

In terms of a comparison with Morrow, the temporal aspects of Wilson's claims are much more interesting than arguments about biological reductivism and determinism per se. For it is here, in the *human spirit*'s necessary *drive* toward *total knowledge*, that Wilson's own quasi-Hegelianism emerges. In his case, the unfolding of evolutionary (natural) "history" is brought to epistemic and social completion through the progressive (and hence teleological) advancement of his positivistic scientific materialism. For the "unity of knowledge" that Wilson later refers to as "consilience," a term he takes from nineteenth-century-logician William Whewall, is understood in a particularly positivistic fashion.[16] Wilson *only* recognizes the epistemic merit of knowledge produced by disciplines that model themselves methodologically on the predictive, law-producing, and quantifiable forms he considers to characterize the natural sciences. This is the "objective" territory on which the unification and completion of knowledge is to be founded.

According to Wilson, "Scientific materialism embodied in biology"[17] will eventually inform and transform every area of the social sciences. Indeed, this predictive materialism will even delineate the future forms society as a whole can take, by providing an "increasingly precise specification of history."[18] Previous attempts "to devise laws of history that can foretell something of the future of mankind" "came to little because their understanding of human nature had no scientific basis."[19] But, as "the social sciences mature into predictive disciplines, the permissible [future] trajectories will not only diminish in number but our descendents will be able to sight further along them."[20] The more complete our evolutionary knowledge, the more ethics, politics, and society will be seen to conform to this knowledge and the more predictable our future social trajectory will be.

It seems, then, to paraphrase Hyams's fictional character Evans, that it is Wilson's "scientific materialism" (let us call it sci-mat) and not a Marxist dialectical materialism (diamat), that will bring about the perfect (quasi-Hegelian) Synthesis. And, as with Morrow's Law, our freedoms must apparently change and narrow as some trajectories (and forms of ethics, politics, and society, such as anarchism) are deemed scientifically *impermissible*. That said, things are more complicated (or perhaps less coherent) than they originally appear, for, at the end of *On Human Nature* Wilson raises a "perhaps final *spiritual* dilemma."[21] For this same knowledge will, he argues, open the possibility of genetically engineering humanity itself. The "human species" could even "choose . . . to imitate the more nearly perfect nuclear family of the white-handed gibbon or the harmonious sisterhood of the honeybees."[22]

Now offering this "choice" of another (genetically engineered) route to Morrow's Hive surely undermines Wilson's argument that our ethics, politics, and society must follow where our genetic constitution leads. For the complete knowledge provided by scientific materialism will, in his terms, enable us to genetically fit ourselves to any kind of politics we might want.[23] Even anarchism might become biologically possible!—at least if anarchists were not likely to be ethically and politically resistant to engineering a solution to our "spiritual" dilemmas so final that even Morrow's system of educational indoctrination would be surplus to requirements.

Contradictions aside, Wilson's reading of "Natural History" clearly takes the form of a secular, science-based eschatology, the *potential* implications of which are hardly less extreme than Morrow's science-fiction schemes. Of course, Wilson, like Morrow, expects resistance to his theories based on what he would consider archaic pre-scientific and/or anti-scientific understandings of ethics, politics, and society. But those sociological and philosophical theorists skeptical of sci-mat's abilities to attain a transcendent truth that will guide humanity to its glorious future are dismissed in *Consilience* as "relativists" who typify "the growing dissolution and increasing irrelevance of the intelligentsia" (46) and (in another surprising turn given his previous comments) they are also said to constitute "a rebel crew milling beneath the black flag of anarchy" (42).

Of course, Wilson's understanding of anarchy is so elastic that this "rebel crew" includes anyone who dares challenge science's overarching authority. For, despite claiming to have read authors like Derrida "with some care" (43), Wilson then lumps him together with all forms of "[u]sually leftist" "postmodern" thought, such as "Afrocentrism, constructivist social anthropology, "critical" (i.e., socialist) science, deep ecology, ecofeminism, Lacanian psychoanalysis, Latourian sociology of science and neo-Marxism" (45), all of which, to be sure, might challenge his positivistic model of the "unity of knowledge," but few of which would actually regard themselves as either "postmodern" or "relativists" or, in most cases, "anarchists." Be that as it may, Wilson has already determined that this dissolute rag-bag of "corybantic" Romantics will inevitably be on the losing side of "Natural History" and as such are of little consequence. Indeed, in a further echo of Hegel, their *disruptive* influences merely provide contradictory tensions that serve to make the arguments for scientific materialism and the new world *order* stronger: For "in the Darwinian contest of ideas, order always wins, because—simply—that is the way the real world works" (46).

> The Transition will proceed at an accelerating rate. Man's destiny is to know, if only because societies with knowledge culturally dominate societies that lack it. Luddites and anti-intellectuals do not master the differential equations of thermodynamics or the biochemical cures of illness. They stay in thatched huts and die young. Cultures with unifying goals will learn more rapidly than those that lack them, and an auto-catalytic growth of learning will follow because scientific materialism is the only mythology that can manufacture great goals from the sustained pursuit of pure knowledge.[24]

The religious overtones of this eschatology are unmistakable.[25] Indeed Wilson is quite explicit about such a reading. He may have lost his Christian faith but, as he remarks in *Consilience*, religious feelings were "bred in me . . ." (4) and he means this both culturally and evolutionarily, for "[w]e are obliged by the deepest drives of the human *spirit* to make ourselves more than animated dust, and we must have a *story* to tell about where we came from, and why we are here." Science "is a continuation [of Holy Writ] on new and better tested ground to attain the same end." "When we have unified enough certain knowledge, we will understand who we are and why we are here" (5; emphasis added). Come the end of days, the practitioners of sci-mat will have ushered in a new and more perfect world order.

> Reason will be advanced to new levels, and emotions played in potentially infinite patterns. The true will be sorted from the

false, and we will understand one another very well, the more quickly because we are all the same species and possess biologically similar brains. (46)

From "Natural History" to Hermeneutics

Morrow and Wilson proffer ethically and politically dubious eschatologies that mix together human history and natural evolution in ways that would be very difficult, *indeed impossible*, to completely untangle. They each present a "*story* . . . about where we came from, and why we are here." These take the form of (quasi-Hegelian) grand narratives where our "natural" past continues to inform and place ethical, political, and social limits on our present and future in ways that are only now beginning to come to our conscious attention (as, for example, through Morrow's Law or Wilson's sociobiology), that is, to become part of "mankind's" self-knowledge. This progressive "self-consciousness" of the sociobiological whole (which seems to parallel certain claims about the emergent intelligence of ant societies) could, ironically, be considered the natural historical equivalent of Dilthey's "historical consciousness" (*geschichtliches Bewusstsein*). Indeed, such a comparison opens up interesting critical hermeneutic perspectives on Morrow and Wilson's claims.

For Dilthey, "man is a historical being"[26] and "historical consciousness" is not just the reflexive awareness of this *historicity* (of human existence necessarily being historically constituted and situated) but proceeds inexorably toward "objectifying the full extent of its [own] historical determination."[27] In other words, our awareness of our inescapable debt to history is precisely what enables us to intellectually grasp the (hermeneutic) limitations on our understanding that this indebtedness imposes and the possibilities it offers, thereby attaining higher, more objective and complete, forms of "self-consciousness."

In Morrow and Wilson's case, humans are natural historical (rather than definitively historical) beings that have reached the stage of being reflexively aware of their evolutionary situation and now proceed inexorably to objectify the ethical, political, and social limits and trajectories this determines. This objectifying "natural historical self-consciousness" itself, in its turn, becomes an integral, vital, and irresistible force in the teleological movement of humanity toward a future where nature, history, and knowledge are completed in each other and where ethics, politics, and society take a final form in accordance with this completion. Like Dilthey, Morrow and

Wilson presume a notion of epistemic progress that carries the whole of humanity along with it. It is the "human species" that attains this progressively "higher" and more complete "self"-consciousness, the "human species" that is the "subject" and "object" of this knowledge, and in their case it is the "human species" that will supposedly make these choices about engineering its future. *The human species, one might say, is already being treated as if it were a superorganism that is, amongst other things, a brain.* After all, from Wilson's perspective, the diverse singularity of individuals and societies, together with their radically different "self-understandings," just mask the underlying (genetic) unity (our human nature) to be carried forward, no matter what, by the tireless scientific vanguard who exemplify the "human spirit" to know.

Now the obvious irony in any such comparison is that Dilthey's entire philosophical project was an attempt to counter exactly the kind of all-inclusive methodological positivism that Wilson's notion of consilience exemplifies. Dilthey's response to Comte's positivism was to argue for two parallel but distinctly different methodologies, one positivistic and descriptive for the natural sciences, and the other interpretative (hermeneutic) for the social/historical sciences. Even today, it is easy to understand why many in the humanities and social sciences might be tempted to take this "bipartisan" line when faced with the colonizing intentions of a discipline where even the language of the "human spirit" is appropriated to serve sociobiological purposes.

But this "bipartisan" line of critique needs to be resisted precisely *because* it depends upon trying to disentangle nature from culture and the natural sciences (on which Morrow and Wilson both claim to base their interpretations and predictions) from those historical and social sciences which Dilthey referred to (in the course of his own attempt to do just this) as the *Geisteswissenschaften* (hereafter the "human sciences").[28] For, while Dilthey's humanism would support the kind of argument that suggests that the natural sciences have no place treading on the humanities' territory, such a bipartisan approach risks artificially (and impossibly) separating knowledge concerning human existence from knowledge of nature.

In fact, the human "sciences" are often at their most enlightening when they reveal just how the stories we tell, like the narrative structures of Morrow and Wilson's works, depend upon more (or less) skillful and more (or less) conscious admixtures of nature and history. What is more, many of the most compelling arguments that counter the positivistic faith in natural scientific methodology exemplified by Wilson[29] also depend crucially on highlighting the messy reality (and diversity) of actual scientific practices and their inescapable *historicity*, that is to say the multifarious ways that natural science, too, is embedded within particular times, places, and cultures. Here one might think, for example, of the pivotal role of Malthus's sociopolitical works in the development of Darwin's theory of natural selection.

Wilson's positivism regards these historical and interpretative complexities as inconsequential. However, historically and socio-logically, much could be written about his own deployment of cybernetic metaphors and models in an age dominated by "informa-tion technology" and much remains to be said about the historicity of biologists' ideas concerning the ethical, political, and social rele-vance of ant "societies." For example, it is surely interesting that one of Wilson's myrmecological predecessors, Caryl P. Haskins, sets his discussion *Of Ants and Men* (1939)[30] squarely in the political context of his own time when asking whether ant colonies exemplify "Fascism or Communism?" And Lustig provides an excellent account of the interplay of social/political mores in the work of late nineteenth- and early twentieth-century myrmecologists Auguste Forel, Erich Wasmann, and William Morton Wheeler.[31]

Such works, which tell singular natural histories that disrupt the metanarrative of a unitary authoritative natural history, are vital because, although even the most ardent sociobiologist recognizes that understandings of "social" insects have been historically and culturally variable, their lack of reflexivity about their own historicity, together with their positivistic faith in progress toward total knowl-edge, soon reasserts itself. For example, Matt Ridley's bestselling sociobiological account of *The Origins of Virtue* notes that Shakespeare used the beehive as a metaphor for "hierarchical Elizabethan society, writ small."[32] But, Ridley argues, a "beehive is a collaborative project on far more levels than first appears,"[33] and so "is not, as Shakespeare thought, a despotism, run from above. It is a *democracy*, in which the *individual wishes* of the many prevail over the *egoism* of each"![34] A sociologically oriented skeptic might suspect that the expanding influence of Chinese-style capitalism might soon elicit myrmecological comparisons that will claim Ridley's conclu-sions about "democracy" are as dated as Shakespeare's.

The human sciences cannot, then, afford to isolate themselves territorially or methodologically. Especially when, as is the case

with Dilthey's (and many other forms of) humanism, this separation
is achieved precisely by deploying (whether explicitly or implicitly)
various idea(l)s of the "human spirit" that, although supposedly
untainted by, or entirely transcending, mere biology, might *also* be
said to treat humanity as a kind of idealist (super)organism, as a
collective (and self-conscious) being analogous to single (and self-
conscious) individual being.[35] This kind of humanism would leave
the ethical, political, and social dangers engendered by Morrow and
Wilson's quasi-Hegelian eschatology entirely unaddressed. For what
is at stake here is *not* the difference between the "nature of ants"
and the "history of humanity." Rather, the issue is the difference
between certain *universalizing* and deterministic readings of
"Natural History" (together with their potentially dire ethical,
political, and ecological consequences) and what Gadamer refers to
as the *universality of hermeneutics*, which I would suggest is the
natural/historical (lowercase) inescapability of interpretation.

In the course of his critique of Dilthey's residual Hegelianism,
Gadamer gives these universalizing formulations a name: "historicism." And Gadamer's critique, which as Grondin notes lies at the
very heart of his hermeneutics and his magnum opus *Truth and
Method*,[36] also applies to Morrow and Wilson's views insofar as they
are indeed the Natural Historical counterparts of Dilthey's "historical
consciousness." Moreover, despite his own residual humanism
(see in the following), Gadamer's criticisms further elucidate the
problems inherent in bipartisan strategies while his anti-historicist
hermeneutics opens the possibility of an alternative *interpretative*
"materialism" that is explicitly not formulated as a meta-science, a
hermeneutic equivalent of sci-mat or diamat, or any synthetic
method reaching a universal truth. Rather, this interpretative materialism (in the form of a hermeneutic ontology) pays (ethical)
attention to diverse understandings and recognizes the necessary
incompleteness of all interpretations that would include myriad
possible "natural histories."

Myrmecological Historicism: Reading the
Book of Natural History

In the foreword to the second edition of *Truth and Method*, Gadamer
remarks that Dilthey "was seduced by its hermeneutic starting-
point into reading history as a book: as one, moreover, intelligible
down to the smallest letter."[37] In doing so, he was merely applying

the "old hermeneutical principle of textual interpretation . . . to the coherence of life insofar as life presupposes a unity of meaning that is expressed in all its parts" (224).

> Dilthey thought he was legitimating the human sciences epistemologically by conceiving the historical world as a text to be deciphered. . . . Everything in history is intelligible for everything is text. [In Dilthey's words . . .] "Life and history make sense like the letters of a word." (240–241)

Dilthey, of course, was hardly alone in this approach. Gadamer also notes historical precedents and similarities between the "self-understandings" of early humanistic sciences like philology and the early natural sciences who also thought "the investigation of nature meant deciphering the 'book of nature'" (182).[38] Indeed, the idea of nature as a book has a long history[39] that is, perhaps, only now beginning to be replaced with cybernetic ideas of nature as "informational" systems. It is easy to understand why this might be so, for having been long accustomed to writing histories of past events and placing them within the covers of actual history books, we are almost compelled to think of the past itself as something available to be read as a coherent bound(ed) whole united by a single narrative thread.

Gadamer recognizes the historical entanglement of hermeneutics with this metaphor but wishes to distance himself from both "Dilthey's entanglement in the aporias of historicism" (218) and what he considers his "speculative idealism" (227). Dilthey's historicism is obviously motivated by his desire to claim that the human sciences, too, proceed toward a truth that is just as complete within its own pre-given boundaries as the natural sciences are in theirs. But since Dilthey makes "historical consciousness" (humanity's historical self-awareness) itself the ground of the human sciences, he also effectively separates this "consciousness" from the actual materiality of history. Dilthey's approach, Gadamer says, harbors the "danger that the actual reality of the event, especially its absurdity and contingency, will be weakened and misperceived by being seen in terms of the experience of meaning" (xxxv), regarding everything as "only language and language event" (xxxiv).

It is worth following Gadamer's argument here in more detail. Historicism and speculative idealism might be considered twin consequences of uncritically reading history (and, presumably, natural history) as a book. The temptation historicism falls prey to is to provide a narrative structure for history as a whole, one that allocates every event a specific role and meaning in the light of an overarching story (e.g., universal progress or decay) that is said to underpin, conjoin, and when grasped, explain, such events. Such universalism is intellectually seductive precisely because it claims to provide an interpretative key with which the secret of the past's relation to present and future events can be laid bare. It takes various teleological forms, such as the inevitable working through of God's plans, the dialectical self-unfolding of the world-spirit (Hegel's *Geist*), the "manifest destiny" of the United States, the inevitable triumph of communism, and so on.

Morrow's and Wilson's readings of natural history clearly succumb to this temptation and Gadamer points out that, although highly critical of Hegel's philosophical claim to have uncovered "reason" in history, Dilthey, too, falls back into a quasi-Hegelian form of historicism. As already mentioned, this is largely due to his attempt to develop a hermeneutic methodology that could serve to distinguish the "human," "social," or "moral" sciences (*Geisteswissenschaften*) from the "natural" sciences. Indeed, Gadamer argues that Dilthey, like Hegel (and ironically like Wilson), actually invokes "a form of spirit" in reference to that "historical consciousness" that is both the concern and *telos* of his humanistic hermeneutics. Indeed, this "historical consciousness," which the social sciences constantly strive to enlarge and improve, "sees all the phenomena of the human, historical world only as objects by means of which the spirit [historical consciousness] knows itself more fully. . . . Hence for historical consciousness the whole of tradition becomes the self-encounter of the human mind" (229).

This, Gadamer claims, explains why Dilthey also falls prey to the second temptation, that of a form of speculative idealism which thinks that only what has, or can be accorded, conscious linguistic meaning *matters*. This seems an absurd position when stated so baldly. Clearly all kinds of events of which we are entirely unaware have occurred in the past and continue to occur and to affect what we often term "the course of history." But, Gadamer claims, Dilthey has (idealistically) defined history as those elements of the past about which we have some historical consciousness and are therefore able to speculate. He has done this because, *beginning from the perspective of human historical consciousness*, if we have no awareness of a past event then we are in no position to interpret or understand it.

Any such event is, quite literally, meaningless (it lacks any meaning for us and, perhaps, for the event's contemporaries, too) and as such cannot possibly affect our *reading* of history. But this, Gadamer argues, still leaves an immense gap between history visualized as a book which we read and the actual *reality of events* and *materiality* of the past.

Despite his overt materialism, there is also a parallel here with Wilson's grandiose claim that science places *all* of nature at the service of "human consciousness," that a theory, like sociobiology, can, through consilience (and given another few years), capture and (re)present the *whole Truth and nothing but the Truth about Humanity and Nature for mankind as a whole.*

The parallel extends further given that Dilthey's response to the charge of idealism would presumably take a similar, historicist form to Wilson's defense of consilience. For Dilthey certainly recognized that the "book of history" is currently incomplete, that events whose significance previously escaped our attention continue to come to light. But if, as historicism might suggest, our knowledge of the past is *progressing* (our historical self-understanding is improving), then this incompleteness is not so much a sign of idealism (of ignoring the materiality of the past) as it is a realistic recognition of a *current* lack concerning our knowledge of that past. Indeed, it is this "epistemo-logical" gap that motivates the human sciences to further develop that historical consciousness by which more and more such events are recognized and woven into the ever more complete fabric of the story presented to us. So, insofar as he is a Hegelian despite himself, Dilthey's historicism suggests that we are indeed gradually working toward comprehending the book of history as a whole, that is, univer-sally, or in his terms, too, "objectively." For all their methodological differences, the parallels between Dilthey's hermeneutic historicism and Wilson's positivistic historicism are striking.

We can see how closely speculative idealism and historicism are intertwined here, and not just for Dilthey but for Morrow and Wilson, too. Despite his scientific materialism, Wilson's teleological account of history (quite at odds with the nonteleological unfolding of an evolutionary dynamic that is *never* complete and that would be described by many evolutionary biologists as fundamentally directionless)[40] is just as indebted to a quasi-Hegelian form of specula-tive idealism as Dilthey, involving as it does the necessary completion of humanity's "historical consciousness" over time. Of course, Wilson, who is so dismissive of his historicist antecedents, would like to regard this progressive species-wide human consciousness as being fundamentally determined by nature and eventually full of time-less truths, but such an idea owes little to any scientific understanding of nature and is virtually inconceivable outside of its particular (post-Hegelian) historical setting. It finds its contemporary ana-logues in Marxist millenarianism or Fukuyama's explicitly Hegelian celebration of the ideological triumph of neoliberal capitalism.[41]

Wilson's supposedly pure scientific ideas thus draw, knowingly and unknowingly, on all kinds of sociohistorical sources; they are affected (and effected, that is, brought about) in all kinds of ways by their historicity. This historicity undermines Wilson's argument that scientific materialism deals in truths that are *not* beholden to any particular historical circumstance, for even to say this presumes an ability to excise, or at least recognize and account for, influences on sci-mat that *are* culturally and historically partial. Ironically, he cannot *know* what *all* those "prejudicial" influences are, or how to take them into account unless he is willing to accept something very like Dilthey's (historicist and idealist) theory of historical hermeneutics. In short, you cannot have a natural history that claims to predict the direction of history or the permissible forms of ethics, politics, and society without actually taking into account history itself and historicity—and this leads us right back into the problems that Dilthey faced. For if Gadamer is right, history can never be completely taken into account, both because of the histo-ricity of those who interpret it, who tell its story from perspectives that are affected/effected by particular times and cultures *and* because of the *materiality* of history itself.

Gadamer's understanding of hermeneutics is quite different from Dilthey's in several ways, all of which touch on this notion of interpretative completeness. From Gadamer's perspective, the "universality" of hermeneutics can never be that of reaching or even approximating completeness or objectivity, precisely because this would suggest the attainment of an ahistorical position, defined as an understanding that stands outside of all social and historical situations. This unreachable, ahistorical position escapes the hermeneutic context, the historicity, that makes understanding possible—for understanding always begins from within historically informed and influenced interpretative horizons. In other words, where Dilthey emphasizes the (methodological) ability to account for, and thereby counter, the effects of historicity on our understanding as part of the process of attaining (or at least approaching) a universally valid understanding, Gadamer insists upon the universality of that historicity, of our always interpreting an event or a text from a particular situation that is, at least in part, historically constituted. This is what understanding is. Interpretative completeness could only occur if, in some self-contradictory way, humanity was not actually, or ceased to be, the "historical being" which Dilthey himself defined it as; that is, if in some Hegelian fashion there was an end to history where everything has, at last, been accounted for and evaluated without residue.

This incompleteness is not just a matter of the inescapability of differing interpretative contexts, of historicity, but also of the impossibility of actually accounting for, and therefore countering, everything about how that historicity affects one's interpretation. "Fundamentally, the experience of historical tradition reaches far beyond those aspects of it that can be objectively investigated."[42] In other words, history often effects (brings about) our consciousness in ways that we are not, and cannot, be entirely conscious of. Historically *effected* consciousness (*wirkungsgeschichtliches Bewusstsein*—the consciousness brought about by and through its temporal precedents) is not the same as being conscious of history's effects, for self-reflection can only reach so far. Such influences are themselves matters of historical interpretation with all the limits that implies (not to mention being subject to chance, speculation, unknown chains of causation, historical erasure, and so on), and these influences are certainly not all in the form of the transmission of *ideas*. To say that history works in mysterious and frequently unpredictable ways is not anti-scientific mysticism; it is, to paraphrase Wilson, "the way the real world works."

Gadamer's point is not simply that self-consciousness is necessarily *historically* incomplete in the sense that we cannot know everything about the past and its effects upon us, but that this incompleteness is a consequence of the *materiality* of history. Dilthey offers a "historical consciousness" that (through hermeneutic reflection) becomes thoroughly "self"-aware. In contrast, Gadamer introduces the notion of "historically effected consciousness" to retain the sense of consciousness effected by (material) history *and* consciousness of being so effected while not treating the two as equivalent or the former as reducible to the latter. In other words, as humans are temporal and historically situated beings this "historicity itself is never resolved in self-knowledge"[43] either at the level of humans as individuals or, more importantly for current purposes envisaged as a species (or a superorganism), at a particular stage of historical "development."

Indeed, to even take complete self-consciousness as an epistemic ideal, as an end, is already to uncritically adopt a historically particular idea of knowledge and truth—and one, as Gadamer points out, that is indebted not only to Hegel but to Descartes. "For Dilthey, a child of the Enlightenment, the Cartesian way of proceeding via doubt to the certain is immediately self-evident."[44] It is a method Dilthey's speculative idealism uncritically assumes and shares with Wilson's positivism. But Gadamer's point is not just that this notion of truth only makes sense when seen from *within* Enlightenment *tradition* but that this Cartesianism treats both history and language itself as *immaterial*, as being entirely transparent to, and in the service of, self-consciousness. It is certainly not just "postmodern" leftists who might have problems with this idea. So Dilthey's notion of "historical consciousness" is not just problematic because it ideal-izes a kind of superorganismic "self"-consciousness where the whole of humanity can objectify the whole of its determining history. It is not just that the whole of humanity is not a self-consciousness in any sense analogous to individual self-consciousness (although Dilthey, like Wilson, and all forms of humanism certainly elides the gap between these levels) but because understanding itself is not to be thought of primarily as a form of self-consciousness (even at the level of the individual). Understanding, Gadamer argues, is an *event*, something that happens to a historically effected consciousness in its encounter with others.

Hermeneutic Ontology

This is not the place for a detailed engagement with Gadamer's hermeneutics, but some attention needs to be given to the nature of the interpretative event in relation to his critique of idealism and historicism, even though Wilson would dismiss such philosophical questions out of hand as, for example, when he states:

> To solve these disturbing problems [the dilemmas Foucault raises about modern identity without faith in God or trans-cendent Reason, the way power subtly corrupts morality, questions of how one is to live, what values one can hold fast to, and so on] let us begin by simply walking away from Foucault and existentialist [sic] despair. Consider this rule of

thumb: To the extent that philosophies confuse and close doors to further inquiry, they are likely to be wrong.[45]

In other words, to the extent that philosophies raise doubts about positivism, challenge the natural sciences' ability to provide the whole story, suggest there might be ethical and political (or for that matter ecological) concerns about the potential outcomes of scientific "progress," or try to address difficult questions about values in an unfamiliar language, Wilson says we should ignore them. One suspects that if Wilson heard about Gadamer, he would simply add him to his long and inaccurate list of anarchistic relativists.

But Gadamer's hermeneutics are not relativist, or subjectivist. Rather, they focus on interpretative incompleteness as the real (material) condition of existing as finite historical beings. To be sure, and despite its title, Gadamer's (1998) key text is critical of any and all claims to attaining a single *truth* or *method* and he is certainly no positivist. But one can be anti-positivist without arguing that all claims to tell the truth are "equally valid"[46] (whatever that might mean) or being anti-scientific.

Gadamer's work concerns understanding itself, that is, what it means to understand something, whether it be a work of history, another person, a work of art, or even, perhaps, a "natural" event. And while Gadamer claims that understanding as such is an experience that is universal (something "we" all experience), he does not think that there is only one way of reaching (or justifying) understanding, or only one kind of understanding. Science is, at best, just one mode of understanding among many and one not fully explicable even in its own terms.

> The phenomenon of understanding not only pervades all human relations to the world. It also has an independent validity within science and it resists any attempt to reinterpret it in terms of scientific method.[47]

Scientific *understanding*, too, is an event—*Eureka!*

The view that only scientific understanding matters arises, Gadamer argues, because "science attempts to become certain about entities by methodically organizing its knowledge of the world. Consequently it condemns as heresy all knowledge that does not allow of this kind of certainty and that therefore cannot serve the growing domination of being" (476) (the kind of domination that

Morrow's Hive and genetically engineering humanity would exemplify). The various modes of understanding relating to, for example, textual interpretation, art, and (natural) science, are not reducible to, or substitutable for, each other (understanding a painting aesthetically is not the same as understanding the chemical composition of the paint, or the genetic constitution of the artist). "Through a work of art a truth is experienced that we cannot attain in any other way" (xxiii). And while Wilson might dismiss this possibility, claiming that *truth* is something attained and verified through, and only through, utilizing a scientific methodology, Gadamer sees the experience of art, philosophy, and so on (and, I would argue, certain other nonscientific ways of experiencing nature) as the "most insistent admonition to scientific consciousness to acknowledge its own limits" (xxiii).

These various modes of understanding are available to us because of the kind of beings we (and by "we" Gadamer means those human beings capable of understanding) are *and* because of our histories. They are not just consequences of employing different methods or, perhaps, of employing any kind of method at all. Rather, they are ways of relating to, and being (existing) in, the world. One might say that these understandings are natural *and* historical possibilities but that Gadamer is not in any sense trying to present a natural history. (Indeed *Truth and Method* is, in large part, written as a critical interpretation of the history of how "understanding" has been wrongly, *but given its post-Hegelian historical context, understandably*, understood philosophically, as reading the Book of History.)

Different modes of understanding are possible because we are, in Dilthey's terms, "historical beings" *through* and *through*. History is not (as Wilson treats it) something analytically detachable from our natural existence. We cannot completely erase history, even via a "thought experiment" that would then reconfigure and predict its course on the basis of knowledge concerning an abstract "prehistoric" "natural" human being (whether this is combined with other quantifiable data). No such being exists or has ever existed, since what we mean by a "human" being is always already social and historical. Nor have we any way (scientific or otherwise) of completely untangling what our modes of thinking (our understandings) owe to nature and what to history. No bipartisan strategy will work. As we have seen, the "we" who engage in this thought experiment are unable to excise the history that intrudes even when attempting to think of "nature" ahistorically, since that thinking is always a "historically effected consciousness."

History is also part of who we are at a very fundamental level. Our self-consciousnesses are composed through the recognition (even if we subsequently deny it) of our own historical finitude as beings. That is to say, following Heidegger, Gadamer recognizes that self-consciousness, understanding one's self *as a self*, is founded in the recognition of that self's *historical* limits, its prior nonexistence and its inevitable mortality. *I was born*—there was a time before *I* existed—*and I will die*—there will be a time after which *I* cease to exist. (And while no one requires evolutionary theory to recognize their mortality, perhaps Wilson's view that individuals are merely loci for reproducing genes also expresses something of this finitude.[48]) This *self*-consciousness, this radical recognition of our own historicity, our inescapable temporal and existential finitude as an individual human being, is the most fundamental form of self-understanding. This is why Heidegger and Gadamer argue that the way that the humans understand and situate themselves in relation to their experiential and existential world, their "being-there" (*Dasein*), is fundamentally a matter for hermeneutic ontology. In this sense then, "[t]raditional hermeneutics [including Dilthey's] has inappropriately narrowed the horizon to which understanding belongs" (260–261), narrowed it to interpreting texts, and to the human sciences' ability to read and interpret history as a text. But "[u]nderstanding is the original characteristic of the being of human life itself" (259).

Hence the importance of understanding "understanding" itself and the universality of hermeneutics in the sense of Heidegger and Gadamer's hermeneutic ontology rather than universality in the Diltheyan sense of attaining interpretative completion and objectivity. Hermeneutics is universal (it is fundamental to human existence) in the sense that understanding as such is (for Heidegger and Gadamer) a specifically human possibility and the ground of individual self-consciousness.

This hermeneutic ontology, which takes as its starting point the inescapable historicity of the self and its self-understandings, cannot be collapsed back into a Cartesian model of a self-consciousness that is fully transparent to itself. And this is important both from the perspective of the limits already discussed in relation to "historically effected consciousness" and the closely related limits (and possibilities) that arise from these self-understandings being "housed" within languages which also have social, historical, and material existences that transcend the consciousness of the finite individual. For Gadamer, like Heidegger, argues that we cannot attain such self-understanding outside of language. Language is the "horizon of hermeneutic ontology."[49] In Heidegger's terms, "Language is the house of being."[50] "We human beings, in order to be who we are, remain within the essence of language to which we have been granted entry. We can therefore never step outside of it in order to look it over circumspectly from some alternate position."[51] The limits and possibilities of language are never entirely transparent to those whose thinking, conversation, and (self)understanding occur within, but can never entirely "master" it (which, of course, is the starting point of Derrida's philosophy although this seems to have entirely escaped Wilson's "careful" reading). So, as *Truth and Method* states, rather than emphasizing

> a false methodologism that infects the concept of objectivity in the human sciences [or . . .] the idealistic spiritualism of a Hegelian metaphysics of infinity . . . [i]f we start from the fact that understanding is verbal, we are emphasising, on the contrary, the finitude of the verbal event in which understanding is always in a process of being concretized. The language that things have—whatever kind of things they may be—is not the *logos ousias*, and it is not fulfilled in the self contemplation of an infinite intellect; it is the language that our finite, historical nature apprehends when we learn to speak. (476)

This again is why Gadamer's hermeneutics might be considered "materialist." The very possibility and scope of understanding and interpretation are *materially* effected because language carries within it a history of (only partially explicable) relations to the world that are never just matters of the transmission of self-evident ideas. These relations are not "subjective" for "understanding is never a subjective relation to a given 'object' but to the history of its effect; in other words understanding belongs to the being of that which is understood."[52]

This, unfortunately, is where Gadamer's own humanism comes into play in a way that threatens to undermine at least some of the ways in which he wants to distinguish himself from Dilthey's historicism and idealism. For although his conception of language is not subjectivist and is *historically* materialist, he nonetheless does privilege human language to the extent that only what appears in it can be understood. Hence his famous insistence that "*Being that can be understood is language*" (474)[53] also comes, as Moran notes, "dangerously close to a kind of linguistic idealism"[54] insofar as it can underemphasize the way that the materiality of nature and natural histories exceeds language. That is, one might be tempted (although as his critique of Dilthey shows, this is *not* Gadamer's position) to think that things *only* exist insofar as they leave their traces in human language and history, that understanding is all about language working on and through itself. The ecologically minded might want to ask, for example, to what extent the understanding we have of ants *belongs* to the ants and not just to the history of human appropriations of ants. (At least the natural sciences *attempt* to say something about ants as beings whose existence *exceeds* their effects in human language and history.) The problem is that science regards this excess only as a matter of the entities "objectivity."

We should also note that Gadamer certainly limits understanding (and language) to humans and only humans in a way that comes close to reconstituting a metaphysical form of humanism, that is, definitively separating humans from all other beings on the universal ground of their possessing language and understanding. Those with an interest in natural histories might actually be much more willing to recognize that "understanding" is ontologically prior to language and self-consciousness, a capacity (although certainly with many varied forms) that is fundamentally important to the existence of many beings that are in no way human; they might also accept that other species, including ants and bees, have various languages.[55] Gadamer's absolute separation thus has some of the same potential effects as Dilthey's bipartisan approach and would also seem to treat humanity (or at least those who have reached self-consciousness through human language) as if it were a kind of "(super)organism" (both superior to all other beings and potentially united as a communicating functional whole).[56]

Hermeneutics, Ethics, and Ants

Gadamer's hermeneutic ontology reveals interpretative incompleteness and finitude as the real (material) condition of human understanding. Understanding (although not subjective) is a fundamental event that enables the self to find its way about in the world. Even the natural sciences depend upon a mode of understanding that, while different from others, still has the form of a kind of practical ability to know one's way around, recognizing, being able to see connections between, and situating oneself in relation to that which is understood; "all such knowledge is ultimately self-understanding

(*Sichverstehen*: knowing one's way around)." "Thus it is true in every case that the person who understands, understands himself (*sich versteht*) projecting himself [*sic*] on his possibilities" (260), but this understanding is *never* fully explicable even to one's self.

Language is, Gadamer argues, what opens this projective possibility as a possibility, offering a kind of freedom (although not absolute freedom) to resituate one's self in relation to others and the world. Language is the medium of an understanding which he memorably describes as the "fusion of horizons" (302–307). For every event of understanding is an experience that challenges and reconstitutes our self-understanding, and alters this limited/limiting horizon; it gathers that which is understood (about another person, a text, etc.) into that self-understanding, but not by making it conform to the contours of that self's preconstituted consciousness.

Now this notion of the fusion of horizons might seem, superficially, to share much with Wilson's notion of consilience, but they are fundamentally different. Wilson's scientific materialism claims to discover and then progressively fit the pre-cut pieces of the world together like a jigsaw puzzle to produce a complete picture where everything slots into its predetermined place. So long as the pieces are correctly positioned, its end result could not be otherwise. This is consilience. But Gadamer envisages the "fusion of horizons" in terms of a dialogue, an incomplete understanding reached through participating in an ongoing historically situated conversation. Here conversational themes arise, are transformed, and transform; they are lost and sometimes recovered; ebb and flow according to specific conjunctions, histories, and ethos; are reconfigured, engaged with tangentially or become, at least for a while, central; they are diluted or concentrated, narrowed or enlarged in their encounters with radically different perspectives. Of course, the direction a conversation takes may become entrenched for a while but can also be radically changed by unexpected or particularly pithy interjections. A conversation is neither ordered nor chaotic; it follows no set rules (not even that of equal time or equal weight being given to all participants!), and there is no method of reaching incontrovertible conclusions. It may (or may not) result in mutual understanding (a fusion of horizons) concerning certain things, but such an understanding is not a compromise, a halfway house, but an appreciation of where other conversational partners are "coming from" and where differences lie. Meaningful conversation is actually quite anarchic and not at all like a board meeting or a formal debate.

In order to flourish, a conversation requires openness to other's opinions, an ability to listen as well as speak because (not despite) of the fact that differences will sometimes be irreconcilable. Understanding, one might say, is inherently a *matter* of ethics, of being open to differences not stamping one's authority. Differences cannot be resolved by forcefully claiming to represent the one true way, since this obviously kills any possibility of a conversation at all. There are no prescribed results to be achieved, not even a consensus, for the conversation is of value in itself, it is enlightening—understanding is an event that opens new (never-before envisaged) possibilities. It is *creative*, bringing about new situations, new relations. Such conversations, in their creativity and their necessary exposure of one's own self-understandings before the wider world, might also be regarded as inherently political, at least in an Arendtian sense, where a political act is an expression of human individuality and freedom, a beginning where "something new is started which cannot be expected from whatever may have happened before."[57]

In other words, ethics and politics *as such* are key constituents of any effective conversation while specific ethical and political values are often constituents of the kind of "prejudices" (or fore-understandings) that constitute the horizon of understanding from which hermeneutics activities begin. These "prejudices" are not simply biases to (supposedly) be eradicated in the name of objectivity but constitute a key aspect of where we come to an understanding from. This, too, is *Truth and Method*'s claim (265–285). This does not mean that all prejudices are somehow "equal," nor does Gadamer want to demean the ideal of objectivity which, he argues, "distinguishes the human sciences from the natural sciences" (285). However, the ordered objectification of the world is a limited mode of understanding, not its paradigm, and one of its consequences is to claim that science per se is ethics and politics free. This is why Wilson is resistant to the idea that ethics/politics might place any limits whatsoever on the acquisition of scientific knowledge. For a hermeneutic ontology, ethics and politics are not just segments of the jigsaw puzzle to be left in place but integral aspects of any ongoing conversation.

Of course, regarding understanding as a dialogue or conversation clearly reflects Gadamer's own disciplinary history, but this only exemplifies his wider points. In any case, conversations are not limited to (or even the dominant paradigm in) philosophy. Rather this offers a concept of understanding that is not reducible to an overarching method of reaching a singular truth. And the way we understand our relations to the world (e.g., as a "conversation" or an "interrogation," as being party to an ongoing discussion or ordering and circumscribing a collection of objects) *matters*—it has material effects. These effects are ethical, political, social, and ecological.

Wilson's sci-mat, like Morrow's Hive, attempts to subject ethics, politics, society, and ecology to a single scientific world order held in "mankind's" consciousness. From this perspective, as Morrow and Wilson's disdain for nonscientific criticism exemplifies, there is nothing left to discuss—a "conversational" (hermeneutic) understanding is superfluous. All we need do is make managerial (cybernetic) decisions on the basis of natural history's irresistible laws. But histories, which are always natural *and* human, do not operate in this way. They are not predictable because they are not just composed of inert objects but of complex material relations *and* various understandings (including various understanding of history) inextricably intertwined. History is not, as Wilson suggests, a machine whose output can be determined or calculated given sufficient quantitative information. History has a materiality that is also affected and effected by the (limited) self-understandings of historical beings—self-understandings that are many and varied and not reducible to a unitary progressive self-understanding of the whole in anything like the (idealist/historicist) super-organismic sense required by Dilthey, Morrow, and Wilson. There is no natural history, no single story to be read and understood in its totality, it is indeed a myth of scientific materialism. Conversely, natural histories are not just stories (fictions) but partially interpretable compositions of historically effected consciousness.

What is more, recognizing that interpretative incompleteness and finitude are the real (material) condition of understanding is a prerequisite for any effective ecological ethics and politics. For even scientific ecology teaches us that managerial decisions made on the basis of natural science alone have frequently caused appalling social *and* ecological damage precisely because they operated on the mistaken assumption that they were dealing with complete understandings of predictable systems of objects. Yet such is the hold of a Wilsonian view of scientific materialism that such failures are never regarded as methodological failures in any overarching sense. Rather than regard them as revealing the limits of "the scientific method" itself in providing what Wilson claims it provides—the complete and necessary grounds for understanding the world—the blame is laid on a failure to apply the method intensively, extensively, and consistently enough (including, of course, the failure of the human sciences to be sufficiently scientific). But so long as the limits of scientific understanding per se are not recognized, the attempts to treat the world as if it was indeed completely understood by natural history can only lead to ever more complete ethical, political, social, and ecological disasters.

This is not to say that a hermeneutic ontology like Gadamer's can provide the perfect solution to our social or ecological problems. It offers no such assurance and no method for achieving any such end. However, its sensitivity to historicity and to understanding interpretative limits, and in the way that ethical openness to diverse forms of existence is an integral aspect of its practices of enlarging self-understandings, holds open some hope for socialities that are creative, relatively free, and have undetermined futures—not at all like the managerialism of Morrow's Hive. The creative possibilities that emerge in such conversations might even help ensure that there is no *end* to history or to nature, at least for a few billion years. But when Wilson asks, "At what point does a society become so nearly perfect that it is no longer a society?,"[58] the answer in human terms might be "when it comes to envisage itself as a superorganism where individuals are understood only in terms of their cybernetic functions in relation to a sociobiological totality." And ethically and politically this model of social "perfection" also has a history—it is known as totalitarianism—and it is not only anarchists like the fictional Evans who find themselves opposed to such an understanding. And the blame for and solution to our current historical predicament does not lie in something called natural history or in human nature but the current managerial "order of things" (to borrow a natural historical phrase from Foucault), one that claims, and strives, to be free of ethics and politics. It is not, as Morrow claims, "freedom [that] destroys itself by destroying the conditions in which it can exist," but those who want to subject even freedom to natural historical laws.

Postscript

Some may suspect that I have been unfair to Wilson in comparing his scientific work with the megalomaniac ideas of Morrow. After all, Wilson is not a fictional character but a real scientist basing his comments on real science. Nonetheless, as we have seen, Hyams grounds Morrow's ideas squarely in the same kind of scientific reading of ant societies that Wilson's work develops and Wilson's scientific materialism is indebted to far more than "pure" science. The difference between them is not that Wilson merely presents "facts," but that Hyams draws and writes scientific "facts" into an ethical, political, and social context where the reader is encouraged to be *critical* of their supposed implications (i.e., of Morrow's interpretation), whereas Wilson presents them as incontrovertible limits on the extent of ethical, political, and social critique (e.g., when claiming that certain sociopolitical forms like anarchism are merely fictions). This difference is, perhaps, best illustrated in Wilson's most recent work. For, in a bizarre twist that further entangles the relations between natural histories, Wilson, the passionate conservationist, has just produced a novel based on sociobiological and environmental themes.

Unfortunately (if not entirely unexpectedly), according to the jacket blurb of *Anthill: A Novel*,[59] the ultimate lesson Raff, his 'environmentalist' protagonist, learns from both his (yes, you guessed it!) prolonged study of ant colonies and his later political involvements is that "'war is a genetic imperative,' not just for ants but for men [sic] as well." Worried by the spread of suburban sprawl, Raff's moment of enlightenment comes as he watches two squirrels "engaged in a territorial dispute."

> The ownership of the land, and the power and security it provided: that was what drove the battles of squirrels. And the cycles of ant colonies. And that was . . . in a tragic sense . . . what runs the world. (261)

The connections Hyams's fictional character draws between ant and human societies appear in Wilson's novel as sociobiological commonsense. While ant colonies are described as a superorganism with a "collective mind" (246), Harvard (where Wilson works) is not only "the world's greatest university" (273) but also "altogether a human anthill, a kaleidoscope of specialists, whose lives are molded to ensure their own well-being through service to the greater good" (275). (Perhaps Morrow would have been better off investing his billions in Harvard!) Here, as in his more scientific works, Wilson offers a managerial (cybernetic) model that moulds individual actions according to their roles in the greater social organism while arguing this can only happen where more fundamental forms of genetic selfishness (i.e., "selfishness" at the level of genes as well as the individual) are satisfied.

This is, perhaps, why Raff, the purportedly highly individualist environmentalist in Wilson's novel, proceeds, not by opposing the corporation that seeks to bulldoze the ecology of his (and Wilson's) Alabama childhood (together with its warring ant colonies) but by going to work for them. This ensures that when an opportune moment eventually arrives, Raff has achieved the executive position and the conservative credentials necessary to present a plan that will make his employers money *and* save some of his favorite swamp. Luckily, since his employment otherwise concerns only *city* lots (and he has even managed to join the Audubon Society) his "ecological principles" apparently remain entirely uncompromised, while his conservatism is in any case quite genuine. He has, by now, become the inspirational leader of Boy Scout Troop No. 43 and an enthusiastic member of the National Rifle Association—"the only things I know that come close to target practice with a .22 rifle are a deep massage and sex" (325).

Raff's "brilliant" plan is to propose an extremely selective up-market housing development for the ecologically minded super-rich combined with a tax-deductible nature reserve. Presumably, as with ants and squirrels, the territorial instincts of this ecologically "gated community" will kick in whenever it becomes necessary to see off further intruders wishing to buy, sell, or develop the land. (That they may have become super-rich by speculating in land or property seems not to have crossed his mind.) And as to whether Alabama's millionaires are likely to be ecologically minded, Raff does not, as one might expect, have recourse to Wilson's arguments about genetic predispositions toward biophilia but reminds his captive audience that "conservation" and "conservative" share the same etymological route and that "green is the new red, white and blue" (349).

The differences between Wilson and Hyams's novels are not matters of using and misusing science but of their very different ethical and political evaluations. And while Hyams, quite realistically, has Morrow's plans gather an economic momentum that ensures they will proceed despite his being assassinated by one of Evan's accomplices, Wilson ensures a "happy ending" by having Raff achieve his (incredibly parochial) ends *and* survive an assassination attempt by anti-environmentalist religious fundamentalists. Indeed, he is saved by the murderous intervention of his childhood acquaintance, the swamp's resident shotgun-toting sociopath who otherwise makes (a no-doubt sustainable) living selling frogs' legs to local garages. The "frogman" also has, in the shape of Old Ben, a fourteen-foot alligator, an eco-friendly way of disposing of the would-be assassins' bodies.

No doubt Wilson's novel, despite its attention to scientific detail, is not a great work of literature, but then nor would Hyams have claimed this status for his own work. He explicitly regarded his novels as being in the tradition of French satire while Wilson (at least as a novelist) just sets himself up to be satirized. But the message of Wilson's novel seems abundantly clear: Environmentalism does not require any kind of ethical, social, or political change at all, for everything can be accomplished *within* the anthill of corporate capitalism. Unfortunately, this is, indeed, a complete fiction, albeit one that will carry a lot more political weight with neoclassical economists and those wielding power than Hyams's work ever could. Luckily, however, the future still remains an open question.

3

Layering: Body, Building, Biography

Robert Mugerauer

Introduction

Among the most challenging issues facing environmental hermeneutics is how to think about person-world relationships in an
integrated manner—not by way of conceptually separated natural
environments and social spheres—as if there were either some "pure
nature" untouched by our interpretations and actions or any human
life apart from environmental dynamics. Rather, the interactions of
the physiochemical and biological, the individuating and communal
dimensions—at all scales—provide our subject matter. For instance,
it makes little sense to carry on studying "sense of place" and
"identity" as we have been, assuming that these phenomena are
stable and that the terms are independently meaningful.[1] When
environmental philosophy seriously took up a study of place in the
1970s and 1980s (joining geography, architecture, planning, and
ecology) the early work focused on describing specific places through
case studies of buildings, landscapes, and settlements and argued for
environmentally appropriate design to overcome "placelessness."[2]
The phenomenological tendency to analyze the atemporal and apoliti-
cal features of traditional and exotic vernacular places simultaneously
limited treatment of differences and changes. This led to the recognition of the need for a greater emphasis on the historical and social-
political dimensions of person-place relationships.[3] Today, a fuller
environmental hermeneutics is emerging with the appreciation of the
role of language and of the co-constitution of organism-environment
dynamics at every level.[4]

As for the role of language, as I argued elsewhere in a Heideggerian
mode when I initially attempted to outline the dimensions of an

environmental hermeneutics, "language enables the environment to come forward into experience."[5] Related theoretical dimensions and empirical case studies continue to be developed, especially in the area of hermeneutics and narrative identity. For example, within this collection, Brian Treanor examines the role of narrativity in understanding place and the natural sphere, Nathan Bell considers the ecological self as an environmental variation based on Ricoeur's hermeneutical ethics, and Martin Drenthen theoretically and practically elaborates on how nature narratives approach the "landscape as a multilayered text in need of interpretation."[6]

Additionally, now that we increasingly appreciate the way organisms and environments mutually specify one another at multiple scales,[7] we need to more carefully consider the diverse ways physical-biological phenomena generate communities within surround-worlds and how, in turn, the emergent individual and social lifeworlds co-generate the Umwelts within which they belong.[8] Many of the authors in this volume take up this second biological dimension as well as that of language. Using cases from the Netherlands, Drenthen explicitly argues for the integration of a historical-regional sense of place and identity with a biological understanding of processes and ecological diversity-integrity. Forrest Clingerman stresses that we need to move beyond understanding persons as stable or closed identities by attending to our complex biocultural experiences of embodiment in space across the temporal rhythms of days and seasons, not only with immediate sensitivity to phenomena such as temperature but to what is given in the modalities of memory and imagination.[9] Bell also joins this broader conversation with his analysis of self in terms of human-animal capacities, including perceptions and judgments.

Thus, though this chapter may seem somewhat unusual in focus compared to my other sorts of investigations such as "Toward a Phenomenological Hermeneutics of Rivers: Integrating Design, Ecology, and Politics,"[10] in fact it simply joins many others in reflecting on the biological-linguistic basis from which self-organizing and built environments, not to mention our own embodied consciousness, emerge. In other words, rather than only investigating the usual phenomenal realm of people operating within buildings and landscapes, the following tries to locate our open character as persons within the interactive and continuous layers of the dynamic unfolding of our micro- and macro-environments.

Layering

Most of us are already familiar with layering from a variety of perspectives, as we readily can see by considering the two major ways in which layering happens. In the first there is a laying down which we experience in natural phenomena. When we witness volcanoes, such as Mt. Etna or Mt. St. Helens, we are seeing the contemporary continuation of the ancient processes by which material comes from deeper in the planetary core and is laid down on top of what was the existing surface. This may happen directly in the form of thick lava (igneous rock) or a blanket of fine ash that covers large areas and even whole cities.

We can also think of the exposed layers of sedimentary rock that gave geologists the first clues as to the actual age of the earth and the processes through which it was formed. While the exposed stratifications resulted from the grinding of the tectonic plates that produced gigantic areas of buckling and mountain formation (a dynamic to which I will return later), many of the stratifications themselves are the hardening of what originally had been millions and millions of years sedimentation in the world's oceans. Here a first phenomenological characteristic shows itself: Layering is sedimentation in both place and time. The stratifications of the ocean floor become a mountainside, which first occurred by the gradual laying down of place, in place, as place. Sedimentation is a laying down in time, as a record of temporality.

This layering is not unlike what Heidegger points out about truth as *aletheia*: It is a dimension of the play in which all building up or presenting of what is given is simultaneously implicated in the covering up or hiding of something else that was there before (or that might come to be disclosed in the future). The disclosure of geological phenomena such as rock formation—which includes the record of the formation of life itself, preserved in the fossils of microorganisms, plants, and swimming creatures—and the attendant opening up of the meaning of historical physical processes normally happens not through a dramatic event but through incremental erosion and our subsequent investigations. In like spirit, we find that Merleau-Ponty seriously explored the notion that language itself is a layering: "Language is sedimentation, the naturalization of the invisible surplus."[11]

Here we can appreciate part of the force of the idea that nature is a book to be read (which has its own complex history in religious and literary hermeneutics). Certainly we are used to considering levels of meaning in texts, a powerful tradition and method derived from medieval and nineteenth-century biblical exegesis and more recently reinvigorated by Northrop Frye. Or, we can think of Anselm Kiefer's series of works that fuse dimensions of painting, sculpture, books, and nature. His famous sculpture installation, *The High Priestess/Land of Two Rivers*, consists of almost 200 books made of lead, presented in two steel bookcases. The books themselves are composites not only of heavy, gray lead but copper wire, glass, clay, earth, vegetables, burlap, human hair, and original and magazine photographs.[12]

The second major way in which layering happens is through folding, in some sense already considered above in regard to the geological buckling that happens under the gigantic pressure of the colliding tectonic plates, eventually exposing the gradually sedimented ocean floor. In his three lectures delivered at the Collège de France from 1956 to 1960 and published as *Nature*, Merleau-Ponty speaks extensively about the phenomena of nature in terms of layers, leaves, folds, and folios. As Robert Vallier (the translator of the courses) notes, "The French term *feuillet* is here translated" both as "leaf" (of a plant or tree) and as "folio leaf, that is, the sheets of paper that printers use [that are] folded and bundled to make a booklet" of pages. More generally, the latter "bookish" leaf would be an "individual fold of a full uncut folio leaf." Specifically, however, Vallier goes on to say:

> But, Merleau-Ponty seems to take *feuillet* as the full folio sheet, not just one fold of it, in the sense that, in being folded into the space for different distinct texts, it is, in the present context, the image of a kind of endlessly productive doubling and redoubling of the basic "stuff/powers/structuring" of "Nature" into many kinds and orders. What is important in this is that the sign "leaf" carries with it significations like "unfolding," "unfurling," and "doubling/redoubling" all at once; the reader must hear both the organicist signification and the "paginational" signification at once.[13]

For Merleau-Ponty, then, the book of nature refers to more than vegetative leaves or a generally useful metaphor. Most seriously, he repeatedly attempts to think through the ontology of nature: "Nature as a leaf or layer of total Being. . . . The problematic of philosophy. Nature: an ontological leaf—the thin leaf of nature—essence is divided in folds, doubled or even tripled."[14]

> [I]n particular, regarding life, the concern was to study the unfolding of the leaf of Nature–regarding the human, the concern is to take him at his point of emergence in Nature. Just as there is an intertwining [*Ineinander*] of life and physiochemistry, i.e., the realization of life as a fold or singularity of physiochemistry—or structure, so too is the human to be taken as the intertwining [*Ineinander*] of life with animality and Nature.[15]

In this regard, Merleau-Ponty contends that the fuller human fold of the human body is "no longer animality" insofar as it is "another manner of being a body" in dynamic interaction with nature.[16] He goes on to explore this intertwining, that is, "what makes these beings (Nature, humans) be—and be 'in one another'—i.e. makes them be together on the side of what is not nothing," by means of images of "a hollow or fold of Being": "a new dimensionality . . . in the milieu of Being, the hollowing out of a singular point where language appears and develops of itself if nothing is opposed to it."[17]

In this chapter, I do not attempt such an ontological investigation, but in the same spirit concentrate instead on exploring three empirical phenomena—biology, building, and biography—that are themselves further layerings constituting the world. Though this is not the place, it certainly would be worthwhile to consider the entire arc of life to see the complex layering that is autopoietically generated by the interactive dynamic that occurs at all levels between organisms and their environments.[18] Here I focus on the human realm in the context of nature, leaves, and layers, the imaginative symbolization of which covers the span from the archetypal image of the "Green Man" (that appears in thousands of years of drawings, paintings, sculptures, and especially wood and rock carvings) to the high-culture exploration of human-environmental relationships and memory as in Kiefer's *Ways of Worldly Wisdom—Arminius's Battle* (1978–1980) (that combines images of important German figures, the forest, and writing).[19]

Biology and the Body

At the cellular level, we find that layering indeed is a basis of life. Cells themselves exhibit the folding over and enclosure that enable life to occur; or better put, cells substantially are the folding and enclosure that we know as "the living." On a macro-scale, we strikingly see the layering resulting from basic cellular life in stromatolites, whether fossilized as in present-day New Mexico (again, the seabed risen to the earth's surface) or still living in Australia, the Bahamas, and Florida. These laminar structures are the oldest fossils, perhaps up to 3.5 billion years old, "resulting from life processes and the rock-building activity of primitive algal and/or bacterial communities."[20] Stromatolites are collectivities in which "the layering reflects a rhythmic growth of the structure due to seasonal periodicities or internally driven oscillations of deposition": an "algal film or mat composed of monospecific or polyspecific communities of blue-green algae and/or bacteria," where detrital particles [grains] are trapped by the mat, after which algae rise though and form a new mat on top, and finally the algal filaments calcify.[21]

At the micro-level of the living as the "arising and abiding" of cells, we can look more closely at membranes and their increasing complexity. Most simply, membranes are what enact life's possibility and actual beginning. In bacteria or algae, the membrane differentiates two realms, thus enabling distinctive chemical processes to take place "inside" the membrane—or within what we could call the minimal living being. Of course, membranes simultaneously allow the passage necessary for two-way transactions with the "outside" (e.g., as energy resources move in, and waste out; or, in connectivity with other cellular phenomena). Though there are other types of "bounding envelopes" in cells (rigid cell walls, secondary membranes, and capsules), the membrane is the only feature they all have in common: "All contemporary cells are surrounded by a unit (biomolecular) membrane, which appears as a triple-layered structure . . . (made of amphiphilic lipids with attached or inserted proteins)." Indeed, "the universal character of bimolecular leaflet membranes appears to be a property of phosphor/lipid molecules and the way in which they aggregate into sheet-like structures."[22]

Given the wondrous manner in which "protein polymers automatically bend themselves into useful shapes, guided only by their affinity for water [a matter of hydrogen bonding as their hydrogen atoms are attracted to nearby oxygen atoms in water] and the social relationships between their constituent amino acids," a brief look at how this works certainly is worthwhile.[23] To set a more adequate definition or description: Membranes are composed of phosphor and lipid molecules—a combination of a *phosphate* portion on the *outside* that is *polar* and *hydrophilic* and of a *lipid* portion on the *inside* that is *nonpolar* and *hydrophobic*. To remind ourselves of our chemistry and biology (and as a simple outline of what follows), the accompanying table shows the two "dimensions" in relation and difference.

Polar	Nonpolar
Hydrophilic	Hydrophobic
Phosphor molecules	Lipid molecules
Outside of membrane	Inside membrane

As Dee Silverthorn explains,

Life as we know it is established on water-based, or aqueous, solutions. . . . It is useful therefore to classify molecules on the basis of their ability to dissolve in water. . . . Molecules that dissolve readily in water are hydrophilic [hydro-, water + philic, loving]. Most are polar or ionic molecules whose positive and negative regions can form hydrogen bonds with water molecules. Molecules that do not dissolve readily in water are hydrophobic [-phobic, hating]. They tend to be nonpolar molecules that are unable to make hydrogen bonds with water molecules.[24]

Polarity, then, is a fundamental feature, operative at the atomic or electrical level, and underlying what does or does not readily dissolve in water. Molecules whose atoms share electrons unequally develop regions that are charged positively (+) and negatively (–). Specifically, because the nucleus of one atom has a stronger attraction to shared electrons than another one does, there develops a slight difference in charge, for example, as happens especially with nitrogen and oxygen. In the case of water (H_2O), the oxygen attracts an electron and generates a negative pole, while the hydrogen generates a positive pole.[25] Thus, because water is highly polar—certainly a critical physiochemical feature essential to life as we know it—and since other molecules that are also polar (with positive and negative differentiation) can easily form bonds with each other, polarity is the marker indicating whether solubility is likely or even possible.

In contrast, lipids (biomolecules composed of carbon, hydrogen, and oxygen) are nonpolar. That is, carbon's and hydrogen's electrons "are distributed so equally [that] there are no regions of positive or negative charge." Unlike oxygen, they neither attract electrons nor have their electrons attracted to unequally charged neighbors, including H_2O. Because they are not able to form hydrogen bonds, they do not easily dissolve in water. They are hydrophobic.[26] To sum up: "Polar molecules generally dissolve easily [in water]. Nonpolar molecules do not."[27]

At a more phenomenal level, "[t]he lipids (fats and oils) are the most hydrophobic group of biological molecules. They cannot interact with water molecules when placed in an aqueous solution, so they tend to form separate layers, like oil floating on vinegar in a bottle of salad dressing."[28] In order for the oil to combine with water, the hydrophobic must join the hydrophilic molecule. When such complex molecular structures do form, they have unique properties or possibilities, as dramatically is the case with phospholipids (most of which are "diglycerides with a phosphate group attached to the single carbon that lacks a fatty acid"—that is, with a hydrophilic head and a hydrophobic tail). In our case, such molecules obviously are "important components of cell membranes."[29]

We can now better see how the micro-physicochemical processes move further out into the macro-scale. As just discussed, since lipids do not readily dissolve in (i.e., form hydrogen bonds with) water, they instead collect in layers. Once formed, these sheets can and do fold over, creating living beings with three dimensions—three dimensionality actualized as far as life is concerned.[30] But, though here we have the first cases of distinct living "individuals," they are not self-enclosed monads. That is because the processes that constitute cell membranes do so by means of the differentiation of the hydrophobic that organizes on/into an inside and the hydrophilic that organizes on/into an outside. Thus, the layering that folds over to establish "an enclosed interior" or cellular body distinguished from its environment simultaneously provides—or is—the site and means of interaction with the surrounding world. That is, membrane dynamics facilitate the movement of chemicals and signals across the layers, so that what emerges is not at all a radical separation but increasingly sophisticated transactions with physical elements and other living beings.

If we can think of cell membranes in the simplest sense as double-layered phospholipid molecules, then we need to remember the elaboration that unfolds beyond the basic layering itself: Most basically or locally, "proteins are inserted into and through the phospholipid bilayer, and carbohydrates bind to proteins and lipids on the external surface."[31] This results, for example, in a complex of receptor-ligand relationships spanning membranes, the receptors projecting outside the outer surface and complementary helices penetrating to the interior and relaying electrochemical changes.[32] In short, here we have the basis for more elaborate interconnections among cells and cellular systems that, at their furthest and richest reach, eventually elaborate into the complexity of functioning organisms—their growth and development, tissue maintenance, auto-regulation, immunity, and consciousness (thought, memory, and action).

Though we obviously cannot trace the full story that goes all the way to whole organisms and social collectives, membranes provide the basis for tracing the trajectory from individual cells to biofilms (e.g., the early algae or bacteria discussed previously to animal cells and epithelial sheets). Not to be overlooked is the amazing and critical phenomenal change that happens by way of membranes: Beyond the micro-scale hitherto considered, here we arrive at the generated three-dimensional configuration at the scale of our world of ordinary perception and daily experience.

On the basis of the same features already discussed, organic sheets could adopt independent shapes in the autopoietic processes of organisms.[33]

> Adrift in the ocean, [eukaryotes, cells with an internal system of multiple membranes including a bounded nucleus] resisted dispersal by clinging tenaciously to each other in a continuous layer known as an epithelial sheet—life's first real tissue. In contrast to a biofilm [for example, early algae or the microbes that form plaque on your teeth], which conforms to the curvature of a rock, the edge of a tooth, or the surface of a pipe, an epithelial sheet could adopt any shape it chose, even curl up and anneal edge to edge to form a tube or sphere. Within such a hermetically sealed compartment, conditions could be specified by the organism. . . .
> . . . Sheets could be fashioned into shapes, populations arranged to form patterns, and tissues assigned specialized tasks. . . .[34]

We intuitively understand that our skin is an outermost layer of our bodies—our selves—and know that our skin is a very elaborate layering. The outer layer, or epidermis, itself is composed of many cell layers; the inner dermis is a complex of tissue and glands, blood vessels, muscles and nerve endings; beneath, in the hypodermis, there is adipose tissue and upward-extending major blood vessels, and nerves. Or, in muscles, we have an even more complex layering, as is seen if we rehearse the constitution down to the cellular level: The skeletal muscle is composed of connective tissues, blood vessels, nerves, and bundles of muscle fibers. To note only the latter's sub-layering, bundles are composed of individual muscle fibers that are a combination of sarcolemma, T-tubules, sarcoplasm, and multiple nuclei, where the T-tubules are functionally linked to the sarcoplasmic reticulum, mitochondria, glycogen granules, and myofibrils (the latter of which are composed of troponin, actin, tropomyosin, myosin, titin, and nebulin, along with thin and thick filaments, organized into sarcomere).[35] But enough, without going beyond tissues to consider all the organs, organ systems (including the immune, nervous, and other systems), or the entire organism, we can acknowledge that the body is constituted by and as layering upon layering.

From Body to Buildings

Humans love to build, and the oldest structures that we have to ponder were constructed precisely by layering that moved higher and higher, from one course of packed earth to another, from one stack of stones to another, from one layer of brick to another, from one course of cut stone to another, from one floor of steel girders and poured concrete to another. Here the layering takes into account the problems of load bearing, whether that is a matter of one hard substance directly placed on top of another or of indirect transfer by way of arches or flying buttresses (opening up large interior spaces) or in contemporary structures where the frame carries the load, freeing us to vary the interiors and exteriors, too, as with curtain windows hung on the steel skeletons. From pyramids the world over to skyscrapers, then, we find that the same process obtains: layering, from the ground up.

At the urban scale, we again find sedimentation as settlements are built on top of what came before. Over the course of time, whether because of military destruction and conquest or natural disasters and the desire to build newly on the same site, most old cities are historical layerings. In fact, "Cities of the Underground" is a popular television program on the History Channel: In each episode, the story of a city—Istanbul, Budapest, Naples, Berlin, London, New York, and so on—and its people is told by exploring the forgotten or otherwise "off limits" underground spaces from the past and those that are operative

today but not "public."[36] For instance, the whole of Cyprus is a case
in point, with thousands of years of accumulation and displacement
on the same sites: the earliest Neolithic places built over by Mycenaeans,
Egyptians, Phoenicians, Assyrians, and Persians; underground cata-
combs created in Hellenic and Roman times, once burial grounds for
prominent citizens in Pafos, became Christian hiding places then
quarries for the structures of medieval crusaders; and the layering
continued more recently through the succession of Turks, Greeks,
British, and Cypriots.[37]

The history of a city's layers tells the enduring story of its inhabitants, their beliefs and practices—their identities. As a colleague of
mine, Laura Saija, pointed out in showing me around Catania,
Sicily, if socially appreciated an accumulation of ancient building
elements can remain a vital dimension of today's lifeworld. Thus,
Henri Lefebvre points out:

> We are confronted not by one social space but many—indeed, a
> multiplicity or unaccountable set of social spaces, which we
> refer to generically as 'social space'. No space disappears in the
> course of growth and development; the worldwide does not
> abolish the local. . . . Newly developed networks do not eradi-
> cate from their social context the earlier ones, superimposed
> upon one other over the years, which constitute the various
> markets. . . . The corresponding buildings, in the towns, bear
> testimony to this evolution. Thus social space, and especially
> urban space, emerged in all its diversity—and with a structure far
> more reminiscent of flaky *mille-feuille* pastry than of the homo-
> geneous and isotropic space of classical (Euclidean-Cartesian)
> mathematics. Social spaces interpenetrate one another and/or
> superimpose themselves upon one another.[38]

Not only from history and archaeology but from mythology and
psychoanalysis we know that the meaning, the symbolic functions,
of layering is incredibly complex and enigmatic—perhaps especially
so in regard to the underground. There is, of course, a wealth of
social-cultural and psychological literature on the subject, ranging
from Bachelard to Carl Jung. As Jung relates one of his dreams:

> I owned and inhabited a very big house in town, and I didn't yet
> know all its different parts. So I took a walk through it and
> discovered, mainly in the cellar, several rooms about which

I knew nothing and even exits leading into other cellars or into subterranean streets. I felt uneasy when I found that several of these exits were not locked and some had no locks at all.[39]

This is interpreted by his colleague Anelia Jaffé as follows:

The maze of strange passages, chambers, and unlocked exits in the cellar recalls the old Egyptian representation of the underworld, which is a well-known symbol of the unconscious with its unknown possibilities. It also shows how one is 'open' to other influences in one's unconscious shadow side, and how uncanny and alien elements can break in. The cellar, one can say, is the basement of the dreamer's psyche.[40]

Or, in his phenomenology of "The House, from Cellar to Garret," during his "topoanalysis" of verticality, Gaston Bachelard explores "the irrationality of the cellar":

It is first and foremost the dark entity of the house, the one that partakes of subterranean forces. When we dream there, we are in harmony with the irrationality of the depths. . . . [F]or the cellar, the inhabitant digs and re-digs, making its very depth active. The fact is not enough, the dream is at work. When it comes to excavated ground, dreams have no limit.[41]

Though he critiques Jung's interpretation as exaggerating trepidation, in stressing how cellar dreams result in a positive "increase in reality," Bachelard clearly agrees on a fundamental point: As an interpretive image, our buildings, with their layers from cellar to attic, articulate that to be human is to be ever-underway, always open. Here we clearly have arrived at the layering inseparable from our human lives in the fullest individual and social sense.

Biography

We come back to the central importance of philosophy, literature, and the arts as modes of reflection, which we can take to mean at least what Martin Heidegger calls "meditative or recollective thinking," or Gadamer and Foucault, "effective history" that attempt to retrieve and to take up again the live impulse that continues to come to us today. Artists and writers from Neolithic times onward have set into works the memories, dreams, and imaginings that form some kind of "deeper" layer—how else to put it?—upon which we move in our ordinary daily lives. Whether thought of in terms of a multilayered hierarchical cosmos as in the ancient Mediterranean, the Mayan world, medieval Christianity or Islam, and many other places, the image of the cosmos (perhaps with the heavens and gods above and the land of the dead below) and the course of human life (often including what is believed to lie before or after it) commonly delineate layerings and our movement across them. Even absent the gods and an "otherworldly" transcendent realm, the same basic spatial schema remains in contemporary secular terms with the subconscious, unconscious, or collective conscious.

Without dwelling on it, it is also clear that the homology of the cosmic model and personal life includes our built world, which extends beyond individual buildings or monuments or even cities to entire landscapes. For example, the Pueblo world bodied forth a unified sacred-natural-personal whole that nonetheless contained its differences as layerings. "The six Tewa levels or categories of being are, from lowest to highest," Dry Food People, that is, common villagers; *Towa é* or keepers of discipline and internal intermediaries; Made People, the society's members and chiefs; then across a spiritual line, Dry Food People Who Are No Longer, the spirits of the common villagers; *Towa é*, that is, *Tsave Yoh*, supernatural whippers and guards; Dry Food People Who Never Did Become, the supernaturals leading the first people. Note that the six layers do not correspond to a range from deep in the earth to high in the heavens. Rather, the first three (the lowest) are at the earth or village level, while the latter three are at spiritual levels, both below and above. At the same time, the six groups do circulate across the territory in a symbolically generated pattern of distribution, populating the underground with its series of connected underground lakes, the middle ground of the village at the earth mother navel, and the mountains with their lakes and shrines.[42] Again, we find the same dynamic phenomenon: the unfolding together of place and life trajectory, each episode, each dimension, each meaning layered upon another.

Of course, most of us are centrally concerned with what bears upon or unfolds as one's own life and life's work, with what and how our own world is delineated. A wonderful literary exploration occurs in the writing of W. G. Sebald, who gives us haunting novels probing memory and identity. In one striking incident in *The Emigrants*, he tells of a painter, Ferber, in his studio in a deserted industrial section of Manchester, working mainly by erasure.

> He applied the paint thickly, and then repeatedly scratched it off the canvas as the work proceeded, . . . [as to his drawings] I often thought that his prime concern was to increase the dust. He drew with a vigorous abandon, frequently going through half a dozen of his willow-wood charcoal sticks in he shortest of time; and that process of drawing and shading on the thick, leathery paper, as well as the concomitant business of constantly erasing what he had drawn with a woolen rag already heavy with charcoal, really amounted to nothing but a steady production of dust, which never ceased except at night. Time and again, at the end of a working day, I marveled to see that Ferber, with the few lines and shadows that had escaped annihilation, had created a portrait of great vividness.[43]

In fact, this character is modeled on actual artist Frank Auerbach, who operates by the artistic equivalent to the processes of geological destruction and sedimentation with which I began this chapter.[44] Not only his studio floor but his whole mode of working testifies to the likelihood that the process of painting and drawing itself and that of trying to delineate a person would appear to be the same. Indeed, in its own scale, Auerbach's description of the way he paints landscapes echoes the language of the geological processes of layering: ". . . the way I work means putting up a whole image, and dismantling it and putting up another whole image, which is actually physically extremely strenuous."[45] The same layering is operative in his portraits, a process that is as laborious and time-consuming as life itself:

> There is no underpainting or outline sketch in Auerbach's paintings; the sitter has to return to the same pose each time. Head of E.O.W. (1955; cat. 45) took two years and three hundred sittings to complete.[46]

> Very rapidly laying down fresh marks into just made marks, angrily or eagerly scraping off areas, leaves an end layer of wet paint perhaps a few millimeters in depth. Sometimes the surface is flattish and packed; sometimes there are blobs, rutted strokes

and marbled blends of colors. The surface's consistency and colour is a result of using primarily bright, viscous Stokes oil paint, thickened with its rather greasy extenders. The brush pokes into piles of few colours, which are rapidly mixed with the finger and with brushes. . . .

   During the course of each session . . . a fresh visual idea catapults itself into an image. . . . If so the surface is scraped down, although as Auerbach admits 'evidence accrues'.[47]

Or, to shift back from the more dramatic equivalents of volcanoes and earthquakes to the more gentle and gradual parallel to sedimentation, art critics inevitably make the connection between Auerbach's mode of working and the bringing-forth of character.

The charcoal drawings made in these years [1953–54] were worked until the surface had a texture slightly reminiscent of suede, with light spreading across the handsome forehead and straight brow, modeling the composed, expressive mouth and running down her throat down the page. More clearly likenesses, they also made transparent her fortitude and doubt.[48]

The finished charcoal or pencil drawings that Auerbach has made of his sitters testify to the daily crises in his art. In some, such as Head of Julia (1960; cat. 98), the paper has been erased so often that it has completely given way and had to be patched. In almost all, the dust of the charcoal or graphite has sunk in too far to be eliminated, softening the light from white to silver, while the faint criss-crossing of unwanted lines brings a vibrant energy to the heads. More visibly than in the paintings, the final image is the summation of the many rejected attempts that made it possible. It is as if the crisis has to be provoked daily to move the art forward so that a new and previously unexpressed meaning can be forced from the subject.

In 1985 or so, Auerbach began to draw with a hard pencil as well as charcoal. While using pencil his working method of many sittings and many rubbings-out remained the same but the surrounding paper became lighter: the heads did not have to struggle with surrounding darkness.[49]

The artists and writers themselves, the characterization of the subject, and the artistic procedure (with its by-products) all move through the difficult layering and continual relayering that go on over a life's course. In Auerbach's and Sebald's own cases, we might suppose that negotiating this triple nexus and instantiating the insight that—more accurately—such constitutes a life's course, are not accidental. Neither were strangers to the destruction of layers of life and the need to rebuild them from scratch, certainly never a smooth or linear process. Auerbach had been born in Berlin in 1931, "son of Max Auerbach, a patent lawyer, and Charlotte Nora Burchardt, who had studied art." In 1939, "after arrangements initiated in 1937, Auerbach and five other children, sponsored by the writer Iris Origo, were sent to school in England. Auerbach's parents took him to Hamburg to board ship," after which they never saw each other again.[50] Sebald was born in 1944 in the small village of Wertach in Southern Germany; but, dissatisfied with life in that part of the world, he moved about Europe, and eventually emigrated to England and taught at the University of East Anglia, where in voluntary exile "he always considered himself a guest in his adoptive homeland."[51]

It certainly is not an exaggeration to say that who we are—or become—"in word and deed" as Hanna Arendt felicitously puts it, is always underway and unavoidable.[52] Hence, just as Kierkegaard, Derrida, and others contend (and implement in what they do), "Auerbach sees the virtue of repetition as that of making statements which nobody else could have made" and Sebald that "a particular instance of human temperament is invested with a cognitive force . . . that unhappiness, that melancholy becomes a profound form of knowing."[53] Our identities, our life stories, are unceasingly made, unmade, remade—by ourselves and more so by others—until our time is over, or hopefully, even afterward if we are remembered.[54] Further, insofar as who we are is a matter of what we say and do, we need a place for this to unfold, a public place within a shared lifeworld.

That leaves us amidst issues not only of who we are but where we are, to which the "answer" at which we have arrived is no less confusing than any other yet discovered: We are they who always and unavoidably are in the midst of the laying out—layering—of our bodies, buildings, and biographies. Here I appreciate that others' explorations and mine resonate with one another. For example, Forrest Clingerman explains how because our core temporal rhythms in "the daily cycle permeated with layers betray an even greater nuance to the temporality of place," overcoming what is obscured by much of our daily built environments requires restorative connections biologically as well as narratively if we are to have healthy transitions between our lifeworld states or a coherent sense of self.[55] Obviously, we find ourselves already, and unavoidably, underway within a very complex, intensely personal environmental hermeneutics. The subject matter itself and the dialogue are well begun, but here further open up—not only among the contributors to this volume, but to the readers who hopefully will continue to conversation as full partners.

Might Nature Be Interpreted as a "Saturated Phenomenon"?

Christina M. Gschwandtner

Could elements of "nature" appear to us as what Jean-Luc Marion calls "saturated phenomena"?[1] And if so, how might that be useful for environmental thinking? While at first glance it might seem obvious that natural phenomena could be experienced as saturated, Marion himself has never employed such phenomena as examples for his notion of the saturated phenomenon. In fact, there is almost no reference to (nonhuman) animals anywhere in his work and a tree is mentioned only once and in that case is listed together with a triangle as a "technical object" and thus a "poor" phenomenon.[2] Marion has never engaged environmental concerns in his writings.[3] He focuses almost exclusively on humans (the self and its encounter with the other) and on the divine. Yet, he does contend that some other phenomena can appear as saturated to us: historical or cultural events, paintings and other works of art, gifts or "objects" of sacrifice, and our own human flesh.[4] Why not animals or trees or ecosystems or planets?[5] In this chapter, I examine the possibility that natural phenomena may indeed be interpreted as "saturated" in Marion's sense despite his own neglect of them. I use the word "interpret" here advisedly, as I will suggest that the hermeneutic element plays an important role in this, one much larger than generally acknowledged by Marion in his phenomenological account.[6] I will begin with a summary of Marion's treatment of the saturated phenomenon and a consideration of the role hermeneutics plays in it. I go on to examine in what ways natural phenomena might fit this description, highlighting the importance of the hermeneutic dimension in identifying a natural (but indeed any) phenomenon as saturated. I conclude by drawing out some possible implications of interpreting natural phenomena as saturated phenomena.

What Is a Saturated Phenomenon?

Marion first develops the notion of the saturated phenomenon in an early article responding to Dominique Janicaud's criticism of his *Reduction and Givenness* (where he had explored the notion of a possible "third reduction" to charity) as exercising a "theological turn."[7] This early articulation of the saturated phenomenon, which Marion later heavily qualifies, is thus primarily in religious terms. Marion provides a much fuller account in his work *Being Given*, which seeks to lay out a phenomenology of givenness more generally. Most fundamentally, his project is to examine phenomena that do not appear as objects.[8] Here Marion describes the saturated phenomenon as a paradox which flips Husserlian understandings of phenomena on its head. *Poor* phenomena (the ones he claims Husserl describes) are those constituted by consciousness which employs intention in order to add to the little intuition given by the poor or "everyday" phenomenon. Such poor phenomena are perceived by an intentional subject who constitutes them as objects by providing whatever is lacking in the intuition received from them by consciousness. It is thus able to reach (and impose on the phenomenon) a clear signification.

Conversely, *rich* or *saturated* phenomena give abundant data to intuition yet any intention that might be directed at them or any attempt to impose signification upon such phenomena always fails or at least falls short:

> Because it shows itself only inasmuch as it gives itself, the phenomenon appears to the degree that it arises, ascends, arrives, comes forward, imposes itself, is accomplished factically and bursts forth—in short, it presses urgently on the gaze more than the gaze presses toward it. The gaze receives its impression from the phenomenon before any attempt at constituting it.[9]

These phenomena give "too much"; their intuitive excess cannot be contained. Marion depicts them as overwhelming and bedazzling. They cannot be grasped or controlled but blind us with their excess. They defy our attempt to analyze them as deriving from a clear cause.[10] Saturated phenomena undo all our usual categories of experience and thus are a "counter-experience" which reverses the usual direction of constitution.[11] Instead of consciousness constituting the phenomenon, the experience of such phenomena constitutes consciousness:

> I cannot have vision of these phenomena, because I cannot constitute them starting from a univocal meaning, and even less produce them as objects. What I see of them, if I see anything of them that *is*, does not result from the constitution I would assign to them in the visible, but from the effect they produce on me. And, in fact, this happens in reverse so that my look is submerged in a counter-intentional manner. Then I am no longer the transcendental *I* but rather the witness, constituted by what happens to me. Hence the para-dox, inverted *doxa*. In this way, the phenomenon that befalls and happens to us reverses the order of visibility in that it no longer results from my intention but from its own counter-intentionality.[12]

Saturated phenomena are identified precisely through the effect they have on the one who witnesses them.

Marion employs the categories of experience, as they are articulated by Kant, in order to show all the ways in which saturated phenomena defy them.[13] There are four (or ultimately five) such ways. First, phenomena can saturate our sense of *quantity* by giving too much information, by overwhelming us with data, by providing an event of such richness and complexity that it cannot possibly be contained. Marion identifies historical or cultural events as such phenomena. Second, phenomena can bedazzle us with *quality* and blind us in their overwhelming visibility. These are works of art for Marion, especially paintings.[14] Third, phenomena can be overwhelming in *relation* inasmuch as they can appear as so immediate that no relation or analogy can be established with them. Here Marion appropriates (fairly uncritically) Henry's analysis of the human flesh. Finally, the phenomenon of the human face is saturated in terms of its *modality*. Marion draws heavily on Levinas' analysis of the face, although he chides Levinas for what he deems his far too exclusive emphasis on ethics. *Any* encounter with the human face can become a saturated phenomenon. Marion's preferred example for this is an erotic instead of an ethical relation with the other. Finally, Marion suggests in *Being Given* that there might also be phenomena saturated to a second degree (what he calls the "paradox of paradoxa") inasmuch as they combine all four aspects of saturation. This is the phenomenon of revelation which Marion posits as a possibility (while making no claims about its historical actuality).[15] He insists that he is merely investigating a structural possibility. If there really were phenomena of revelation they would appear as saturated in these four respects.[16] Marion also suggests that while the "simply saturated" phenomena push to the very edge of the phenomenal horizon, the phenomenon of revelation transcends any horizon whatsoever.[17]

Marion has also analyzed other phenomena in terms that would suggest that they are saturated or has even explicitly identified them as such. Thus the erotic phenomenon is clearly saturated in many respects.[18] The same claim can certainly be made of the various religious depictions Marion provides throughout his work, especially his analyses of the Eucharist.[19] This also seems to apply to his more recent explorations of the gift (qualifying some of his earlier analysis in Part II of *Being Given*) and of the phenomenon of sacrifice.[20] In all these contexts, Marion no longer seems to employ the quasi-Kantian categories as strictly.[21] Thus he seems happy now to speak of saturation as richness and then to qualify in the examination of the phenomenon the ways in which such richness overwhelms us. What matters to the saturated phenomenon is its bedazzling excess, its overwhelming splendor, its giving more than we can possibly bear, and the impossibility (or at least inadequacy) of reducing it to a mere object. These are the characteristics shared by all saturated phenomena.[22]

Hermeneutic Dimensions of the Saturated Phenomenon

Marion engages in further analysis of these phenomena in his book *In Excess* which devotes one chapter to each of the phenomena outlined in *Being Given* (although the final chapter examines less a phenomenon of revelation than the possibility of mystical theology, that is, appropriate language about the divine).[23] Already in this book, Marion admits that some of the saturated phenomena might include a hermeneutic dimension. He briefly outlines such a possibility of an "endless" or "infinite" hermeneutics in the chapters on the event and that on the face of the other.[24] He sees little need for hermeneutics in the case of the phenomena of art, of the flesh, or of the divine. In chapter 2 on the event, Marion engages in a close phenomenological analysis of the space in which he and his audience find themselves: the room (the *Salle des Actes* in the *Institut catholique* in Paris) in which the lecture is conducted. Yet he goes on to show that this and other phenomena (his second example is the friendship between Montaigne and La Boétie but he also speaks of the phenomena of human birth and death), if rightly considered, actually are saturated. Here Marion explicitly admits a hermeneutic dimension of the event. It must always be interpreted and interpretation can never come to a close, the phenomenon can never be fully grasped: "Such a hermeneutic would have to be deployed without end and in an infinite network. No constitution of an object, exhaustive and repeatable, would be able to take place."[25] Marion admits a similar need for interpretation in the chapter on the icon or the face of the other (chapter 5):

> All that I would perceive of the other person as regards significations and intentions will remain always and by definition in the background and in deficit in relation to his or her face, a saturated phenomenon. And, therefore, I will only be able to bear this paradox and do it justice in consecrating myself to its infinite hermeneutic according to space, and especially time. For as I have already observed, even after the death of this face, hermeneutics must be pursued, in a memory no less demanding than the present vision. And it will be pursued—or at least should be—after my own death, this time entrusted to others. The face of the other person requires in this way an infinite hermeneutic, equivalent to the "progress toward the infinite" of morality according to Kant. Thus, every face demands immortality—if not its own, at least that of the one who envisages it.[26]

Shane Mackinlay is right to point out that this is a hermeneutic that comes only *after* the event, an interpretation of what has already taken place.[27] Marion does not consider the more contextual hermeneutic dimension that might precede and situate the event and thus shape our experience of it.

It appears, then, that saturated phenomena do require interpretation of some sort. This is closely linked to Marion's account of the recipient of the saturated phenomenon whom he describes as someone called, interlocuted, witnessing, or finally devoted, given over and even addicted to the phenomenon. Marion consistently attempts to maintain a fragile balance between the priority of the phenomenon (which he tends to emphasize more strongly) and the recipient's response to it, which alone makes the phenomenon visible. (A phenomenon is not a phenomenon if it is not experienced by someone.)

Marion contends that the devoted "remains in the end the sole master and servant of the given."[28] In this paradoxical balance, it is the phenomenon that has the initiative for Marion. It "gives itself"[29] freely without any limits imposed upon it, especially in the case of the saturated phenomenon. The saturated phenomenon is hence always anonymous. It does not announce its identity. Marion stresses this dimension of anonymity (at least in *Being Given*, considerably less so in his more recent work). We do not know the origin of the call: Does it proceed from Being (Heidegger), the other (Levinas), the flesh (Henry), or God? The phenomenon, Marion contends, is only identified in the response. The recipient serves as the screen upon which the phenomenon crashes and thereby makes it visible.[30] Yet, if it is the recipient's role to identify the phenomenon, to determine it, then intentionality and interpretation have not been completely suspended.[31] Marion is careful to insist that the recipient can never "grasp" the phenomenon fully, that it always gives too much, is always an overwhelming experience. And yet, the recipient identifies and describes, makes choices in light of the phenomenon, and maybe even creates a great work of art. Thus even if the *adonné* is not a Cartesian subject in total control of its world, it does have an important hermeneutic dimension. It tells the phenomenon's story.[32] While the story may well remain fragmented due to the incredible weight and overwhelming nature of the phenomenon, it is still a story or a vision or an experience. It is thus hermeneutic.

More recently, Marion has qualified his proposal even further in a way that appears to give much more space to hermeneutics (though that continues to remain largely unacknowledged). In "The Banality of Saturation," he responds to the criticism that saturated phenomena are rare and only seen by very few people (which might imply that they do not actually exist).[33] Drawing on the medieval term of the vassal [*ban*], he claims that they are actually "banal" and quite accessible to anyone. In this context, he suggests that a phenomenon might actually move from "poor" to "saturated" or that the same phenomenon might be interpreted as either poor or saturated:

> The banality of the saturated phenomenon suggests that *the majority of phenomena, if not all* can undergo saturation by the excess of intuition over the concept or signification in them. In other words, the majority of phenomena that appear at first glance poor in intuition could be described not only as

objects but also as phenomena that intuition saturates and therefore exceed any univocal concept.[34]

He goes on to appeal explicitly to the role of interpretation in this transition:

> Before the majority of phenomena, even the most simple (the majority of objects produced technically and reproduced industrially), opens *the possibility of a doubled interpretation*, which depends upon the demands of my ever-changing relation to them. Or rather, when the *description* demands it, *I have the possibility of passing from one interpretation to the other*, from a poor or common phenomenality to a saturated phenomenality. [35]

Hence here the difference between whether something appears as a poor or a saturated phenomenon seems to depend on the recipient's interpretation of it as either common or rich. While that is a fairly strong hermeneutic claim, it is not clear in the rest of the essay that Marion actually intends to go that far. To show that saturated phenomena are "banal" or easily accessible, he gives examples of all five senses: seeing, hearing, smelling, tasting, touching, in each case contrasting a poor with a saturated experience. Yet it is not really the *same* phenomenon that is experienced in one case as poor (e.g., the female voice over the loudspeaker at the airport or train station which only conveys information) in the other as rich (e.g., the voice of the diva to which no critic can do justice). The voice over the loudspeaker (even if it is pleasant) obviously *is not* the voice of the diva. It is thus not as clear as in the initial claim that it is the interpretation that matters here: To listen in rapture to the loudspeaker's voice or to critique the diva's performance is to identify the phenomena incorrectly, but not to move the *same* phenomenon from poverty to saturation through the interpretation.

Yet although this is not highlighted by Marion, it does appear that both phenomena require a hermeneutic context or horizon. It is not only the voice (the phenomenon) that differs but also the context. It matters for our identification of the phenomenon whether we stand at a train station and the platform for our outgoing train has just been changed or whether we sit in the Met or the Lyric expecting to hear Renée Fleming or Anna Netrebko. Marion does not seem to acknowledge this context as much as appears required. Indeed, to push this a bit further, presumably one would have a more saturated experience if one had listened to many operas and many divas and were able to hear the particular nuances of the performance in a way not possible to the neophyte (or someone who is not an opera lover). Marion suggests as much when he speaks of the person who can distinguish the smell of Chanel from that of Guerlain or the vintner who "knows what he or she has tasted and can discuss it precisely with an equal, though without employing any concept, or else with an endless series of quasi-concepts, which take on meaning only after and only according to the intuition that is the sole and definitive authority."[36] Despite Marion's insistence that the intuition "is the sole and definitive authority," these cases do suggest that one could be "trained" to some extent to perceive and identify saturated phenomena, that a hermeneutic circle might be established in which greater understanding and appreciation is gained through further exposure and greater awareness of the larger context. Marion here also clarifies that he did not mean to suspend horizons entirely: "The hypothesis of saturated phenomena never consisted in annulling or overcoming the conditions for the possibility of experience, but rather sought to examine whether certain phenomena contradict or exceed those conditions yet nevertheless still appear, precisely by exceeding or contradicting them."[37] Later in the essay, he suggests similarly that the appropriate reception depends on the phenomenon (i.e., whether it can be constituted as an object or has to be described as saturated), but he adds: "This affair is not decided abstractly or arbitrarily. In each case, attentiveness, discernment, time, and hermeneutics are necessary."[38]

The hermeneutic dimension comes to the forefront again in the final section of the essay where Marion examines the role of the witness (to the saturated phenomenon) who is no longer a subject. This witness "develops *his* vision of things, *his* story, *his* details, and *his* information—in short, he tells *his* story, which never achieves the rank of history."[39] Hence this clearly seems an admission of a hermeneutic dimension in the reception of the phenomenon (Marion actually calls it an "infinite hermeneutic"), even if (or maybe precisely because) it is never a definitive version (no final constitution is possible). The response of the witness is precisely a hermeneutic one, although in Marion's view it is not a hermeneutic horizon which contextualizes the event but rather an attempt to articulate its effect after the experience. Marion concludes by reiterating his conviction that "real" phenomenology has as its task the "making visible" of what is previously "unseen," which he implies is a much more worthwhile task than mere interpretation of what is already visible.[40]

This goes back to a claim that Marion had made early on. In an early essay exploring the possibility of a "Christian philosophy" (an identification he usually rejects because of its Thomistic connotations), he is very dismissive of hermeneutics, which he considers a mere interpretation after the fact, rendering such interpretation utterly relative and therefore ultimately meaningless.[41] Instead he suggests Christian philosophy should be "heuristic" in character in that it would discover previously unseen phenomena (found in the realm of theology), formulate them philosophically (presumably via phenomenology), and then "abandon" them to strictly philosophical investigation. Marion makes similar claims about the role of the artist in chapter 3 of *In Excess*.[42] This idea that phenomenology "discovers" phenomena which did not previously exist and thus is not merely interpretive in character is reiterated in his most recent book, *Certitudes négatives*:

> This enlargement here does not simply consist in a hermeneutics of already visible and received phenomena (moving them from objectivity to eventness), but in *discovering* saturated phenomena so far misunderstood by virtue of the very excess of their evidence.[43]

In this book, he portrays hermeneutics as a first enlargement of phenomenology whose role it is to move the phenomenon from the horizon of the object or being to that of givenness. The analysis of saturated phenomena goes far beyond this initial (hermeneutic) enlargement and finally culminates in a "third enlargement," that of "negative certainty." With this notion, Marion seeks to express the fact that we can know "with certainty" that such phenomena cannot be fully known and that Cartesian parameters for certainty do not apply to them. He argues that such phenomena are the human person (self or other), the divine, gift, sacrifice, art (and basically anything else that is saturated). Negative certainty applies to these phenomena not because they are *not yet* fully known, but because they *can never* be fully known. Their negative certainty is constitutive of their very phenomenality. While Marion here also does not engage natural phenomena, he is somewhat more explicit in his criticism of technology and reductive science, which is evident to some extent already in his designation of "technical objects" as "poor" and the contention that nothing is really produced in them but that they are only applications of concepts. In *Certitudes néga-tives*, he combats what he calls a nihilistic view of the world, fueled by technology, which reduces humans (and the divine) to mere objects. Although no explicit environmental concerns emerge here, one may wonder whether the current destruction of the environment is not also a result of such reductive, purely technological, thinking.

Might Natural Phenomena Be Given as Saturated?

Although Marion himself does not make this application, I see little reason why nature may not appear or give itself (or be given) as satu-rated at least in certain contexts. Several essays in the present volume illustrate this.[44] If the primary characteristic of a saturated phenomenon is its overwhelming nature, the fact that it cannot be constituted, that it provides too much information, then many natural phenomena would qualify. Clearly a biotic system such as a wetland, a tidal pool, a region of rainforest, even an anthill are complex phenomena that cannot be completely grasped by human consciousness. Although we can certainly *try* to impose concepts upon them or turn them into an object (which obviously happens in much scientific study), these always remain inadequate.[45] Natural phenomena, whether singular (a particular animal or individual tree) or more complex (an entire ecosystem), are *not* technical objects. Since they cannot be fully constituted by human consciousness, or as McGrath says do "not depend on a subjective synthesis," they must be saturated phenomena on Marion's terms. This is evident in Clingerman's depictions of the effect dawn, dusk, and the seasons have on our experience of "temporal depth." Feeling "the traces of autumn, winter or spring" surely is not about constituting them as objects. In fact, Clingerman explicitly refers to the "complexity" of places and the relationships they establish.[46] All these suggest con-notations of saturation as Marion has depicted them. Drenthen's analysis of human interaction with the landscape show how land-scape is experienced as saturated, how it can even affect a people's "soul." Uncovering the rich layering of the legible landscape can "recognize the many alternative ways in which one can articulate the meaning of a landscape that is worth protecting."[47] This is a recognition that the landscape is not merely an object which we can constitute at will but a much richer phenomenon which is multilayered and much closer to something like Marion's events than to an object. The North Sea or the Waal River is recognized as unpredictable and as not possible to be controlled. Similarly, Donohoe explores rich connotations of monuments which try to "bring that which is immemorial, other, and which surpasses us" to experience, resulting in "a kind of humility." A monument is saturated inasmuch as it makes "us feel the death of the other in its particularity that brings us face to face with our own death, but also face to face with our own life."[48] All these are characteristics of saturated phenomena as Marion has outlined them.

This becomes particularly evident in the depictions of encounters with nature in much nature writing that certainly qualifies as phenomenological, even when it is not explicitly identified as such. Whether it is Annie Dillard's moving meditations on her experiences at Tinker Creek, Bernd Heinrich's careful observations in the Maine woods,[49] Edward Abbey's gripping accounts in *Desert Solitaire*, John Muir's experiences in the Sierra Nevada, Aldo Leopold's meticulous descriptions that precede and ground the more theoretical final chapter of the *Land Ethic*,[50] or any of the many similar descriptions of "nature experiences," natural phenomena are consistently described as saturated phenomena in these writings. To give just one example:

> The secret of seeing is, then, the pearl of great prize. . . . The secret of seeing is to sail on solar wind. Hone and spread your spirit till you yourself are a sail, whetted, translucent, broadside to the merest puff. When her doctor took her bandages off and led her into the garden, the girl who was no longer blind saw "the tree with the lights in it." It was for this tree I searched through the peach orchards of summer, in the forests of fall and down winter and spring for years. Then one day I was walking along Tinker Creek thinking of nothing at all and I saw the tree with the lights in it. I saw the backyard cedar where the mourning doves roost charged and transfigured, each cell buzzing with flame. I stood on the grass with the lights in it, grass that was wholly fire, utterly focused and utterly dreamed. It was less like seeing than like being for the first time seen, knocked breathless by a powerful glance. The flood of fire abated, but I'm still spending the power. Gradually the lights went out in the cedar, the colors died, the cells unflamed and disappeared. I was still ringing. I had been my whole life a bell, and never knew it until at that moment I was lifted and struck. I have

since only rarely seen the tree with the lights in it. The vision comes and goes, mostly goes, but I live for it, for the moment when the mountains open and a new light roars in spate through the crack, and the mountains slam.[51]

Dillard's encounter with the tree overwhelms and bedazzles her, she finds herself envisioned by it instead of imposing her vision on it. It impacts her when she did not expect it or prepare for it—it has all the connotations of saturation and counter-experience Marion outlines. In general, these books contain bedazzling, overwhelming, unsettling, at times breathtakingly beautiful and deeply moving accounts, calling us to cease our objectifying of nature and to experience it differently, precisely as saturated with meaning and as valuable on its own terms. The very power of their impact on us is precisely this dimension of excess or saturation that either overwhelms us and makes us feel small or maybe even makes us permanently devoted and even "addicted" to natural phenomena. Many environmentalists are committed to ecological actions and policies precisely because of their own profound experiences of nature as saturated or because they were moved by reading such accounts. As Treanor points out, "Muir's narratives about the High Sierra, and what it did for and to him, moved his readers to value mountains they would never climb and for whose scientific history they cared not a whit."[52] To receive something as a saturated phenomenon is to assign a value to it that cannot be measured or objectified.

Yet not all people experience nature in this fashion. One may well argue that nature does not have the same impact on everyone. Apparently in the nineteenth century, some people covered their eyes when they had to travel through the Alps since their lack of geometry was perceived as ugly. Many tourists that go to national parks spend a few minutes there and don't see much more than the visitor center. Increasingly, children no longer have any sense that food comes from anywhere but the grocery store. Here apparently a way of "seeing" (as Dillard stresses in her account) and thus a hermeneutic dimension, is necessary in order to be sensitive to nature as saturated. This seems very similar to the accusation made by some that where Marion experiences saturation they see nothing. Marion responds that this does not mean that there is nothing there: "[T]he fact or the pretense of not seeing does not prove that there is nothing to see. It can simply suggest that there is indeed something to see, but that in order to see it, it is necessary to learn to see otherwise because it could be a question of a phenomenality different from the one that manifests objects."[53] One requires "eyes" to see, a kind of hermeneutic attunement to the phenomena. Some see more than others. Artists see particularly well—they are gifted (the only context in which that translation for *adonné* actually works):

> The painter renders visible as a phenomenon what no one had ever seen before, because he or she manages, being the first to do that every time, to resist the given enough to get it to show itself—and then in a phenomenon accessible to all. A great painter never invents anything, as if the given were missing; he or she suffers on the contrary a resistance to this excess, to the point of making it render its visibility. . . . Genius only consists in a great resistance to the impact of the given revealing itself.[54]

In fact, one of Marion's most recent books is called *Le croire pour le voir* which asserts that believing is required in order to see.[55] This, however, clearly seems to imply a hermeneutic dimension to such phenomenal "seeing." And this is not the hermeneutics Marion disdains, the "mere" interpretation which renders the phenomenon utterly relative, dependent solely upon the recipient's (by implication capricious) "interpretation." Rather, it is the hermeneutic context (the horizon Marion so often wants to do without) which makes seeing possible.[56]

One might ask, then, how people might be inspired to experience nature as saturated. Can one be guided into having a saturated experience of something? Or, as Treanor, wonders, "Can narrative elicit intimacy, love, and concern?" Can nature be experienced "vicariously"? How do we best understand the "significant difference between narrative experience and real experience"?[57] Might a description of a natural phenomenon as excessive—as happens in the writers mentioned previously—open others' eyes to see differently? Can good nature narrative highlight the saturation of the phenomenon we might have missed in our own experience? This implies an important hermeneutic dimension for any application of Marion's thought to nature, namely an insistence that while sometimes saturated phenomena overwhelm us utterly without any apparent prior preparation, at other times we require hermeneutic preparation—that includes an important role for the imagination as McGrath suggests—in order to experience a phenomenon as saturated.

Yet the idea of the saturated phenomenon might apply not only to such more "positive" encounters with nature. Clingerman is right that "this is the case not only for wildness as paradigmatically 'natural,' but also for built environments, destroyed or toxic ecosystems."[58] In light of Marion's account of the event, we might well speak of environmental destruction, especially global climate change, as saturated phenomena. The exorbitant release of carbon into the atmosphere has unleashed forces that seem utterly overwhelming and indeed often blinding in their complexity. Not only do we still know very little about the real consequences climate change will have on the planet but we realize increasingly that it is such a complicated phenomenon that it might never be fully understood. Climate change is so complex that we cannot possibly make secure predictions about how it will affect the planet, since even the weather as a chaos system cannot be predicted securely. Certainly the 2011 earthquake and tsunami in Japan (and the earlier one that caused such destruction in Indonesia, Sri Lanka, and various other countries on the day after Christmas 2004) are beyond all description, overwhelming in magnitude or even in "quality," though certainly not in the positive sense usually described by Marion. We have no way of being sure of such things as the runaway impact of feedback loops or of the capacity of the ocean to continue to absorb carbon. In many of these cases, it is not merely a matter of not having sufficient information but of the very complexity of the event and its countless interactions with other phenomena. These events cannot be understood as effects not because they do not have a cause but because there is an abundance of possible causes, not because there is too little information but because there is too much of it.[59]

Here the more recent notion of "negative certainty" might also prove helpful. As Smith seeks to articulate in his essay, it is impossible to provide "a narrative structure for history as a whole, one that allocates every event a specific role and meaning in light of an overarching story (e.g., universal progress or decay) that is said to underpin, conjoin, and when grasped, explain such events." Smith insists that Dilthey, Morrow, and Wilson operate with reductive idealist versions of history that do not sufficiently recognize its materiality and complexity. Such "myths" of natural history that interpret it as a "single story to be read and understood in its totality" operate "on the mistaken assumption that they [are] dealing with complete understandings of predictable systems of objects." We will never have "complete understandings" of saturated phenomena, precisely because they are the kinds of phenomena that escape definitions of knowledge in terms of certainty. Marion's depiction of the event seems to fit this complexity outlined by Smith and the incommensurability of climate change well: "In effect, what qualifies it as event stems from the fact that these causes themselves all result from an arising with which they are incommensurable."[60] Instead, Smith suggests, this

> opens the possibility of an alternative interpretative materialism that is explicitly not formulated as a meta-science, a hermeneutic equivalent of sci-mat or diamat, a synthetic method of reaching a universal Truth. Rather, this interpretative materialism (in the form of a hermeneutic ontology) pays (ethical) attention to diverse understandings and recognizes the necessary incompleteness of all interpretations that would include myriad possible "natural histories."[61]

Similarly, Marion admits that the saturated phenomena which impose on us "negative certainty" are best depicted via an "infinite hermeneutics" that provides multiple—even endless—accounts of such experiences.

Implications of Interpreting Natural Phenomena as Saturated

What difference would it make if natural phenomena were regarded as saturated? First, it might make us more reluctant to think that we can give quick answers to environmental problems or solve them with advances in technology. In fact, for Marion the opposite of the saturated phenomenon is precisely the "technical" object.[62] Although at times he contrasts what he calls "common-law" phenomena (as poor) to saturated (rich) phenomena, he often employs the language of "technical object" and comments precisely on this link with technology.[63] Indeed, many of his examples (besides mathematical concepts) are products of technology. He contends that technology first imposes a concept and that whatever it "manufactures" as an object is a mere result of this concept.[64] It should thus not even be called the "production" of an object. Real production or creation happens only in the case of saturated phenomena. At one point he comments on "the monstrous commercial city, almost unlimited and without form, oozing its own vulgarity, awash in items for sale."[65] Although his critique of technology is not as trenchant as that of Michel Henry, on whom he largely seems to rely here,[66] he certainly agrees that technology reduces humans to objects and is quite critical of its pernicious effects, especially in his more recent work *Certitudes négative*. In a Lenten address on the topic of "Faith and Reason" at Notre Dame de Paris, he condemns "the dehumanization of humans to improve humanity, the systematic sapping of nature to develop the economy, injustice to render society more efficient, the absolute empire of information-distraction to escape the constraints of the true."[67] In *Certitudes négative*, he repeatedly refers in derisory fashion to "technoscience." This is a theme that could profitably be pursued further.

If a phenomenon such as global climate change is recognized as a saturated phenomenon (of the event), it is also admitted that it is overwhelming in its complexity and requires deep and thorough engagement (and a kind of humility before its enormity). It might enable us to deal more fully with the issue, raised by several of the essays in this volume, why a relation of concern with the environment is a better interpretation than one that is indifferent to nature or exploits it. For example, Bell presents various interpretations of the "good life" and how these impact our relation to the environment. How would we judge between these interpretations? Recognizing natural phenomena as saturated might help answer Bell's question: "How can we argue and genuinely persuade others that interpretations that are more positive towards, or inclusive of, the environment are more valid? How do we argue that people *should* include nonhuman others?"[68] The proposal of the saturated phenomenon insists that there is something rich about the phenomenon even when it is not recognized. Although there is no phenomenon without a recipient or witness, some ways of receiving or witnessing are better than others, more appropriate for the phenomenon. Experiencing nature as saturated is a better reception than reducing it to an object. This also calls, even on Marion's own account, for hermeneutic endeavors that would give a variety of analyses of the event, recognizing that one version alone will never do full justice to its complexity. We hence need good narratives about nature as saturated phenomenon, accounts that are "rich, complex, or highly developed"[69] and hence able to convey some of the nonconstitutable excess of natural phenomena. Providing such narratives makes us more aware of our environment, integrates the natural and the social more fully, and thus maybe can forge a path to experience both as saturated and thus not reducible to an object. Maybe there is also a possible connection here to Forrest Clingerman's focus on remembering and imagining as sensitivity to time within place which gives a richer (more saturated?) understanding of nature, one in which we "engage and find meaning in the world." If imagination truly helps us to "see otherwise,"[70] maybe it can help us see nature as a saturated phenomenon instead of as an object and can help us recognize the complexity of the layers of the legible landscape (Drenthen).

Thus, speaking of individual natural phenomena as saturated, especially as such phenomena are frequently depicted by nature writers, might give us greater sensitivity for the other and might even make it possible to contemplate what it might mean to respond to the (animal) other in more ethical ways.[71] This could be clearly pushed beyond encounters with other animals to encounters with trees or mountains. No tree is merely an object for consumption, no mountain merely a possible site for future strip mining. To treat trees, rivers, springs (and even gas deposits in the Marcellus Shale) merely as resources is to reduce their saturation to that of an object and to believe that we know them with certainty (and completely) when we treat them as such. It is to ignore the complexity and excess of the saturated phenomenon: Instead of being blinded by it, we turn a blind eye to it. Much environmental destruction is precisely a result of having reduced complex ecosystems to mere objects of exploitation without recognizing the excess of life present in them (this is particularly true of the rainforest which contains more species than we can presently imagine, most of which remain undocumented and unexamined). Scientists involved in botanical, zoological, ethological, ecological, and geological research frequently witness to this sense of abundance or excess and acknowledge the limitations of their research (which indicates a need for further research not for ceasing it). One might suggest again (but differently) an affinity with Marion's account of the artist (see earlier discussion). Saturation here precisely does not function as a way to stop our examination of the phenomenon. As the artist has a special genius for depicting the unseen (or invisible) on canvas, so the scientist has special talents for investigating the complexity of natural phenomena. This might help resolve to some extent the ambivalence Treanor notes in regard to strictly scientific and more narrative accounts. Smith describes this as "a bipartisan approach" that "risks artificially (and impossibly) separating knowledge concerning human existence from knowledge of nature." Instead, we must recognize "the messy reality (and diversity) of actual scientific practices and their inescapable historicity, that is to say the multifarious ways that natural science too is embedded within particular times, places and cultures."[72] Scientific and narrative accounts can come together, or, as Treanor suggests (relying on Rolston), "scientific understanding can, in some cases, enrich personal experience of nature." This possible parallel between the artist and the scientist might indeed be a way to recognize "that these two ways [ecological knowledge and imaginative engagement] of relating to nature mutually inform, support, and correct each other in a complex and symbiotic relationship," as Treanor requests.[73] Maybe this also is a kind of "giftedness" that comes with a heavy responsibility.

Finally, we might wonder whether this helps us rethink what it means to be human. Marion consistently stresses that his analysis of the saturated phenomenon has deep implications for the notion of the self which is no longer a Cartesian self-sufficient subject. Instead, the one who responds to the saturated phenomenon is the witness, the one devoted, given over, even addicted to the phenomenon. The recipient of the "self" of the phenomenon only becomes a self of his or her own in this encounter with and response to the "self" of the phenomenon:

> Therefore, the *self* of what shows *itself* and gives *itself* can never be verified through inference or constitution, which would collapse it equally into the in-itself of the object (or the thing without phenomenality). But it could be through the impression, or rather through the pressure that it exerts over the gaze (and, of course, over the other modes of perception). This pressure bears down in such a manner that it makes us feel not only its weight, but also the fact that we cannot in any way master it, that it imposes itself without our having it available to us—we do not trigger it any more than we suspend it. The *self* of the phenomenon is marked in its determination as event.[74]

Thus, if natural phenomena can be recognized as saturated in this fashion, this may well help us address some of the anthropocentric dilemmas that continue to haunt environmental thinking. The aporia of how to move away from anthropocentric thinking that values the human as superior to all other species and focuses exclusively (or primarily) on human concerns toward a more biocentric or ecocentric position that values other forms of life for their own sake always is complicated by the fact that it is the *human* who continues to speak and to articulate even the biocentric or ecocentric position. How would a human really know what an animal felt or what the world might look like from the view of a plant?

This aporia seems addressed at least to some extent in the paradox Marion continually articulates in the notion of the counter-experience of the witness. While all phenomenology is about human consciousness of phenomena and thus recognizes that it is contextual and cannot literally climb into the consciousness of an other (whether it is the other person, the divine, or indeed animals, plants, or stars), Marion (following Levinas) articulates a phenomenology that is primarily a response to the other. Here the self only becomes possible as it responds to and receives the incoming of the other who has the initiative. And unlike Levinas, he does not insist on the other as human but envisions the possibility that any phenomenon could have such an initiating "self." If an encounter with a painting or the experience of a historical event can individuate me, why not an encounter with a (nonhuman) animal or the experience of natural (rather than historical) phenomena? Peter Steeves describes such an encounter:

> And then one night while doing dishes in the back washroom a black lizard the size of my thumb crawled in through the barred window above the sink and sat on the wet cement near the faucet. He moved impossibly fast; his toes were spread impossibly wide in a graceful fan at the end of each foot; he stuck impossibly to the dripping wall of the sink, cranked one eye in my direction and turned to face me. Self-conscious, I froze. The moment was pregnant with possibility. . . . But the lizard. . . . His motions, his thoughts were unscripted. The relationship was open on both ends. We could become anything together.[75]

Treanor envisions the possibility that "narratives can open us to other, non-human experiences and worlds,"[76] while Bell[77] asks:

> What if I see myself as a human animal, or just as another animal, as just another earth creature, a plain citizen, a fellow voyager . . . what if I essentially interpret myself specifically and humanity generally as, more than anything else, just another animal, or at least just another mammal? Well, in that case other animals, at the least other mammals, would count as other-than-humans in something more like a categorical

sense. He is other-than-human. I am other-than-lion? Regardless, under such a view animal others are other selves. The animal other is a self to which I am an other. And in this case, the animal other can place an injunction upon me.[78]

Clingerman also recognizes that "the self cannot be fully known or manifested without being distinguished from the other." Thus we might envision place as "an other-than-self as self that envelops the self." The self responds to the other. Consequently, "by re-placing the self and remembering place, nature becomes more than a backdrop; it is a participant in the narrative, an other that embodies memory, an other that locates imagination, and thereby an other that provides a constitutive element of selfhood."[79] (Maybe Marion's focus on the self-giving of phenomena can strengthen this sense of nature as participant? Maybe the "embodied text of the world" can be interpreted as a phenomenon of the event?)

I wonder whether the saturated phenomenon can also help distinguish between the healthy and unhealthy environmental identities Utsler mentions. Does "interaction with nature" "provide what we might call a hermeneutic reconfiguration of the self"[80] precisely when it is experienced as a saturated phenomenon? Then it might become possible, as Bell suggests, that the self "has experienced a change, not just in his or her perception of the world but in his or her own self-understanding."[81] Or, as Drenthen insists, "the notion of landscape legibility can also help us to understand the relation between landscapes and human identity." He suggests that this might precisely help us to "adjust the anthropocentric place narrative of heritage landscape protectionists and to broaden our sense of human place-history (landscape biography) and our ethics of place."[82] Approaching nature as a saturated phenomenon opens the possibility of understanding ourselves differently and experiencing ourselves envisioned (transformed and constituted) by the nonhuman other. And this might also open up Marion's account to an important dimension lacking in his phenomenological analysis. As Derrida suggests in *The Animal That Therefore I Am* (mostly in response to Levinas), a phenomenological account that ignores our animal nature and does not recognize nonhuman others is no longer viable today. We cannot extract ourselves—even philosophically—from the ecosystems which sustain our life or ignore the destructive impact we have on them. Nature places an injunction upon us which we ignore at our peril.

5

Must Environmental Philosophy Relinquish the Concept of Nature? A Hermeneutic Reply to Steven Vogel

W. S. K. Cameron

In a series of astute papers over the last fifteen years, Steven Vogel has developed a remarkably compelling social constructivist critique of "nature." Having drawn on several well-known and widely accepted postmodern worries, he might have appeared vulnerable to the traditional environmentalist's equally well-known counterthrust: doesn't the reduction of nature to culture simply efface nature in one last, hubristic and utterly anthropocentric gesture? Yet Vogel's argument is carefully constructed, the fruit of a thoughtful and sober mind. He carefully avoids the overstatement to which some have been prone. Notwithstanding many excellent arguments, however, Vogel appears not to have carried the day: Few have responded, and the dominant tone of those who have has been impatiently negative. While it is hazardous to conclude anything from silence, I suspect the silence of embarrassment in the face of Vogel's stunningly coun-terintuitive conclusion: that the failure of Cartesian dualism and post-structuralist suspicions challenge us to develop an environmental philosophy *without* the concept of nature.

In this chapter, I will concede most of Vogel's arguments while attempting to avoid his conclusion. Vogel has given excellent reasons to jettison the *familiar* concept of nature, yet we still need a suitably modified version to illuminate the shared world that is—in some sense—really independent of us. Indeed a concept of nature is necessary not only to make sense of ordinary scientific facts and common intuitions but of claims that Vogel himself must make. But having accepted Vogel's case, how will I escape his conclusion? We'll make a detour through Hans-Georg Gadamer's philosophical hermeneutics, which contests the traditional opposition between language and world. As Gadamer shows with critical concepts like "language" and "world," so we will also see with "nature": a sufficiently concrete look at the way that concepts are continually reconstructed *can* capture the way that words present the world "in itself."

Our itinerary follows from these goals. Having outlined Vogel's powerful critique, we'll consider his reconstruction of environmental philosophy without the concept of nature. Vogel's proposal, though well-grounded, nevertheless raises other problems. To circumvent this, I'll invoke Gadamer's hermeneutic account of the world, since the same argument permits a parallel recovery of "nature." The received view of nature may be historically particular and conceptually confused, but we can, should, and indeed must redeem it.

Vogel on the Manifold Problems with our Concept of "Nature"

Vogel published the heart of his case in three seminal papers.[1] The first surveys several strands of generically continental and specifically poststructuralist reaction against the traditional understanding of nature. Yet as Vogel also shows, the post-structuralist critique of nature "as origin" and "as difference" is itself vulnerable in other ways. Thus Vogel develops a fourth, quasi-Marxist conception of nature as "practice" that begins from our interactions with a world that could neither be conceived nor exist without us. Vogel's second paper develops a new environmental ethics that can function after the death of (the traditional concept of) nature. Finally, his third paper blocks a possible countermove by showing that unless we invoke a philosophically and biologically suspect dualism, we cannot distinguish human activities and artifacts from other putatively "natural" processes. Furthermore, frankly admitting the naturalness of *all* human artifacts—from human stool to babies to organic produce to superfund sites—encourages the environmental virtues of self-knowledge and humility. Rejecting the traditional concept of nature—as surprising as that may be to environmentalists— thus produces not only a more consistent but a less hubristic account of the human-nature relation. To convey how compelling Vogel's arguments are, we must take them up in detail; but to avoid repetition, I'll collect first his critical, and then his constructive claims.

Vogel's primary target is the romantic view of nature that most environmentalists presume: nature as a complex, interlocking whole within which humans should and once did fit, but that they have now betrayed through hubristic efforts at calculation and control. The "fall" is variously identified and dated by competing strands of this tradition,[2] but all see humans as behaving naturally "when they act in accordance with 'natural processes' (i.e., those that would take place anyway in their absence), while their actions are harmful and unnatural (and immoral)" when, transfixed by "the hubristic dream that our actions could fundamentally transform (indeed master) nature," they produce technologies "whose ultimate consequence is environmental disaster."[3]

Vogel starts with Bill McKibben because his widely read book[4] sums up the core concerns of contemporary environmentalists in an accessible, powerful, timely, and terrible image of humans as having brought about "the end of nature." Whether we consider the jet stream's worldwide dispersal of pollutants, the hole that industrial chemicals have eaten through the ozone, or carbon-caused changes to the climate itself: Human beings can no longer regard nature as independent of us, for they are now reshaping it. McKibben, to be sure, concedes some obvious limits of his metaphor. He knows that his home's apparently wild setting is second-growth forest reclaiming land denuded a century and more ago by foresters and farmers. Yet virtually all of North America—let alone other more deeply humanized parts of the planet—has been shaped and reshaped under human pressure. And as Vogel notes, this raises a fundamental question: If nature has ended, when and how did we bring that about? We face a regress: European settlers cleared trees for lumber, fuel, and farms, but before that the land had been cultivated for centuries by Native Americans, who in turn inherited a land newly stripped of megafauna by their forebears, the first immigrants. Has nature "ended," or did it ever really exist?

And even if we could answer that question in a sensible way, the "end of nature" thesis has impractical consequences and presupposes an incoherent human-nature relation. Imagine, Vogel asks, that McKibben were right: What would we do then? There would be nothing else *to* do but mourn and perhaps offer penance. Moreover, like many contemporary environmentalists, McKibben incoherently presupposes two conflicting views of nature: nature as what's other than us, that is, as the nonhuman; and nature as everything, including us. The problem is not that he has two views, since both make sense; it's that he incoherently oscillates between them.[5] Vogel demonstrates this in two dilemmas.

First, many environmentalists accept Lynn White's argument that the ecological crisis developed under the baleful influence of Christianity, "the most anthropocentric religion the world has ever seen."[6] In response, we must change our role "from conqueror of the land-community to plain member and citizen of it."[7] Yet Vogel demurs: For if we *are* just part of nature, why do *our* special qualities, *our* creations and waste products, "end" nature? Earth has suffered several major extinctions at least one of which was driven by the toxic waste (oxygen) produced by a new organism; more commonly, small ecosystems are dramatically reshaped by new colonists. If so, how are our effects any different from those of cyanobacteria, kudzu, or beavers? And if they are different, why resist those who preserve a special place for humans? Would not our uniquely toxic effects, if not our special powers, distinguish us from "plain members and citizens"?

One might save the contrast between humans and nature by drawing the line not between one species and all others but *within* one special species—human beings. On this view, we are just like other species in many respects—thus our common origin as evidenced by the overlaps in our biological makeup—but some of our products and activities have allowed us to push beyond natural bounds. Yet as Vogel shows, this view is also unsustainable—or at least, it's unsustainable without resurrecting dualism in another form. We cannot argue with Robert Elliot that humans have somehow escaped the natural bounds of their ecological niche, for

> this claim is entirely circular. . . . [Y]ou can't distinguish between species that disrupt the stability of "naturally evolved checks and balances" and those that remain within "naturally constrained ecological niches" unless you've already decided that natural systems are all marked by stability, by checks and balances, by constraints and brakes—but if humans are natural and nonetheless do what Elliot says they do, then clearly natural systems are *not* all so marked.[8]

Eric Katz draws the line elsewhere, arguing that some but not all human activities are unnatural insofar as they push beyond our biological or evolutionary capacities. But this argument, too, is circular: "[W]hy is the building and operation of an engine not an expression of humans' natural capacities unless we have previously decided,

on other grounds, that technological action is outside of nature?"[9]
And when Elliot suggests that certain rational capacities place us
"outside the natural order," Vogel replies,

> Such a remark is the tip-off that we've entered Cartesian terri-
> tory here, where the dichotomy between humanity and nature
> turns out to depend on one between mind and body—which is
> to say that the dichotomy has now been introduced directly
> into human beings themselves. But to enter Cartesian territory
> is to leave behind, on the one hand, Darwin, and, on the other
> hand, the last two hundred years of philosophical discussion. . . .
> [T]his view commits itself to a form of metaphysical dualism
> that is both philosophically and biologically suspect at best,
> and that furthermore itself sounds suspiciously anthropocentric
> (how come our special skills get to transcend nature?).[10]

In fact, not just we ourselves but *all* our artifacts—from excrement
to compost to traditional crafts to nuclear waste—are natural: made
from natural materials by beings who arose in nature and behave
according to natural laws.

Indeed as the previous block quote reminds us, not only does the
Darwinian view of our biological origin ground us and our products
in nature, in addition, we cannot anchor the human-nature divide
philosophically. Vogel cites the post-Hegelian and poststructuralist
critiques of the given; and the same problem arises in the Anglo-
American approaches of the pragmatists, Wittgenstein, and Quine.
In a secular age, it's tempting to take nature as the "origin or
foundation on which everything else is built," but contemporary
poststructuralism impugns the "very notion of nature as origin,"
revealing "the complex processes of linguistic and social construction
required to produce that appearance" as natural, original, or foundational.[11] On reflection, neither "nature" nor "wilderness" exists:
both have been socially and historically constituted.

Consider, as brief indications, how variously "wilderness"
appears: In North America, we use a sophisticated plumbing system
to manage the apparently (but not actually) untamed power of
Niagara Falls; in longer-settled environments in Europe, Africa, and
Asia, the idea of "untouched wilderness" has still less plausibility;
and even in North America, that notion arose largely because early
European settlers didn't recognize the sophisticated and significant
means by which earlier Indian occupants had cultivated and managed

the land.[12] We've already noted the historical regress set off by the search for "untouched" nature; on the poststructuralist view, this continual deferral of wilderness and nature, and thus of either as origin, undermines the idea altogether.

Nor is this an unusual case, for parallel problems have emerged in science studies. Here, too, efforts to resolve the problem of the given have resulted in politically and socially thick descriptions of science-in-practice. Not only does science proceed, as Karl Popper thought, through a series of "conjectures and refutations," but

> the objects that scientific theories "describe" are themselves in a certain very real sense artifacts, produced in laboratories by complex linguistically mediated social practices. Thinkers such as Latour, Hacking, Pickering, Crease, and Rouse have emphasized the importance of coming to understand that science is above all a matter of practice—that the laboratory is a *labora*tory, as Rouse puts it—and that the entities it studies therefore have to be viewed again in a certain sense as constructed ones.[13]

What does this mean? In place of the discredited idea of nature as origin, many poststructuralists understand nature as "difference"—that is, "the name we might give to the otherness of the world, to that which is always left out of any attempt to grasp the world as a whole and bring it entirely into the light." We must "attend, in every language or conceptual scheme, to what that scheme occludes, excludes, inhibits—more, . . . attend to the crucial fact that every such scheme does occlude, exclude, inhibit something, and does so essentially, because this is what a scheme *is*." Nature stands "for the gap between frameworks, for that which is left out," a gap we experience in the resistance to and unexpected consequences and side effects of all human activities.[14]

Poststructuralism has the salutary effect of chastening the hubris of human technology and demanding a deeper modesty in our dealings with the earth. Yet we cannot rest here, Vogel continues, for poststructuralism itself faces several unresolved challenges. First, it appears unreasonably idealistic. Historians have shown that nature has meant different things at different times,[15] but is it as plastic a concept as poststructuralism suggests? Has it no objective, or at least intersubjective, trans- (or deeper-than-) historical content? If nature is just a name for the gaps in our conceptual structures, how can we even *intend* it, much less identify what we want to protect from our domination? Yet to replace nature as kaleidoscopic and unnamable "difference" with the seemingly more adequate notion of nature as that Other that perpetually eludes both full comprehension and complete domination does not solve, but only exacerbates the poststructuralist's problem. To identify nature with a hypostatized, eternally inaccessible Origin was precisely what poststructuralists had resisted in the traditional concept of nature.

For would-be poststructuralist environmentalists, however, the theoretical challenge of idealism pales in comparison to the much more immediate and practical problem of relativism. If nature merely stands for the difference between thought and thing, we cannot re-reify it as a Thing especially deserving of honor, protection, or preservation. In particular, claims about nature as a functioning whole or aspirations to grasp and act in accordance with its telos are in principle out of reach. Yet if nature is no thing, but merely what in things escapes our grasp, it's unclear why it should be any more accessible in wild places than in our waste products or machines, unclear how it could ever have been under threat, and unclear what sort of action could or should be taken to "save" it. Since it is in principle beyond conception, nature no longer has any determinate meaning. Can there be any point speaking, much less acting, on behalf of what we cannot understand?

And this, finally, brings us to Vogel's constructive claim. The romantic tradition we inherited impractically and incoherently severs us from nature while poststructuralism falls to the challenges of idealism and relativism. Abandoning nature as always-deferred origin, beyond-our-grasp difference or (re-) hypostatized Other, Vogel focuses on the mutually transformative *relationship* we have with nature. We grasp nature, in short, primarily through practice. While he concedes that his view is a form of "social constructivism," Vogel intends the metaphor of construction quite literally—and thus he can answer the charge of idealism. Hegel had argued that we know the world in itself because we continually constitute it; Marxists identified this constitution more concretely with human labor. Vogel agrees, "It is through our practices, . . . [and] above all laboring practices, that the world around us is shaped into the world it is."[16] For Vogel, practice is not ontologically secondary, the activity of sovereign subjects working on objects, or the vain attempt to grasp always-inaccessible things in themselves: Practices are rather prior to both subject and object poles. As Vogel notes, I am always *in media res*, "finding both myself and the world in which I act to be the products in turn of earlier practices."[17] Practices just are the tie between us and the world and the medium of our mutual interaction.

Vogel's focus on practice also permits a reply to the charge of relativism. Unlike the naïve romantic and the poststructuralist paradigms, we need no outside standard—neither nature identified as origin and authority, nor nature as some inaccessible (and thus unhelpful) difference or Other. All we need is the standard of self-knowledge inherent in practice itself. Negatively, we must identify and overcome the alienation inherent in our practices of construction—that is, our regarding as "natural" what is in fact the effect of our own activity. Nature as hypostatized Other in both romantic and poststructuralist versions is one obvious target, but so are nature as moral, metaphysical, or aesthetic authority; nature as passive field of experimentation and manipulation; and nature as economic resource. Having recognized ourselves and our world in our practices, we can adjudicate what practices we will continue to affirm by the standard implicit in practice itself: viz., "[w]orld-constituting practices that acknowledge themselves as such, that know their implications and take their responsibilities seriously, are to be preferred over those that do not."[18] By challenging us to "acknowledge our own entanglement in and indeed responsibility for" constructing the world in which we find ourselves, Vogel's proposal is inherently liberatory: For to admit the many signs of poor ecological health in the world around us is implicitly to accept the call to construct a world we'd more happily acknowledge as our own.[19]

Is the Concept of Nature Hopelessly Compromised?

Is Vogel right? Can we, ought we, indeed must we relinquish the concept of nature as hopelessly confused? Perhaps—but my aim is to redeem it. Our general concept of nature is certainly ambiguous: Concrete descriptions are socially, historically, and culturally dated; and we cannot secure them either by studying nature "in itself" or by distinguishing "natural" from "unnatural" activities, artifacts or wastes. The problems Vogel highlights have been acknowledged far too rarely and reluctantly. Yet if I am sympathetic to Vogel's philosophical reservations, I'm also chary of relinquishing an old and resonant word. Given our desire to preserve a global ecosystem long under attack and now vulnerable to collapse, it is impolitic to abandon a word with such positive metaphysical, theological, and aesthetic connotations.[20] Fortunately, the concept of nature *can* be redeemed—not as something isolated from the human, but as an essential aspect of our human experience of the world.

Moreover, it would not only be politically useful to save the concept of nature, Vogel himself needs either it or an analogue both to express scientific commonplaces and to carry out any critical reflection on practice. For first, in one obvious sense we must conceive the world as independent of us—and hence of our linguistic constructions. Humans of roughly our physical form have existed for at least 100,000 years; have had the linguistic and cultural capacities for adventurous exploration, colonization, and adaptation for 50–60,000 years; have husbanded land and animals, modifying environments on a landscape scale with fire, irrigation, and farming for 10–12,000 years; and have dominated ever larger areas in ever more intensive, industrially mediated ways for the last 200 years. "Untouched" nature has indeed been gone for a long time or for very much longer, depending on what we take as the threshold of "significant" influence. Yet on a geological scale—let alone that of the universe—we're clearly latecomers, so we need a word for the ordinary insight that nature has existed without us for most of its history. Moreover, our imaginative and scientific capacities lend us the freedom not only to conceive but to think concretely about the world beyond us and our effects. If the History Channel's documentary series "Life after People" is admittedly speculative, it nonetheless meaningfully considers how our artifacts will decay after we can no longer maintain them.[21]

The second reason to preserve the concept of nature is less obvious but more critical. Vogel summarizes what I accept as indisputable evidence that we have conceived nature in historically and culturally particular ways that have changed over time. If, in addition, we accept the linguistic turn as the common fruit of the last century in both continental and Anglo-American philosophy, then we cannot naively check our changing conceptions of nature against some prelinguistic given. Vogel offers a good first answer to the worries about idealism and relativism that follow on those developments by focusing on practices as reciprocally producing us and our world, and by asking what practices we would acknowledge and affirm. But evaluating practices presumes a (contemporary) interpretation of what they are and of how they might be good, and either interpretation may be shortsighted or wrong. The question of truth is thus deferred but not eliminated. We can and do evaluate practices, and over time our practices can improve[22]—but we need to reflect more concretely on *how* this happens. For if we can gain better insights into the world in general, we can also improve our knowledge of nature in particular. Our challenge, then, is to understand how we can conceive the world as transcending any current interpretation and how we modify that interpretation in the light of our new insights. To do this, I will introduce Gadamer's philosophical hermeneutics, which was created to bridge the untenable dualisms of romantic aesthetics, historicist historiography, and post-Kantian constructivism. Although he is famous—or infamous—for highlighting our indebtedness to tradition, Gadamer also reveals the limited transcendence we can achieve through reflection on concrete experience. As we will see, a sufficiently concrete consideration of how this happens demonstrates that we need not do away with nature, even though we must (and must continually) reconceive it.

Hermeneutic Reflections on How Language Reveals the World—and Nature

Turning to Gadamer for help in gaining access to the world in itself may appear quixotic. He accepts Hegel's critique of enlightenment and romantic appeals to immediate intuition and also Heidegger's postmodern insistence that Dasein cannot secure its awareness of self or world since it is always already Being-in-the-world. Worse still, Gadamer develops Heidegger's early insight that linguistically mediated presuppositions preshape our experience of the world into a version of the linguistic turn. And if we cannot see the world either naked or whole, we must begin with inherited interpretations— thus Gadamer's "rehabilitation of tradition." Yet his defense of the authority of tradition offends both modern and postmodern aspirations to escape the dead hand of the past. Worse still in an environmental context, does it not also condemn us to understand nature in traditional categories that have already proven toxic?

   This view of Gadamer's work is widespread, understandable, and deeply mistaken.[23] I cannot now defend Gadamer in detail, but in what follows, I will elaborate his hermeneutic analysis of experience— the core of his reply to the charge that we are condemned either to repeat or irrationally to reject the views we inherit. The charge against Gadamer is clearly plausible, since a parallel problem haunts every attempt to think through the implications of the linguistic turn: thus, for example, Quine's worries about the indeterminacy of translation,[24] Peter Winch's Wittgensteinian qualms about cross-cultural communication,[25] Kuhn's difficulties explaining the rationality of paradigm shifts,[26] and the performative problems implicit in the Sapir-Whorf hypothesis that language structures our experience of the world. Yet if the challenge Gadamer faces is both common and critical, his virtue is to trace the problem to its root—our long-standing tendency to hypostatize both concepts and the world—and then to resolve it by revealing the continuous, dynamic, historical, and hermeneutic interaction of concept and world in concrete experience.

Gadamer addresses the challenge posed by the linguistic turn in three basic moves.[27] Since we cannot secure our perspective in imme-diate evidence or a grasp of the whole, we must rely on inherited presuppositions. The modern, of course, takes offense: Might not our presuppositions be wrong? Gadamer's first response paraphrases Hegel's retort to the foundationalists: Might not our error be the fear of error itself?[28] If we experience our presuppositions as illuminating the world in the first place, why assume that they're wrong? On Gadamer's view, our trust of tradition, like our trust in expert authorities, is sufficiently well grounded in experience: We accept expert advice because we so often find it right.[29] Like Hegel too, however, Gadamer realizes that once the skeptical challenge has been raised, it cannot be answered merely by shifting the burden of proof.[30] Gadamer thus elaborates the language-world relation and, even more importantly, shows how we evaluate our linguistically embedded presuppositions even though we lack access to the world "in itself." We must consider these second and third moves in turn.

On the traditional view—widely accepted from Plato to the early Wittgenstein—concepts are mere labels used to grasp ontologically prior things. Poststructuralists, in contrast, typically reverse the relation of priority: They take the failure of ancient realism, rationalist and empiricist foundationalisms, romantic intuitionism, and Hegelian holism as showing that language preforms our experience of the world. In contrast to both, Gadamer resists prioritizing either concept or object, arguing that they are, and are essentially, correlative.

To show this, Gadamer starts from Heidegger's startling observa-tion that among all animals, only humans have a world.[31] All animals live in and react to their immediate environments, but language alone makes possible our conceiving the world *as world*.[32] As Hegel has shown, sensation may claim access to our immediate environ-ment but it clearly cannot reveal the world as a whole,[33] for that includes not just what's immediately present but also what's far away, the deep past, the distant future, as well as what is actually true despite our current misperceptions. None of these are accessible, let alone combinable, except by means of language alone. Yet while language necessarily conditions our experience of the world, Gadamer joins Hegel in resisting the analysis of language either as a tool or as a medium.[34] Through careful phenomenological reflection, Gadamer reveals the limpidity of language. Concepts never thematize themselves but disappear in bringing the world into view. The language-world relation, again, is correlative: "[L]anguage has no independent life apart from the world that comes to language within it. Not only is the world world only insofar as it comes into language, but language, too, has its real being only in the fact that the world is presented in it."[35]

So far so good—but if language always presented the world as it was, we could not explain the evident fact of error. Gadamer certainly intends no naïve realism. And if he denies the traditional correspondence view that language mirrors the world in favor of the expressivist view that the world exists in, and only in, language, it is obviously true that language sometimes gets the world wrong. In fact, he not only acknowledges the possibility of error but distinguishes "normal" expectation-confirming experiences from surprising ones, noting that the "latter—'experience' in the genuine sense—is always negative."[36] Paradoxically, however, the negativity of experience is curiously productive: For it is the disappointment of our expectations, and not their confirmation, that leads most directly to a better understanding. Yet while Gadamer values this dialectical element of experience, he denies the Hegelian hope that it could end in full comprehension. The truly experienced person has better judgment than the rest of us, but she is also the least dogmatic: Her richer experience has revealed how readily she may yet be surprised.[37] Gadamer thus testifies with Aeschylus that we learn through suffering, and not just by suffering this or that particular thing but by discovering the limits of humanity. In this sense, experience is ultimately the "experience of human finitude."[38]

If the world comes to be through language but language can get the world wrong, Gadamer must show how we learn through linguistically constituted experience. He does so through a concrete, supple, historical account of concept formation.

The key to this, Gadamer's third move, is to relinquish our tendency to hypostatize concepts. Concepts, of course, have meanings that we usually use unproblematically to refer to a world that appears equally obvious. But just because our conceptually constituted expectations should capture the world as it is, their disappointment poses a critical choice: We may continue in denial of our disappointment, applying the old concept but subtly shift our expectations by extending or reinterpreting it, or turn to another concept we had not expected to need. Just as (and because) our experience is historical, our concepts must be historical, too—stretching, adapting, and multiplying to capture the world of our actual experience.[39]

The closer we attend to experience, moreover, the more obvious this dynamic becomes. Though my spouse is the occupational therapist, she defers to me to explain her profession to new acquaintances since my distance from the field frees me to oversimplify ruthlessly. Conversely, the people closest to us are the most difficult to describe—and not because we don't know them but because we know them too well to be entirely happy with any finite description. The frustration this evokes may tempt us to wish to communicate wordlessly—but since that clearly cannot help another understand, we slog on to explain why what we've said so far is not adequate. Tact often limits how long we may continue, so human speech "brings a totality of meaning into play, without being able to express it totally." But that offers no permanent barrier to adequacy, for "every word, as the event of a moment, carries with it the unsaid, to which it is related by responding and summoning"—and thus when it's necessary, we find the time and energy to converse further.[40]

Gadamer concludes that the concept of the world in itself is deeply ambiguous. In one sense, it is both inaccessible and unnecessary:

> It is true that those who are brought up in a particular linguistic and cultural tradition see the world in a different way from those who belong to other traditions. It is true that the historical 'worlds' that succeed one another in the course of history are different from one another . . .; but in whatever tradition we consider it, it is always a human—that is, verbally constituted— world that presents itself to us. As verbally constituted, every such world is of itself always open to every possible insight . . . and is accordingly available to others.
>
>   This is of fundamental importance, for it makes the expression *"world-in-itself"* problematical.[41]

While God may have access to the world in itself, we do not: We always perceive it through the perspective inherited from our language and culture. Yet that is no final barrier to insight. Language itself alerts us that our view may be partial: "No one doubts that the world can exist without man and perhaps will do so. This is part of the meaning in which every human, linguistically constituted view of the world lives. In every worldview, the world-in-itself is intended."[42] Through language, the world first appears, and through the dynamic of actual experience, we learn over time where we must adapt our concepts in the light of new insights. We cannot tell in the abstract or in advance whether or how we are mistaken; that insight must await the disappointed expectations that first reveal the limits of our view. But over time, disappointed expectations inevitably occur—and thus provoke the further explorations by which we refine our insights. Finite and historical though our linguistically constituted perspective may be, it provides not only the structure by means of which we understand the world, but by means of which we recognize misunderstanding.

And thus at last we come to the point of this digression through hermeneutics: Gadamer illuminates the way we can and do reconceive nature. What's true of our concept of the world is also true of nature: The view we inherit is historically and culturally particular, but over time it can and will be reinterpreted in the light of new experiences. Some reinterpretations are especially dramatic, as when the romantics reimagined mountain "wastes" as natural cathedrals where one might venture to experience the sublime. In the light of McKibben's concerns about the death of nature, reservations about the now-traditional romantic view, and poststructuralist worries about our access to nature "in itself," we are wrestling with a similarly fundamental reinterpretation. But we must not throw the baby out with the bathwater. Our concepts shape our expectations, and thus our experience, and do so largely behind our backs. But language not only preinterprets nature; it also alerts us that nature may not be as we take it, demands that we distinguish between what we have expected and what we now see, and gives us the freedom to modify expectations in the light of our experience. And by such means we comprehend *nature* more adequately. We will never capture it naked or complete. But because new and surprising experiences inevitably occur, we cannot help developing a more adequate view.

Objections

I hope I have shown why we need not and cannot rid environmental philosophy of the concept of nature—though we must certainly take it in a historically, culturally, and hermeneutically more sophisticated way. We need the concept of nature to make sense of the biologically obvious fact that the world has and will yet exist without us. More importantly, Vogel cannot express the reason for critiquing prior practices without some way of identifying what that critique is meant to help preserve. Even though Gadamer accepts an expressivist account of language, his concrete and historical description of concept formation shows how we can recognize, identify, and correct the flaws in the traditional notion of nature. Quite reasonably, Vogel worries that traditional connotations will continue to mislead us, thus rendering the concept worse than useless. But as I suggested in the introduction, the concept of nature still has enormous political appeal; there's no better way to draw attention to its limits than by thematizing them explicitly; and conversely without that concept we lack a critical tool for drawing attention to what we want to preserve. Yet even if my argument has been compelling so far, I cannot close without considering three critical objections:

Have We Escaped or Just Reaffirmed the Negative Infinity of the Always-Deferred Difference?

We have certainly affirmed what Hegel would consider a "bad" infinity, since there is no way of closing the gap between the full truth we intend and the partial truth we have now. By the careful analysis of clear and distinct ideas or repeatedly accessible impressions of sensation, rationalists and empiricists had hoped to narrow that gap; romantics aspired to close it in a flash of aesthetic or religious intuition; and Hegel believed that a historically articulated, developmental learning process had finally uncovered the global structure of truth. Enlightenment thinkers thereby aspired to a quasi- (and in Hegel's case explicitly) divine grasp of the truth—and against that hope, Gadamer insists, with Kant and Kierkegaard, on the finitude of human insight.

Yet if he thus restricts us to a bad infinity, it's one that all now accept. Who would deny the gap between what we now think and what we may later believe in the light of some new experience? Certainly Vogel must presume it. As we have seen, his standard for adjudicating practices presumes an ideal (self-knowledge) that we will never fully achieve, so our perspective may and hopefully will improve in ways we cannot anticipate. He, too, must intend an always-deferred fuller truth.

But didn't Vogel rightly worry about the endless deferral implicit in both the traditional romantic and poststructuralist perspectives on nature? He did—but we must distinguish different types of deferral. On foundationalist assumptions, the endless deferral of origin means that we'll never make certifiable progress to the truth. Romantics and poststructuralists, on the other hand, refer us to a full truth—an experience of the Absolute, some Wholly Other, or friendship or justice or democracy "to come"—that is endlessly deferred insofar as it can never be conveyed in language. For Gadamer, on the other hand, deferral is necessary; what's problematic is *abstract* deferral. We must admit and indeed insist on the likelihood of a gap between what we think we know and what's really so, but we cannot simply gesture to full truth *jenseits*, "over there." How does Gadamer avoid such an abstract deferral?

First, he draws attention to the speculative structure of language as intending the whole truth even though every formulation is finite and particular. For the dialogical and processual structure of language does not prevent our *intending* a greater truth than we can say in any finite locution. More important, we may indeed have it. That I struggle to describe my spouse shows how well I actually know her. Full comprehension always outruns our capacity to speak the truth all at once but does not always outrun our capacity to intend it. To be sure, using a finite form of words to capture the fuller truth we intend but cannot express is dangerous. Such bare intentions appear abstract and are certainly prone to overconfidence and error. The solution, however, is not to give up the intention but to see how it is realized concretely.

For the homely truth on which Gadamer insists is that every language—created through the long experience of entire cultures—*does* embody important insights into our world. We're certainly fallible, but negative experiences continually prod us to richer (if still finite and fallible) perspectives. And this constitutes Gadamer's second key insight. We do inescapably intend a fuller truth than we now have—as we demonstrate concretely when we're presented with evidence that our old view is mistaken and find ourselves forced either to contest the new evidence or revise our view. And as we do that, the greater truth to which we have deferred is in a finite degree further *realized*. The process of determining and mitigating concrete weaknesses in our former perspective is what keeps Gadamer from both an abstract deferral of presence and a bare assertion of absence.

Even in Gadamer's view, then, two kinds of deferral remain: First, we cannot speak the truth all at once, but may truly intend it; and second, no matter how adequate we think our current view, we realize that it may yet become better. But neither deferral is deeply problematic. Insofar as we really *have* the truth, the first infinity is not bad, since it can be redressed by as much further elaboration as people have patience to offer or hear. And the second deferral—to an eventual better view—is not deeply problematic either. We always discover that we were wrong through concrete, negative experiences, in other words, through the disappointment of particular expectations. But every such disappointment is really a step to a better view. Absolute knowledge remains out of bounds, for we are not and will never be gods. But discovering the concrete limits of some former perspective involves not the abstract deferral of some ultimate truth but the concrete realization of a finite truth.

Isn't This Idealism Redux? What Happened to Vogel's Salutary Emphasis on Material Practice?

A second objection might recall a worry raised previously concerning post-structuralism. Does Gadamer's interest in the way language structures experience not constitute a radical form of idealism?

Were this charge well-founded, it would indeed be serious: I, at least, deeply appreciate Vogel's focus on concrete practices. Yet the worry is completely unfounded. Following Heidegger—and in this respect like Marx and Vogel—Gadamer understands Dasein as Being-in-the-world, that is, as "thrown" to a particular location from which it must "project." Dasein, in short, is concerned with nothing more urgently than its situation and the practices that its situation allows or demands. Indeed, the quickest retort to the charge that Gadamer ignores practice would be that he thematizes the most characteristically human practice, the linguistification of the world. But Gadamer goes further: Language use is not merely the characteristically human practice but plays a pivotal role, since only through language can we conceive, recognize, and evaluate any one of our practices in the first place.[43] Gadamer never denies or minimizes the practices by which we are formed and form our world; he just highlights the way language enables our practices, as also everything else, to emerge into view.

Have I Not Merely Proposed a New Dualism Distinguishing Language from Nonlanguage Users?

In one obvious sense, I must answer "yes." This will not be a problem for Vogel, who rejects dualism but argues that we rightly distinguish language-using humans from nonhumans whose inability to communicate determinately excludes them from participating as agents in the moral community.[44] On both our accounts,[45] however, non-agent status need not disqualify nonhumans from consideration as moral patients. But that answer will not satisfy others, since Vogel and I may both be inconsistent. Doesn't our evolutionary and physical continuity with the natural world—a world from which we emerged and to which we'll return both phylo- and ontogenetically—block any radical split between language and nonlanguage users?

Against the charge of inconsistency, I have three quick responses. First, there are countless differences between us and every other species. Not all can be stigmatized as problematic dualisms, for many are obviously morally relevant. Our digestions determine the finite range of things we can eat—and thus I do the koala no better favor sharing my cheese sandwich than she does by sharing eucalyptus leaves. Similarly, it would be abstract and inconsequential to extend full membership in a moral community to species that lacked the capacity to participate in it.

Second, while translation between human cultures is already difficult and translation between species would be exponentially harder, the distinction between language and nonlanguage user is a defeasibly empirical. If Douglas Adams' Babel fish were discovered and we thereby realized that whales, chimps, blackbirds, or aliens were talking after all,[46] we certainly could welcome one another to a correspondingly broadened moral world.

Third and much more immediately relevant, nonhuman animals surely exhibit different (and differently recognizable) communicative capacities. A few animals use learned signals and many use natural signs to make their needs known in a virtually unambiguous way. Just as with cognitively compromised humans, we may recognize limited rights as agents and we certainly have grounds to advocate for many animals' needs as moral patients.

And finally even if we could one day confidently translate animal expressions, we would necessarily still interpret them from our point of view. With both romantic and historicist historians, Gadamer agrees that we can understand past ages; but unlike both, he denies that we can leap directly into the mind of the original author or into the "closed horizon" of some past age. On Gadamer's view, we enter the perspective of the other from within our own since we grasp another's experience by analogy with our own. Thus the duality of human and animal will remain even if we leave old dualisms behind. And again, if language use is critical to moral agency, the capacity to speak would presumably bind agents of whatever species more closely than bodily similarities to nonverbal species. Not every duality marks a problematic dualism.

Conclusion

Since we have traversed much ground, I'll close with a quick summary. Having noted many grounds for skepticism about the traditional concept of nature, Vogel argues that environmental philosophy abandon the concept entirely. He is clearly right to recognize the limits of the traditional concept. But the political potency of the concept of nature, its necessity for expressing common scientific truths, and its critical role in indicating what we want to preserve when we're evaluating practices we've come to question prevent us from abandoning the concept altogether. Admittedly, its limits will continue to cloud our vision into the indefinite future. But only by preserving the concept of nature (even as we articulate its limits) can we indicate what we cherish in the home that we can and must work together to protect.

PART

II

Situating the Self

Environmental Hermeneutics and Environmental/ Eco-Psychology: Explorations in Environmental Identity

David Utsler

Introduction

Environmental hermeneutics is, as the subtitle of this book claims, an "emerging field." It is not the case that philosophical hermeneutics and environmental discourse have not been thought together before. But a "field" suggests a body of knowledge that is at once diverse yet coherent: Diverse, in that there are multiple perspectives and applications; coherent, in that there are recognizable characteristics that make environmental hermeneutics identifiable as a particular way of engaging environmental philosophy. Hermeneutics itself is widely recognized and understood to have multiple applications across a wide variety of disciplines and themes. Indeed, "environmental hermeneutics" might be simply defined as the elaboration of techniques useful for interpreting "the environment" in environmentally related disciplines.

While interpreting environments as a form of a text is certainly one facet of hermeneutical thinking, a broader, more nuanced description is necessary and springs from the purpose of philosophical hermeneutics itself. Richard Palmer writes,

> . . . hermeneutics achieves its most authentic dimensions when it moves away from being a conglomeration of devices and techniques for text explication and attempts to see the hermeneutical problem within the horizon of a general account of interpretation itself.[1]

Palmer insisted on a broad conception of interpretation, by which he meant that in addition to the interpretation of texts, we seek to understand what is happening in the very act of interpreting itself. We interpret the world around us from the moment of waking until sleeping. "Interpretation is, then, perhaps the most basic act of human thinking; indeed existing itself may be said to be a constant process of interpretation."[2] Hermeneutics is an ongoing dialogical process between the interpreter and her world. This simply means that hermeneutics is not content with simple subject-object dichotomies but operates in the space of a dynamic complex of dialogical relations.

Like philosophical hermeneutics itself, an emerging environmental hermeneutics can and should likewise engage a wide variety of disciplines in a mutual reciprocity wherein environmental hermeneutics both contributes to and gains from such dialogue. The focus of this chapter is to realize this mutual reciprocity between environmental hermeneutics and environmental psychology/ecopsychology.[3] The reciprocal link this chapter will focus upon is the concept of "environmental identity."[4] I will follow the hermeneutical principle that all interpretation entails self-interpretation (a theme prominent in both Gadamer and Ricoeur, among others). Any environmental hermeneutics will likewise entail, explicitly or implicitly, self-interpretation in relation to environments—we can call this self-interpretation the environmental self. Likewise, environmental psychology is explicitly concerned with self-understanding in relation to environments and the ways in which such a relation shapes the psyche. I will proceed by first looking at a few ways that environmental identity has been conceived previously. I will then work out a more specifically hermeneutical understanding of environmental identity, drawing on the hermeneutics of Paul Ricoeur. In the section that follows, I will discuss the discipline of environmental psychology and the related thinking of those working in ecopsychology looking at how environmental identity is conceived therein. In the concluding section, I will argue that environmental identity in environmental psychology and ecopsychology is a psychological "environmental hermeneutics of the self." While it is valid to compare environmental hermeneutics and environmental psychology as distinct and complementary accounts of environmental identity, I will argue that there is also an instance of multidisciplinary accounts of the same reality.

Different Concepts of Environmental Identity

A concept of personal identity and the constitution of selfhood in relation to one's physical environment are not unique to hermeneutics. Some have termed it "ecological identity."[5] While various versions of ecological or environmental identity[6] have been defined and explicated somewhat differently, the unifying thread entails some form of self-understanding that derives from our relationship to the environment, whether "environment" is understood as "the earth," social or cultural environments, or built environments. For Mitchell Thomashow, ecological identity

> refers to all the different ways people construe themselves in relationship to the earth as manifested in personality, values, and sense of self. Nature becomes an object of identification. For the individual, this has extraordinary conceptual ramifications. The interpretation of life experience transcends social and cultural interactions. It also includes a person's connection to the earth, perception of the ecosystem, and direct experience of nature.[7]

Thomashow's definition shares features with a hermeneutical conception of environmental identity. In this quotation, he even notes "the interpretation of life experience" in terms of how we identify ourselves. While not an explicitly developed hermeneutical understanding, Thomashow recognizes that what he calls "life experience"— whether social, cultural, or in nature—is a process of interpretation.

Thomashow's definition of ecological identity pertains primarily to the individual. An individual and personal dimension to any conception of identity will be no less true than one derived from philosophical hermeneutics as well as other disciplines. But collective identities, even those pertaining to the environment, should also be considered. Insightful aspects of a collective environmental identity can be found in environmental justice literature in the work of Robert Melchior Figueroa.

Figueroa's development of the concept of environmental identity links identity to how individuals and communities understand themselves in terms of social location. For statistical purposes, communities are often defined in terms of abstractions, such as zip code or census data. In contrast, Figueroa argues that communities understand themselves in terms of "social location defined by cultural identity." This understanding is one way in which he defines environmental identity.[8] Figueroa expands upon his conception of environmental identity with the following:

> Recognition justice demands that we fully account for the situational aspects of group mobilization for environmental justice by understanding the individual and community environmental identities and environmental heritages at stake. An environmental identity is the amalgamation of cultural identities, ways of life, and self-perceptions that are connected to a given group's physical environment.[9]

For Figueroa, individual and community environmental identities are linked. This parallels very well with a hermeneutical conception of environmental identity that construes identity simultaneously in terms of the same and the other.

What is interesting in Figueroa's conception of environmental identity is that it is not only defined in terms of a relationship to the environment but it gathers all aspects of personal and communal identity that are connected to the physical environment of a person or group. This suggests a more dialogical understanding of the components of identity and self-understanding and (rightly, I think) assumes an integrated connection among them all. Such a conception, I would argue, is inherently hermeneutical in that it encompasses the play (or interplay) of multidimensional realities of existence. This aspect of a "given group's physical environment" would likewise make for an interesting dialogue between environmental justice studies and a "hermeneutics of place" as found in Forrest Clingerman's chapter in this volume. "Place" or "emplacement"[10] and social justice are linked together, which indicates an important dialogue between environmental hermeneutics and environmental justice that needs to take place.

A final hermeneutically interesting observation I would like to make concerning Figueroa's definition of environmental identity is how it is linked with what he calls "environmental heritage." He writes,

> Environmental identity is closely related to environmental heritage, where the meanings and symbols of the past frame values, practices, and places we wish to preserve for ourselves as *members of a community*. In other words, our environmental

heritage is our environmental identity in relation to the community viewed over time.[11]

What is immediately striking is how environmental heritage here is clearly predicated on some form or another of narrative theory.[12] How else could Figueroa's definition of environmental heritage obtain if not in narrative form? Environmental heritage is not merely the recounting of bare historical data but clearly involves some sense of meaning in relation to place for the people who possess the particular heritage. Moreover, if environmental heritage is environmental identity narratively configured over time it also brings to the fore (as does narrative) the notion of memory.[13] Again, Clingerman's chapter in this volume takes up a dialectic of memory of the place of time and the time of place. Although Clingerman and Figueroa are doing very different things, it is clear that similar philosophical ideas form the underpinning of their work.

What is evident in just the two foregoing examples of the concept of environmental or ecological identity is that whether explicitly thought of in terms of hermeneutics, self-understanding, self-perception, and all the components that make for grasping identity and constituting selfhood in relationship to the environment involve interpretation. All forms of environmental identity are inherently hermeneutical or, at the very least, are best conceptually explained with the help of hermeneutics. Environmental identity is always an interpretation predicated upon a dialectic between the self and the other than self of one's (and/or a community's) environment. This claim will likewise show itself to be true of environmental identity as it is understood in environmental psychology and ecopsychology. Before turning to that discussion, I will continue this exploration in environmental identity by developing the concept from philosophical hermeneutics.

Nature as One's Other Self/Other than Self: A Hermeneutics of Environmental Identity

My conception of environmental identity will be developed primarily, and nearly exclusively, from the philosophy of Paul Ricoeur. I will occasionally make reference to Hans-Georg Gadamer, whose own hermeneutics provides a fruitful field for environmental identity. I have in other places used hermeneutics to construct an initial understanding of environmental identity.[14] The present chapter develops environmental identity somewhat more completely, before placing it into conversation with environmental psychology in the next section.

Identity and selfhood for Ricoeur are not static categories but are constituted in a complex, multidimensional way.[15] This will be important for our understanding of environmental identity. If identity is not formed dialogically in relation with others, but is actually opposed or contrasted with otherness (we identify *this* one precisely because it is not *that* one), then there can be no such thing as an environmental identity—that is, an identity that can be understood as originating in dialogue with an environment. Let us then establish that identity is not reducible to mere subjectivity or the psychic sphere, but is multidimensional. Ricoeur opens up his work devoted to the "hermeneutics of the self," *Oneself as Another*, with these words: "By the title *Oneself as Another*, I wish to designate the point of convergence between the three major philosophical intentions that influenced the preparation of the studies that make up this book."[16] Notice that the three philosophical intentions (which shall be discussed in a moment) converge at a point from which Ricoeur's conception of selfhood emerges. One's self is oneself "as another," not an isolated autonomous subject. The three "intentions" that converge are "the detour of reflection by way of analysis, the dialectic of selfhood and sameness, and finally the dialectic of selfhood and otherness."[17] Let us look at each of these features and how they reveal an environmental identity in turn.

The main point of the first intention is that we come to the self reflexively rather than in the "immediate positing of the subject."[18] The reflexive aspect is a series of detours back to the self, which simply means that the self does not know itself with immediacy. I can say "I am" but "who" is doing the saying? "Who speaks?" is a fundamental category of reflective analysis for Ricoeur. The answer to that question can only be answered through detours, which can in the case of any given person be limitless. Reflection can be upon where I am from, my race, my religion or lack of religion, my family, the people I choose as my friends, how and what I work at, how and what I play at, and so on infinitely. There is no immediate access to an "inner core" or an "I" apart from reflective analysis through these many detours that answer the question of who is speaking and, Ricoeur adds, acting, recounting (remembering) him- or herself, and who is responsible.[19] What the path of detours reveals is that one is identified in a complex of relations to and through the other (mostly at this point by way of contrast and comparison, of other people, other places, and other things).

Can the physical world, the "natural environment," of which today there is so much concern, be counted among detours of reflective analysis by which we identify ourselves? Does the question "who?" apply here? Are problems related to global climate change, pollution, clean water, agriculture, sustainability, and so forth matters to which identity are relevant? Certainly, each and every aspect of our physical environment will not become a source of identity. Yet, there are places or experiences of places or physical entities that we find meaningful in one way or another, and this meaningfulness gives us a sense of who we are or by which we identify ourselves. Some are more direct or intentional than others. For example, I have fond and vivid memories of my grandparents' farmhouse that had a large stream that ran through the land behind it and a large wood. I spent countless hours walking through or jumping over different parts of the water. In the summer when it was hot enough to leave an almost dry stream bed, I would find wet muddy places in the shade and play with the crawdads. I caught snakes (much to my mother's chagrin and perhaps terror), set traps in which I caught little to nothing, and generally explored again and again this place of solitude and natural richness. To this day, I count part of who I am in relation to these experiences. What I learned about the world and how it works in the multiple little ecosystems I would watch for hours ("all the busy little creatures chasing out their destinies"[20]) formed aspects of my personality and worldview. These experiences were detours that answered the question "who?" for me.

Other detours are less intentional. For example, we cannot exactly dictate where we are born and grow up. Yet, aspects of our physical environment (and other related ones) shape our sense of self. This can be so negatively as well. I think of classmates from high school that these many years later talk about "how glad I was to get out of that place! I'm never going back!" In part their identity is influenced by self-consciously *not* identifying with a particular place. This, too, is a detour by way of reflective analysis, even if the analysis resulted in a negative reaction to a place. A detour by way of negation is no less a detour. The point is that environments, including "natural" or biotic ones, can be among the detours from which we identify ourselves.

The second intention underlying *Oneself as Another* is the dialectic between sameness and selfhood. Ricoeur uses the Latin *idem* to refer to the former and *ipse* that latter. *Idem* as identity is sameness in both a numerical and qualitative sense.[21] Applied to the person, I am myself and not another. Over time, I am the same person. The young boy who once traversed the streams and woods now sits in front of his computer writing this chapter and all the years between these two events do not change that identification. The qualitative sense is simply that two things can be "interchangeable with no noticeable difference."[22] *Ipse* is identity that "implies no assertion concerning some unchanging core of the personality."[23] In contrast to sameness, *ipseity*, or selfhood, accounts for the dynamic, changing aspects of personality in the same person.

This dialectic between *idem* and *ipse* offers little to our understanding of environmental identity, but there is one insight worth mentioning that will figure into the later discussion of ecopsychology, especially as it relates to therapy. Sameness and selfhood, as Ricoeur defines them, imply that a single person can change. Why is this important? Many environmental identities (ways of interpreting oneself in relation to nature) are not always necessarily good. In the case of those who interpret nature as an endless supply of resources to be exploited, for example, it would be unfortunate if identity was nothing more than an inner core to which we had unmediated access, unrelated to anything external, and fixed. That identity is arrived at through detours is revelatory in that it matters what detours are taken and the disposition in which they are taken. New detours can refigure the self in ways more environmentally benign. And for those who experience violence, conflict, or any number of psycho-social troubles, nature can be healing (more on this later).

The third and final intention is the dialectic of selfhood with otherness that is opened up by *ipseity*. Ricoeur notes that remaining "within the circle of sameness-identity, the otherness of the other than self offers nothing original . . ." but when otherness and selfhood are paired together it offers a way of being with

> otherness of a kind that can be constitutive of selfhood as such. *Oneself as Another* suggests from the outset that the selfhood of oneself implies otherness to such an intimate degree that one cannot be thought of without the other, that instead one passes into the other. . . (oneself inasmuch as being other).[24]

Unlike the first intention where the other is a detour back to the self, keeping self and other distinct, the third intention refers to a more profound intimacy where one is oneself by virtue of being other. Can nature be such an other?

Beginning with Aldo Leopold at least, there is no shortage of writing that proclaims human oneness with the nonhuman world. Much nature writing has emphasized oneself as another (our unity with nature) at the expense of the self (our uniqueness and difference from all other things). The self, however, is prior to the "one's other self." Otherwise what would the other than self be other to? Even though it is the case that we may discover the self more profoundly through experiencing the other, the self was always already there to have the experience. The tension between separation and intimacy must not be forgotten. Retaining tensions such as this one is a hallmark of hermeneutical thinking. After noting these issues, we should not hesitate to understand nature in terms of the third intention. I have elsewhere considered this intimacy in terms of vital being and corporeality.[25] Tradition and culture also join the human and nonhuman together creatively:

> Relationships with the land are generated from and enhanced through cultural histories, stories, and songs. Through my family, I encountered and gathered my Rarámuri cultural history through morals, ecological lessons and observations. Cultural history is more than a story; it's a way of perceiving ourselves as part of an extended ecological family of all species with whom we share ancestry, origins, and breath; a way of acknowledging that life in our environment is viable only when we view the life surrounding us as family. This family, or kin, includes all the natural elements of the ecosystem.[26]

In this passage, Enrique Salmon indicates the relation of his family and culture to the land while also extending his family to include the land. This is no doubt the experience of many indigenous peoples: that life is only "viable" when one's environment is understood as family in terms of what humans share with nonhumans. Salmon says in the same essay that while the Earth has only one voice, it speaks in many languages. Yet, many of these languages are not heard. "The unheard includes not only plants and animals, places or open spaces, streamsides or oceanscapes: they include people. More specifically, the cultures of people who maintain a sustainable and enhancing relationship with their land are at risk."[27] This notion of "many languages" calls to mind Gadamer's famous phrase, "*Being that can be understood is language."* But it is what follows this phrase, typically not commented on, that is of interest here:

> The hermeneutical phenomenon here projects its own universality back onto the ontological constitution of what is understood, determining it in a universal sense as *language* and determining its own relation to beings as interpretation. Thus we speak not only of a language of art but also of a language of nature—in short, of any language that things have.[28]

Gadamer goes on to speak of the "book of nature"[29] and if we are to hear the unheard languages Salmon speaks of, we may say that the problem is one of translation. Just as Salmon extends the notion of family beyond blood to "all natural elements of the ecosystem" (all the while one can realize the different senses of family without conflating or confusing the two), Gadamer extends the notion of language from human language to the way that any being speaks or communicates itself (all the while one can realize the different senses of language without conflating or confusing the two). Interpreting nature as "kin," extending a sense of family to nature awakens the need and desire to understand the language of nature. But why the metaphor of language? Simply, the purpose of language is communication and the purpose of communication is mutual recognition and mutual understanding. Thus, living together with others, human and nonhuman, is a function of language. Gadamer says that "language is a medium where I and world meet or, rather, manifest their original belonging together."[30]

The purpose of the foregoing analysis is to show that the human relationship to nature falls under the scope of what Gadamer calls the universality of hermeneutics. Recognition of this relation leads one back to how nature can be understood as one's other self. Nature, indeed, can be thought of in terms of the third philosophical intention of *Oneself as Another*. We can experience nature as a dialectic of selfhood (*ipseity*) and otherness so that nature is understood in terms of the intimacy Ricoeur highlights. Truly, one cannot think of oneself without the other of nature.

It is evident that the three philosophical intentions in Ricoeur's hermeneutics of the self lend themselves to the understanding (interpretation) of the human relationship to nature and how one's identity can be constituted or understood in this relationship. Environmental identity as a hermeneutical enterprise fosters a way of thinking about and encountering nature. I would also contend that the hermeneutics of the self, shown to apply to the human/nature relationship, demonstrates that an environmental identity is not a component of identity one chooses to have or not. The question is whether I will be self-aware of that component of my identity and what kind of identity or sense of self it will be.

The value of a hermeneutically constructed theory of environmental identity will be further manifested as we explore environmental identity through the lens of environmental psychology and ecopsychology.

Environmental Psychology and Ecopsychology

Environmental psychology and ecopsychology are two different but complementary fields. Each is relevant for our explorations in environmental identity. Before continuing this exploration through these fields, a few words should be said concerning their differences.

Environmental psychology is a branch of psychology concerned with the interaction between humans and their environments. It considers human behavior within environments, the effect of people on their environments, and, reciprocally, the effect of environments on humans. Environmental psychology considers environments as relevant data to understand human behavior—how and why humans act.[31] Like environmental hermeneutics, environmental psychology does not limit its conception of environments but considers all forms of environments whether natural, cultural, and so on. The literature in environmental psychology continues to grow, such as in the *Journal of Environmental Psychology* and textbooks such as Robert Gifford's *Environmental Psychology Principles and Practice*.[32] I will focus on one book in particular, *Identity and the Natural Environment: The Psychological Significance of Nature*,[33] as I continue this exploration in environmental identity. My choice to concentrate on this single book is due to the explicit focus of the editors and many of the authors on "environmental identity." Hence the dialogue at hand is between two conceptions of environmental identity, one drawn from philosophical hermeneutics and another from environmental psychology.

Distinct from environmental psychology, ecopsychology is defined by one of its most prominent voices, Theodore Roszak, as the joining together of the "psychological and the ecological, to see the needs of the planet and the person as a continuum."[34] There has been a strained relationship between environmental psychology and ecopsychology. An article critical, although not totally dismissive, of ecopsychology from an environmental psychology perspective appeared in the *Journal of Environmental Psychology* in 1995 by Joseph P. Reser.[35] As recently as 2009, Reser gave an interview in the journal *Ecopsychology* in which he expressed his admiration for ecopsychology and his hope for it to become more conceptually clear.[36] This mixture of caution and acceptance seems to characterize the attitudes of other environmental psychologists. For example,

> [d]espite the occasional theoretical insights and individual observations to be found in its writings, ecopsychology is not so much a descriptive or empirical psychology as it is an ethical and practical outlook in response to the present environmental crisis; like deep ecology, ecopsychology constitutes a self-transformative practice, and indeed has been formulated as a therapeutic approach. At the same time, these normative philosophical perspectives are by no means irrelevant for more conventional scholarly or scientific approaches, and the packaging of ecopsychology as "psychology" may help it serve as a moral guide or inspiration to the field of environmental psychology proper.[37]

The primary and essential difference between environmental psychology and ecopsychology it seems is that the former is a recognized field of psychology, properly and academically speaking, and the latter is a response to the environmental crisis and a dialogue with psychology, but lacking a full conceptual and theoretical development. That is not to say that environmental psychology is not concerned with normative outcomes or that one cannot draw out normative implications from the findings of environmental psychology. My observation is not that environmental psychology is merely descriptive and empirical whereas ecopsychology is concerned with environmental and ecological ethics. The difference, as I understand it, has to do with origins. Environmental psychology is rooted in psychology with all its methods, practices, and theoretical foundations. Based on the data, environmental psychology can certainly bring one to draw normative conclusions.[38] Ecopsychology, on the other hand, arose as a response to the environmental crisis, not out of a psychological science later applied to such a crisis. And just as environmental psychology's theoretical origins do not preclude a normative aspect, ecopsychology's roots in ecological crisis do not preclude the development of stronger conceptual foundations. Perhaps both would be best served by dialogue where environmental psychology sees its normative possibilities realized, in part, in the efforts of ecopsychology, and ecopsychology would strive for great conceptual clarity with the aid of environmental psychology. For the purposes of the analysis here, my interest is that for both the self is constituted and interpreted in various ways in relationship to environments. The differences, however, are necessary to point out. For purposes of clarity, I will look at environmental identity with environmental psychology and ecopsychology in separate sections.

Environmental Identity and Environmental Psychology

Susan Clayton and Susan Opotow have edited an insightful volume, bringing together a diverse group of scholars, to explore the "psychological significance of nature."[39] What joins the essays in the volume together is that each investigates how it is that our understanding of the self affects how we understand (and behave toward) nature and how the experience of nature also shapes and develops who we are. If nature and the human psyche are connected in any way, then there are going to be ramifications concerning that connection. "Environmental identity" conceived in terms of psychology is going to matter for personal identity and for social identity in relationship to the natural world and, as such, will also entail social consequences. The editors and authors hope insight into the connection between the psyche and nature can result, personally and socially, in more positive environmental outcomes.

Clayton and Opotow recognize that both "identity" and "nature" are not simply defined univocal terms. Each is complex and multivalent. Hence they "propose an integrative construct of environmental identity that encompasses multiple meanings as well as a recognition of the dynamic nature of identity."[40] While they do not use the term, Ricoeur's *ipseity* comes to mind as a way to describe their conception of identity. The recognition that environmental identity is not static and across individuals can be very different based on numerous factors and influences reveals that identity in relation to environments is a matter of interpretation, a polysemic "surplus of meaning." They acknowledge that different environmental identities can lead to different types of action, some good and others not so good. Although their aim is to understand the psychological significance of nature, the essays in the volume comprise a psychological environmental hermeneutics of the self.

Following William James, Susan Clayton agrees that persons have many "selves." So for Clayton

> an environmental identity is one part of the way in which people form their self-concept; a sense of connection to some part of the nonhuman natural environment, based on history, emotional attachment, and/or similarity, that affects the ways in which we perceive and act toward the worlds; a belief that the environment is important to us and an important part of who we are. An environmental identity can be similar to another collective identity . . . in providing us with a sense of connection, of being part of a larger whole, and with a recognition of similarity between ourselves and others.[41]

Environmental identity for Clayton is both a way we orient ourselves to the environment as well as being a product of interactions with our environment. This definition is similar to what was argued for previously. After all, it is not much of a leap to say that Clayton's account of environmental identity involves detour by way of reflective analysis, a dialectic of sameness and selfhood, and, especially in terms of the emotional attachment, the dialectic of selfhood and otherness—that is, all three features of Ricoeur's hermeneutics of the self.

Clayton's and Opotow's book contains a wide variety of essays, some more theoretical and others related to particular case studies. The thread that runs through them all is that nature is psychologically significant (as the subtitle of the book says). One final point here that I will mention (because it will come up again in ecopsychology) is the healing capacity of nature for psychological and social well-being. Two chapters, in particular, the first by Robert Sommer and the latter by Maureen E. Austin and Rachel Kaplan, raise this point in terms of trees, especially in the city.[42] Trees, it turns out, are important to those who dwell in cities and are not lost among the street lights, fire hydrants, and so on. Trees make places in the city meaningful. Sommer's chapter indicates the grief that some people in the city feel when they lose a tree. Austin's and Kaplan's chapter discusses the transformative character of introducing trees and other means of beautification to vacant lots. The effects are numerous from reducing crime to building stronger community identities.

Environmental Identity and Ecopsychology

James Hillman argues, "There is only one core issue for all psychology. *Where is the 'me'?* Where does the 'me' begin? Where does the 'me' stop? Where does the 'other' begin?"[43] For Hillman and other practitioners of ecopsychology, the "other" of nature is also a part of the "me." Beyond simple subject-object dichotomies, a chief insight of ecopsychology is its grasp of the complexities of embodied identity and that identity is not reducible to psyches and externals. He refers to "projective identification" meaning moments in which we attach ourselves to the other ("distant objects") so "fiercely that [we] believe [we] cannot live without them"[44] (this can obviously have both healthy and unhealthy connotations). The similarity of Hillman's observations with Ricoeur's third intention in the hermeneutics of the self is unmistakable.

The healing capacity of nature is a central theme in ecopsychology. Restoring the earth is to heal the mind, to borrow from the title of the book, *Ecopsychology: Restoring the Earth, Healing the Mind*.[45] In the area of environmental philosophy, J. Baird Callicott writes of "environmental wellness."[46] Callicott points out the interrelation of the wellness movement to environmental concerns. In contrast to other aspects of life over which we have a certain degree of autonomy (e.g., diet and exercise), Callicott further argues that because the environment is a "commons" it will take social and political means to make the environment a space for human wellness. He takes it as self-evident that the wellness of the environment is without negotiation necessary for human wellness. He writes,

> I don't think that you need to be a rocket scientist to figure out the environmental quality . . . is essential to overall human health and well-being. The sight of a person jogging through smog-clogged city streets is as much of a visual oxymoron as the sight of a person jogging along smoking a cigarette. An ersatz environment of metal, glass, concrete, and asphalt; congested automobile traffic; drugs, poverty, homelessness, and the street crime they breed—all of these seriously compromise our physical, emotional, and spiritual wellness. Environmental wellness, in short, is a necessary condition for human health

and well-being generally. We cannot pursue personal wellness unless we also work collectively and cooperatively to ensure improvement in our natural and fabricated environments.[47]

It is important here not to have one's attention diverted by the emphasis on physical health away from "emotional and spiritual wellness." In the previous quotation, Callicott points out that mental and emotional health is not just about having a positive outlook, but such health depends on several external factors as well.

By placing people in direct contact with nature in various ways, therapists have discovered that such contact leads to greater emotional and social well-being. The Gnome Project is one such enterprise that, among other things, has shown that children traumatized in war-torn or high-conflict areas are helped by direct contact with nature.[48] Interaction with nature seems to provide what we might call a hermeneutic reconfiguration of the self.

This foregoing discussion on health and wellness serves the purpose of highlighting the importance of developing environmental identity and, further, why a hermeneutics of environmental identity is relevant. Consider it the following way. It is reasonable to say that sensible people agree that health and wellness are better than being sick. Whatever people's varied and diverse conceptions of the "good life" might be, it is doubtful that many of them include mental or emotional sickness and distress. So if environmental hermeneutics can show us a theory of the self through environmental identity and ecopsychology can show us that identity is connected to nature, then we can argue for self-interpretation and self-development and growth in environmentally positive ways. Although in social life there is room for many different understandings and practices that constitute the "good life," there are others that do not. Environmental hermeneutics aided by the hermeneutically oriented conceptions of identity in ecopsychology (and this is just as true of environmental psychology) provides tools for critically understanding what the good life may be, to quote Ricoeur, "with and for others in just institutions."[49]

Conclusion

This exploration into different conceptions of environmental identity is intended to highlight the complementary nature of them all. Moreover, I am arguing that environmental psychology and ecopsychology and their accounts of environmental identity and self-understanding are instances of an environmental hermeneutics of the self. One of the things that makes environmental psychology and ecopsychology of interest to environmental hermeneutics is that, although there are many case studies and empirical data that undergird their conclusions, what remains inescapable is the inherent hermeneutical dimensions. Ricoeur, for example, has amply demonstrated the hermeneutical dimensions of psychoanalysis.[50] People interpret themselves in relationship to their environments and the way they do so is relevant. Environmental psychology and ecopsychology are hermeneutical in that they show how people understand themselves and how that understanding shapes and motivates action, which results in a rich consideration of what is ethical. As such, this essay provides another argument in support of the final chapter of this volume that wonders whether hermeneutics might save environmental ethics.[51] Because environmental hermeneutics can engage a wide variety of disciplines, it can only offer enrichment to environmental philosophy generally and environmental ethics specifically.

Environmental identity, I contend, is a genuine component of human identity. However, it is not understood in some direct and unmediated way. We cannot speak of some direct, intuitive access into the *eidetics* of environmental identity. Indeed, since it is a component of identity characterized by being "environmental," it in its very constitution situated! Identity in general—and environmental identity in particular—is hermeneutical, that is, identity involves a process of interpretation to come into its own. To say that environmental identity is hermeneutical is to say two things: 1) All identity, following Ricoeur, is not grounded in the immediacy of the subject. An environmental identity is a component of identity that obtains in an indirect mediated manner on various levels of our relation and interrelation to the natural world. 2) Likewise, environmental identity is not perceived in a simple dichotomy of the human subject and the natural object. Rather, it is manifest in a complex set of meaningful relations that result in multiple forms of self-understanding. If Palmer is correct, interpreting nature and all environments is a fundamental action of human existence. And if we follow Gadamer that all hermeneutical understanding is self-understanding[52] and Ricoeur that the self is constituted in the act of interpretation,[53] then interpreting environments yields an environmental identity. Any and all other forms of environmental identity are at their root hermeneutical—that is, are fully grounded and more deeply explained in terms of a hermeneutic mentality.

Gadamer observes that hermeneutics operates in an intermediary place such that "its work is not to develop a procedure of understanding but to clarify the conditions in which understanding takes place."[54] So, for example, when environmental psychology highlights the impact of nature on personality and offers definitions of environmental identity, it is hermeneutical principles that make such activity possible. The various ways that environmental psychology construes environmental identity are conceptually undergirded by the hermeneutics of the self. On one hand, environmental psychology and ecopsychology are clearly distinct disciplines complementary to environmental hermeneutics offering insights unique to their respective fields. On the other hand, environmental psychology and ecopsychology constitute a particular kind or instantiation of an environmental hermeneutics of the self, offering an environmentally relevant interpretation of human behavior in relation to and in response to environments. I do not mean to say that environmental hermeneutics encompasses environmental psychology or ecopsychology. But insofar as hermeneutics clarifies the conditions within which understanding is possible generally, environmental psychology and ecopsychology manifests this reality particularly.

My conclusion is that environmental hermeneutics and environmental/ecopsychology have a good deal to offer one another and would mutually benefit from dialogue. A Gadamerian "fusion" of these respective horizons would not only be mutually beneficial on the level of each discipline but would ultimately be beneficial to the environment in which we dwell (and, as such, as Callicott would say, is beneficial to us). In other words, it is characteristic of all three of the areas that comprise the subject matter of this chapter to be most fully realized in application. Gadamer forcefully argues that concrete real-world factors belong to the domain of hermeneutics.[55] Hermeneutics applies not to an abstract world but is operative right in the midst of the concerns of society. Both environmental and ecopsychology are concerned, as are all forms of psychology, with concrete results in people's lives. Therefore, the dialogue between these areas is not intended just for their academic enrichment inside the walls of the academy but ultimately to impact and improve our world. The link discussed here of environmental identity has the power to transform our conceptions of self, such that our dependence upon the natural environment becomes abundantly and unequivocally clear. A rightly construed environmental identity on personal and collective levels may be one of many tools employed as a sensible and effective response to the present-day environmental crisis and the debates over how best to address it.

Environmental Hermeneutics with and for Others: Ricoeur's Ethics and the Ecological Self

Nathan M. Bell

Narrative identity has been a significant area of focus in environmental hermeneutics. This concept of identity builds on Paul Ricoeur's formulation of the ethical intention of the self: *"aiming at the 'good life' with and for others, in just institutions."*[1] While Ricoeur's ethical intention has come up in relation to environmental identity, a full formulation of an environmentally focused version of it has not yet been developed. My intention in this essay is to examine Ricoeur's ethic as it applies to environmental ethics and philosophy. This application is a fresh way to address several major issues of environmental philosophy, while also contributing to Ricoeur scholarship by presenting a new application of his thought.

Environmental philosophy is very broad, seeking to formulate and address a wide range of questions regarding the natural environment. As a subdiscipline of philosophy, it draws on a number of philosophical sources to examine nature. Ricoeur's work is important to environmental philosophy because it provides a rich ground for reformulating and readdressing these questions. Regarding Ricoeur's ethical intention specifically, there are several important questions in environmental philosophy that can be addressed. One is the issue of the human-nature relationship: How do selfhood and identity relate to different conceptions of the environment? This question is related to the important question of what is right environmental action. A second question is how we should extend ethics to nonhuman others, particularly animals. In what ways, and to what extent, might we include nonhuman animals in ethical questions? A third and final question relates to the environment and social justice: In what ways and to what extent do the benefits taken from and the burdens placed on the environment relate to questions of social justice? As I show in this chapter, Ricoeur's ethical intention can address these questions in ways that are both new and complementary to current scholarship in environmental philosophy.

To answer these questions, I develop an environmental version of Ricoeur's ethic as it relates to environmental identity. I define environmental identity as an individual's understanding of her self related dialectically with her understanding of environmental aspects of human and nonhuman others, as well as the natural environment as a whole. Every person has an environmental identity (whether it is concerned, apathetic, or antagonistic) that is a necessary part of her full identity. A foundational aspect of identity, as we examine Ricoeur's ethic, is what one sees as the good life. For a person with a more engaged environmental identity, the good life involves interaction with and preservation of the natural world. Insofar as self-understanding is important for considering one's view of the world and the good life, it is necessary to look at various ways self-understanding might be affected. Because a person judges her actions against the good life, the interpretation of nature in relation to the good life is important to environmental identity. And environmental identity is important to one's actions. How we think about ourselves in relation to nature is the source of our actions in relation to nature—environmental identity is the source of environmental action. And it is our conception of the good life in relation to nature that will determine both.

Following this is the idea of being with and for others. For Ricoeur, this is a reciprocal ethics of the recognition of esteem. This ethical relationship becomes difficult when turned toward nonhuman others. And yet, against the interpretive framework of hermeneutic identity, such a relationship is possible. Relating environmental identity to animal others, the self interprets the animal other as a kind of self, and as such recognizes oneself as an other. This allows for an ethical recognition of nonhuman others (thus addressing the extension of ethics to nonhuman animals). Finally, there is the notion of being in just institutions. Outside of individual interpersonal relationships, people also live in societal institutions. The just institution extends interpersonal ethics to the anonymous many. For a person with a more inclusive environmental identity, it is important to strive for environmentally just institutions (thus addressing social justice questions relating to environmental benefits and burdens). This means striving for institutional recognition of the environmental aspects of others' lives, as well as environmental justice for other cultures. Through the consideration of these three aspects (the good life, being with and for others, and just institutions) in regard to environmental concern, I will show how Ricoeur's ethical intention interacts with a hermeneutic environmental identity and can address the aforementioned questions in environmental philosophy.

The Good Life . . .

I begin with the good life. Ricoeur notes that the good life is not an abstract universal, but an Aristotelian good that is "the good for us."[2] It is, significantly, "the nebulous of ideals and dreams of achievements with regard to which a life is held to be more or less fulfilled or unfulfilled."[3] The good life, then, is the ultimate end of all of one's actions. And yet these actions are also ends in themselves.[4] This creates a hermeneutic circle between the more important decisions one makes and one's conception of the good life, toward which these decisions are ends—an unending interpretation between the parts and the whole. In addition to this, one's very concept of self is expanded by the circular interpretation between self-understanding and action. In terms of ethics, this interpretation of self becomes self-esteem. The validity of this interpretation is a conviction, based on experience, that one can judge and act well in moments, which approximates living well.[5] One interprets her own sense of self against her actions, and her actions against the good life.

The interpretive element of one's view of the good life also relates to the natural environment and one's view of the environmental implications of action. Ultimately, how one views one's own actions (and presumably action in general) rests on the view one has of the good life. Take, for example, a person whose concept of the good life is wrapped up mainly in possession of material goods. This could include owning land, technological devices, vehicles, particular foods, and so on. For this person, actions that otherwise might be considered destructive to the environment are not necessarily seen in this way—they are, rather, just satisfying needs. Certainly there is evidence that this kind of interpretation is common among twenty-first-century inhabitants of the United States, and likely those in most of the major industrialized nations in the world!

This interpretive dynamic explains early U.S. history regarding the environment: The good life is reshaping the land, expanding and growing as a society, and so on. The European settlers of the "New World" saw vast, unsettled wilderness. Given this interpretive framework, their religious background, and the fact that they were escaping a society they saw as unfavorable, these people saw the open land as an opportunity and the wilderness as something to conquer. They interpreted the natural world as material out of which to create a new paradise for themselves.[6] This was an interpretation of the natural world, in relation to the self-interpretation of those people, which led to particular kinds of action. Their action then becomes an object of interpretation for us, whereby we can surmise what kind of environmental identity they possessed and how they interpreted the good life.

In contrast to such views, however, the good life could be interpreted as a harmony with nature. Under this view, the good life is not one without material enjoyments but is primarily focused on a lifestyle and kinds of enjoyments that are less or not ecologically destructive. Under such an interpretation, actions would be viewed differently, and in some cases negatively: Too much driving is polluting, eating meat is wasteful and destructive, and so on. The view of the good life shapes interpretation of one's actions.

We can see something of this view in more recent U.S. history. Take the rise of transcendentalism, for example: Humans were seen as existing simultaneously in a higher realm of spiritual truths (via the soul) and a lower material realm (via the body). This condition gave different value to wilderness: Natural objects were more reflective of higher spiritual truths, and so wilderness was the part of the material world where spiritual truths were most accessible. This interpretation of wilderness, as a place to reach spiritual truths, was dependent on humans being trapped in the material world but able to access the spiritual one. And so, for those who held the transcendentalist view (with its implications for selfhood), the good life involved achieving spiritual truths and moral development, which could be done in the wilderness. And so right action was not destroying wilderness to develop the land but preserving wilderness and going into wild areas without destroying them.[7] This is just one example among many of how changes in views of humanity and selfhood corresponded to new interpretations of the natural world and of right action in regard to the environment.

The dominion view and the harmony with nature view of the good life are opposite extremes, as are the early settler view of American wilderness and the transcendentalist view. Between such polar opposites there are many other potential interpretive viewpoints. For example, a person may have a view of the good life that is focused on material simplicity or outdoor enjoyments, but with a view toward what this person perceives as a more physically and emotionally healthy lifestyle, without a necessarily strong inclination toward environmentalism. In such a case, actions may not be viewed as ecologically good or bad, but the actions one views as good are, incidentally, less ecologically destructive. This is just one alternative example. The point is that one's view of the good life, as it relates (or does not relate) to the environment, will certainly shape how one views actions and, correspondingly, how one should act.

Therefore, if one wishes to prompt others to more environmentally responsible actions, then the relationship between the sense of self and the good life must be considered. To better understand how one's self-understanding might change, I wish to consider how environmental identity, action, and the good life are all hermeneutically interrelated. This is accomplished through a brief detour through Ricoeur's more general theory of interpretation, as it is found in *Hermeneutics and the Human Sciences* and other works.[8]

Ricoeur's theory deals with an interpretation of texts, which can be linked to our understanding of the world. An encounter with a text is a circular interpretation: We presuppose a meaning of the whole, which shapes how we discern the parts, and we encounter the parts as we read and build up/reinterpret the whole. In this interpretation, we encounter a world. This world is made up of the references of the text, which we encounter in interpretation.[9] If we are open to it, a possible mode of being-in-the-world can be encountered via the world of the text. Therefore, in being truly open to the world of the text, we see a way in which we could be, not just in that world but in our own world as well. This also involves self-identity; to understand the world differently involves understanding the self differently—it is a new self-understanding that causes a reinterpretation of the world in which we live. For Ricoeur, the hermeneutic circle is a circle between an understanding of the text and understanding of one's self.[10]

Ricoeur emphasizes that it is important to take what is unfamiliar or alien in the text and to make it one's own, which is the world of the text itself. Appropriating this world for oneself is to understand not just the text but also, again, to understand one's self; it is "to let the work and its world enlarge the horizon of the understanding" which a person has of herself.[11] The point is that we take the unfamiliar world of the text and make it our own, and in doing so we understand ourselves in a different way.

Through this detour, we can see how different encounters with texts can change one's self-understanding. The text of a nature writer, for example, makes distinct references to nature in a way that creates a world in which nature is recognized and has significance and value. The reader of the text, following the references, encounters this world. If the reader does not already recognize nature, and recognize it as having significance and value, then the world of the text (in cases where nature in the text clearly has significance and value) appears unfamiliar. This person makes this unfamiliar world her own—appropriates it—by opening up and letting it speak; appropriation is the key to understanding the text. The reader has experienced a change, not just in her perception of the world but also in her own self-understanding. Through the text, the reader now lives in a world *with* nature—which is to say, in a world in which nature is recognized, and is recognized as having value and significance. Prior to the appropriation of the world of the text, the value and significance of nature did not exist in the reader's world. The reader has, as Ricoeur would say, enlarged her horizon. Through the text, this person may now understand his or her self as having commonality with nature, or may just view the environment in a more positive way.

This is admittedly an optimistic view of what can happen to one reading nature writing. While this does happen in some cases, it would be incorrect to say it does in every case or that it will necessarily occur. A person could read environmentally friendly texts critically rather than sympathetically. It may be difficult to say environmentally sensitive interpretations are more valid (in a hermeneutic sense) than other ones. It must also be noted that the change in view or identity through literature applies just as readily to text that may have anti-environmental references; one could gain what we might call a negative environmental identity through such interpretation. It is nonetheless important to look at how various encounters fit in with the self-understanding that is integral to understanding the good life. This could apply, as well, beyond the text as a written work (e.g., art, film, music, and television), this likewise being dependent on the self-interpretation and the background of the individual interpreter.

Action itself is something that is, like a text, interpreted. Meaningful action is separated from its intention, separated from the agent, and open to a range of interpretations—these interpretations likewise being open to the interpretive model of the hermeneutic circle.[12] It is not only our own actions that are interpreted in regard to the good life, but action is generally interpreted. This explains why others' actions can seem so wrong to us but not to them; I may not understand why people drive fuel-intensive vehicles, while they do not understand why I care what anyone else drives (or eats, or wears). Further, since we interpret others' actions, these actions may influence us as other texts do. We interpret action—which may have its own effect on how we interpret the good life, our own actions, and ourselves.

Ultimately, one's view of the good life exists relationally with our own self-understanding, our own actions, and the various encounters that shape such understanding. This is true of one's identity generally as well as of one's environmental identity. When we do specifically examine the environmental aspect of identity, or when we relate identity to environmental problems, our conflicting views of the good life help to explain why people view the environment and actions differently, both in their own actions and those of others. This dynamic also provides a new way of examining the self-nature relationship. A question that emerges from the self-nature relationship, and one that is of crucial importance to environmental philosophy, is the question of the relationship between human and nonhuman animals, specifically whether this relationship can be classified as an ethical one. This takes us to the second part of Ricoeur's ethical intention, the "with and for others" from which to examine the human and animal relationship.

. . . With and for (Human and Nonhuman) Others . . .

Turning to "with and for others" brings us to the issue of self-esteem, which is the ethical dimension of the interpretation between the self and the sense of the good life. The self is worthy of esteem because of its capacity for ethical judgment; esteem is the recognition of the self as being able to judge.[13] However, self-esteem cannot be understood apart from solicitude, which itself cannot be understood apart from self-esteem.

The notion of solicitude is the heart of what it means to be with and for others. Solicitude begins with Ricoeur's take on the

Levinasian idea of the face-to-face; the other, in the face-to-face encounter, places an injunction on the self, calling it to responsibility, and the self recognizes the authority of the other to make this command. This recognition of authority is what makes the injunction (which was asymmetrical for Levinas) become symmetrical for Ricoeur.[14] The importance of this symmetry is that it explains how solicitude and self-esteem show that the self *needs* others, for now "the self perceives itself as another among others."[15] There are several elements to this idea, but the most important is the idea of similitude—an "exchange between esteem for the self and solicitude for the other."[16]

One's self-esteem is the result of her ability to judge ethically, as a self that sets goals and sees ends. If the self is to recognize the authority of the other, this other must also be able to judge, to set goals, and see ends. And yet, in first making an injunction on the self, the other has shown esteem for me, which solicitude is a response to.[17] The esteem of the self is affirmed through the recognition that the other has esteem *as a self*, or one with the same capacities of the self. "The esteem of another as myself and the esteem of oneself as an other," as Ricoeur says, showing that one has esteem for the capacity to judge in an exchange in which the other affirms esteem for the self, and the self affirms esteem for the other.[18]

The difficult question is whether we can incorporate other-than-human others, or at least animals, into solicitude, and thus into Ricoeur's ethical intention. Ricoeur's work has been applicable, so far, to considering environmental concern more generally, but can we move specifically beyond human environmental concerns? To what extent could one say that a nonhuman other recognizes one's esteem, and to what extent could one say that a nonhuman other has esteem? This is implausible, especially if we take esteem as the ability to judge and act well.

Something similar might be possible if we regard, rather than esteem as such, some kind of agency as the boundary for reciprocation. In other words, we might be able to say that the animal other is a self; not that the animal other is a self in the same way a human is, but that, insofar as the animal has some kind of agency, it is a kind of other self. The reciprocity with the animal other, including its agency or selfhood, is an interpretive measure. After all, it would be strange if, given the hermeneutic and narrative background of Ricoeur's project, we considered reciprocity to be a kind of objective recognition by two agents in a factual world. Rather, we could say that the self interprets an agency in the nonhuman other, and likewise interprets the others' judgment of the self; in other words, the self interprets the animal other as a kind of other self. Interpretation is at the heart of our ethical encounters with human and other-than-human others.

Recall that an important aspect of solicitude is that the self *recognizes* the authority of the other to make this command.[19] This seems to go back, ultimately, to an interpretation of the self. As what does one see one's self? If I see myself as a human in a very rational, culture-focused, anthropocentric way, then I most likely see the animal as being nonhuman in the sense of being less-than-human. The animal other, having less-than-human rationality, could not have significant agency or be an other self, and thus could never place an injunction on me.

On the other hand, what if I see myself as a human animal, or as just another earth creature, a plain citizen, a fellow voyager? The metaphors go on, but what if I essentially interpret myself specifically and humanity generally as (more than anything else) just another animal, or at least just another mammal? In that case, other animals, at the least other mammals, would count as nonhuman in a mere categorical sense. She is other-than-human, I am other-than-lion, we both are animal others. Regardless, under such a view animal others are other selves. The animal other is a self to which I am an other. And in this case, the animal other can place an injunction upon me.

As stated previously, difficulty comes in with the notion of esteem: Do animals have the kind of agency that might allow for recognizing the esteem of the other? Not moral agency, certainly, but what about a simpler notion of agent as seeking intentional action? Animals certainly do seek better and avoid worse states of affairs. Animals also certainly have better or worse conditions to be in; not just that we would say are better or worse, but that animals feel (in some way) is better or worse. If we interpret humans as just animals or just mammals, then it would seem easy to consider some level of agency of animals. But how might we consider the reciprocity of solicitude?

We can gain support for this idea from the work of Val Plumwood. Plumwood would disagree with the idea that intentional agency is necessary for moral consideration; she does, in fact, attempt to refute this idea. In Ricoeur's framework, however, it is unavoidable—for there must be some kind of mutual recognition in order for this kind of ethical relationship to occur. However, Plumwood argues for "our openness to the non-human other's potential for intentionality, including their openness for communicative exchange and agency." Plumwood states, "[O]ur willingness and ability to recognise the other as a potentially intentional being tells us whether we are open to potentially rich forms of interaction and relationship which have an ethical dimension."[20] Thus it is our willingness to recognize the other (I would say our interpretation of the other) which defines whether we can have an ethical relationship, rather than an ability or characteristic that the other does or does not have. To recognize the other as a "potentially intentional being," as Plumwood puts it, corresponds well with what I mean by interpreting agency in animal others, and what it means to call an animal other a self. For both Ricoeur and Plumwood, it is crucial that it is not just the self or just the other, but both, in relation, that defines the ethical situation.

An example of such a relationship might be fruitful here. In his essay "Thinking like a Mountain," Aldo Leopold recounts, as a younger man, hunting wolves with the idea that killing predators means higher numbers of game animal for sport hunters. After shooting at the pack, Leopold reaches the mother wolf just in time to watch the "fierce green fire dying" in her eyes, and he feels, based on this, that "neither the wolf nor the mountain" agreed with his then-view of predator elimination.[21] The dying green fire seems to form an implicit injunction against Leopold, the asymmetrical call to responsibility.

When Leopold says that neither the wolf nor the mountain agrees, however, he makes it symmetrical in his recognition, based on his apparent regret and his later change in view, of the authority of the other to make a judgment. Although this is not the same as a rational human judgment, it is Leopold's perception and recognition of the animal's ability to judge that matters. In Leopold's recognition of the authority of the wolf to make implicit injunctions, he is affirming the esteem of the wolf. But also, since the injunction by the other is an affirmation of the esteem of the self, the wolf is affirming the esteem of Leopold, or at least, since Leopold seems to read a kind of injunction in the dying green fire, for Leopold this is an affirmation of his esteem. Leopold was wrong, according to the wolf, because the wolf is able to judge.

This example shows how similitude can include animal others. For anyone who ascribes to animal others some degree of seeking or agency, while not necessarily human-level judgment, it is possible to interpret of the animal other a kind of judgment against the self, to which the self responds. There is thus similitude and solicitude: Between my interpretation of my self and my interpretation of the animal other I see an other self, and I respond when this other calls me to responsibility, acknowledging the others' judgment of me, which holds me responsible.

We could turn also to the idea of suffering. This is a more common notion in animal ethics—indeed, a large activist movement has been built specifically around ending animal suffering.[22] For Ricoeur, suffering is on the opposite end of the spectrum from solicitude. Suffering is not physical or mental pain but the reduction or destruction of the capability for action, experienced in the other as a violation against the self. Ricoeur argues that "here initiative, precisely in terms of being-able-to-act, seems to belong exclusively to the self who *gives* his sympathy, his compassion, these terms being taken in the strong sense of the wish to share someone else's pain."[23] As with solicitude, suffering comes down to interpretation. I might interpret animal others as incapable of mental pain, consider their physical pain to be irrelevant, or not consider animal action to be worth anything—and thus feel no pain in their reduction. On the other hand, I may interpret the animal's suffering as like mine, or consider the animal enough of an acting agent that the suffering or reduction of the other is something to be lamented, enough so to be felt as pain by me.

Given these considerations of ethics between humans and animals, what are we to say then about inter-animal ethics? If we ascribe to animals a level of seeking or agency, essentially some degree of selfhood, then what is the implication of this for cases where animals harm each other, such as natural predation? It would seem at first that this may be a conflict of ends—for example, either the predator goes without food or the potential prey gets eaten, and either way an animal suffers and is harmed, perhaps even killed. On the other hand, animals exist in various natural relationships, and in such relationships animals exist in accordance to their own seeking or agency, in accordance to their selfhood, up to the end—which does not seem to indicate any kind of ethical error on the parts of animals. It would also seem intuitively inaccurate to call animal predators in nature unethical. Either way, the question cannot be fully answered in the scope of the present work. Animals exist relationally, exhibiting their own forms and degrees of selfhood; this may have interesting implications, but it does not seem to, theoretically or practically, be as much of an issue as human harm to animals and human-animal ethics.

To sum up thus far, we might put the argument as follows: A person's self-interpretation and corresponding view of the good life shapes one's interpretation of both actions and others. In turn, one's interpretation of others influences ethical consideration. Every person has an environmental identity (whether it is caring, apathetic, or antagonistic to the natural world), which will shape that person's interpretation of right action. Further, this view will define a person's interpretation of other-than-human others, including the openness (or lack thereof) of that person to an ethical relationship with other-than-human others. This provides a new way of thinking about the inclusion of animals in ethics: In addition to ethical relationships with other humans, we can potentially have ethical relationships with nonhuman animals. If we move beyond personal ethics, however, this brings us to the question of human social justice and the environment. A major issue in environmental philosophy is this very question, which we can complement and address with the final part of Ricoeur's ethic: being in just institutions.

. . . in (Environmentally) Just Institutions

Addressing just institutions acknowledges that ethical life goes beyond interpersonal encounters. For Ricoeur, the ethical content of justice is equality, and involves the determination of the rights of each person. "Institution" is the "structure of living together" in a historical community, which is bonded by common mores rather than constraining rules.[24] When dealing with institutions and justice, we are dealing with the inclusion of the anonymous, the third party to interpersonal encounters, the inclusion of which goes beyond the instantaneous moment of interpersonal relationships to a span of time. This span of time has less to do with the past than with the future, with "the ambition to last—that is, not to pass but to remain."[25] So "in just institutions" refers to the inclusion of the community, aiming for equality over time.

The crucial notion in equality is distributive equality, both of goods and of roles, tasks, and advantages. This equality is not to be confused with straight egalitarianism. For Ricoeur, the point is to create a link between justice and equality because equality is to the institution what solicitude is to the interpersonal encounter. Justice turns the "other" into the "each" and extends solicitude to all humanity. Whereas earlier we spoke of recognizing the authority and esteem of the other, we now recognize the authority and esteem of all other persons.[26]

Ricoeur's thinking on just institutions is applicable to the environment if we align some of Ricoeur's thought to the major elements of environmental justice theory. First, a major aspect of environmental justice itself is the idea of distributive justice—the question, as Robert Figueroa and Claudia Mills put it, of "how are environmental benefits and burdens distributed?"[27] As David Schlosberg notes, "environmental inequality occurs when the costs of environmental risk, and the benefits of good environmental policy, are not shared across the demographic and geographic spectrum."[28] So a primary concern is the distribution of both environmental benefits and burdens.

Ricoeur notes first that there is an element of distribution in considerations of justice, with regard to which institutions have a necessary role: "external and precarious goods . . . goods to be shared, burdens to be shared."[29] Furthermore, tied to the idea of being part of an institution, there is the "distinction of shares assigned to each individual."[30] Following my conclusion from the previous section, the argument would go something like this: If the natural environment is relevant to a person's conception of the good life, then we can have ethical obligations to others insofar as the environment is concerned (e.g., if someone is harmed by pollution). Because justice is the extension of ethics to rights, it follows that the environment, insofar as it affects people, can be a relevant dimension of justice. We see this first in distribution of both goods and burdens. So, under a Ricoeurian environmental ethic, we can find an easy fit to the idea of distribution of environmental goods/benefits and burdens: both goods and burdens must be shared.

The previously defined concept is a necessary step; however, environmental justice goes further, raising the issue of participation. Figueroa and Mills pose this question: "[H]ow are these distributive decisions made? Who participates in their making?"[31] Schlosberg likewise notes, "[D]emands for individual and community voice and self-empowerment have become a central part of the environmental justice movement."[32] Beyond the problem of distribution, there is the problem of participation: Who decides the distribution? Often those who make the decisions are those who get more of the benefits, and those left out of decision making often get more of the burdens. However, even if distribution was equitable, it remains a question of whether people have a say in the distributions that affect them, as individuals or as a community.

A similar, though distinct, issue comes in the form of recognition. "Recognition" as an environmental justice issue deals with misrecognition of certain peoples. Schlosberg notes that at first "descriptions and critiques of misrecognition focused most explicitly around issues of race and racism."[33] Misrecognition can also relate to class or other cultural aspects, as both Schlosberg and Figueroa note.[34]

The question then is whether Ricoeur partly or fully accounts for some of these further aspects. He does say that distribution "must not be limited to the economic plane," meaning merely to the distribution of material goods (in this case, clean air, water, and usable land, on the one hand, and pollutions or toxins, on the other, would constitute the material nature of this issue).[35] Rather, institutions "govern the apportionment of roles, tasks, and advantages or disadvantages between the members of society," so distribution would include these things also.[36] Roles and tasks could easily be conceived to refer to political roles and tasks; in other words, the roles and tasks of decision making, of participation. Likewise, advantages and disadvantages could refer to the advantage of having, or the disadvantage of not having, a power position or even a voice. While Ricoeur does not speak explicitly of recognition in this work, having a voice would mean some level of recognition. Indeed, Schlosberg notes the relationship between participation and recognition—a lack of recognition will certainly lead to a lack of participation, and participation will foster recognition.

While I have no intention of disrespecting either the nuances or the ramifications of the distinction between participation and recognition, for now we will treat them as parallel and linked issues. This presents a potential problem: Does reducing participation/ recognition to distribution undermine the potential of using Ricoeur's work for environmental justice? Schlosberg, drawing on several thinkers, discusses several objections to treating recognition as distribution. Some claim recognition and distribution go hand-in-hand to the extent that gaining the latter means gaining the former, but this may not work in the real world; without recognition first, distribution may not happen. Further, recognition is not, or at least not only, a good to be distributed.[37]

Ricoeur does partially address participation, noting a link between participation and distribution: "it is to the extent that the shares distributed are coordinated among themselves that those who have them can be said to participate in the society considered, to use Rawls' expression, as a cooperative enterprise."[38] In other words, there is an element lacking if we only consider distribution—people are considered to be a *part* of society if they have a *say* in distribution. As Ricoeur says, "Being part is one thing; receiving a share is something else again."[39] There is a relationship between distribution and participation, which cannot be reduced to one or the other.[40]

So far, it seems that Ricoeur gives us a basis for developing a theory of environmental justice out of an environmental variant of his ethic. What I have shown, thus far, is that an environmental ethical intention based on Ricoeur's ethical intention would include environmental justice, and that what we can get from it is consistent with much of the theory on environmental justice. I do not mean to say, at this point, that Ricoeur has given us something *new* for environmental justice. While he gives us another approach to environmental justice, he does not cover new ground. However, we may get something new out of the second aspect of Ricoeur that relates to environmental justice: the idea of remaining.

Recall that for Ricoeur there is a temporal dimension to justice that is aimed toward the future: with "the ambition to last—that is, not to pass but to remain."[41] Justice has to deal with the mortality of humans and the temporal dimensions that go with it, as well as the temporal finitude of institutions. Further, temporality has to do with power, which "exists only as long as people act together and vanishes when they disperse."[42] While this ties to institutions as such, I believe it is also applicable to communities—communities achieve a voice, achieve power only while they are together, and this voice disappears if they are dispersed.

The aspect of remaining over time relates to environmental justice in several ways. One such way deals with the idea of cultural identity: What does it mean to remain as a culture? Figueroa and Mills note that in the paradigm of development, controlled by the industrialized global North, "long-standing cultural traditions of indigenous groups . . . and many environmental values are sacrificed."[43] Cultural traditions, and thus the cultural identity of people, are forcibly changed. However, for people to truly remain, as justice indicates, their own interpretations and their own ideas of the good life must remain. For a group of peoples to remain means to preserve their culture—their practices, identity, language, and so forth. This ties in to recognition. The misrecognition of cultural groups, itself an injustice, can lead to the disruption of traditions and essentially the erasure of that culture. In addition to the inherent injustice of misrecognition, such an erasure of cultures both permanently eradicates the necessary conditions (in the logical sense) for that group to have power, creating a permanent misrecognition, and eradicates that group itself. This violates the rights of that community or culture to remain, and the rights of individuals of that group to remain insofar as their identities are tied to being a part of that group or community.

David Kaplan, in reference to development ethics, points out how imagination is related to the ability to remain:

> Imagination, so conceived, is fundamental to human experience. It is essentially for cognition and volition. As a result, any notion of human rights, typically understood as derivative of the basic right to think, choose, and act for oneself, implies the right to imagination.[44]

While Kaplan develops broad implications of this, the particular point that I wish to highlight is that culture is fundamental to how people think, choose, and act—in other words, culture is important to the imagination. Therefore, cultural preservation is integral, and the right to remain means, beyond just biological survival, the right to remain as a culture, with a distinctive way of thinking. This does not mean that all cultural practices are free from criticism (as Kaplan likewise notes), but to assimilate or forcibly distort other cultures is to take away that group of peoples' right to remain, as the cultural peoples that they are. As noted previously, this violation of the right to remain operates on various levels.

The right to remain could, moreover, be taken in a more literal sense: the right to remain on the land. Regarding the cultural traditions mentioned earlier in the Figueroa and Mills quote, such traditions often involve a certain kind of relationship to the land (which is why land is an environmental justice issue). The right of people to remain as a culture or community might be tied to their right to remain on their land. It could be a particular piece of land, with spiritual-cultural landmarks, or simply access to a certain kind of land, water, flora, and fauna. Both are disrupted by development and forced displacement.

This could also apply to particular groups within countries of the Global North. While their situations may not be as dire as those of indigenous peoples (and perhaps, in that sense, not as unjust), there is nonetheless a justice issue. Some people have ties to the land based on familial identity, coming from or belonging to a family of multigenerational landowners. Also, though it is less common in the Global North, there are people (often the same multigenerational landowners) who still "live off the land" in a very real sense. To displace such people through development is to disrupt their family-based identity and/or their way of life; again, to violate their right to remain. While this might appear as more of a recognition/participation issue, it is also a distributive issue. The access to certain environmental goods (in this case land) is inequitably distributed, or rather, in this case, unfairly redistributed.[45] This is problematic, however, as land cannot always be reasonably redistributed (if it has been greatly altered). Further, consequent harm to identity done through land misappropriation cannot be undone; this does not mean it is not a distribution issue, just that it is not exclusively a distribution issue.

As we have seen, it is possible to draw environmental justice out of the third part of Ricoeur's ethical intention. An idea of environmental justice must be a part of an environmental variant of this ethic. Ricoeur gives us material to develop concepts of distributive and participatory/recognition justice, giving us the potential for a Ricoeurian environmental justice that can fit well with current theories of environmental justice. Furthermore, with the notion of remaining, Ricoeur gives us ground to develop new ways of thinking about certain aspects of environmental justice and injustice.[46]

Another way in which Ricoeur adds to this issue is the reminder that this is likewise an issue of interpretation. Many would disagree that justice is an "interpretation"; is not justice universal? I would agree that some general notion of justice may well be universal, but the particulars are not. This would account for why there is unfair distribution and misrecognition: because of the variations of interpretation. This ultimately relates back to identity. One's own identity shapes her view of the good life, for herself and others—this determines what one could consider others to need to have a good life, and thus set the limits of distributive justice.

Furthermore, one's own identity will shape what she sees as the constitutive elements of identity generally, and thus shape recognition. For one whose idea of identity and the good life includes the environment, this means that the recognition of others includes the recognition of the others' environmental identity and environmental good life. Insofar as justice is the extension of ethics (according to Ricoeur), it would follow that justice must incorporate environmental aspects. As I have shown previously, Ricoeur's thinking on justice, put in an environmental context, is commensurable with and complementary to current philosophy on environmental justice. More to the point of the present work, we can also draw environmental justice out of the third part of Ricoeur's ethical intention, completing the environmental ethical intention and further addressing issues in environmental philosophy.

Conclusion

In dialogue with Ricoeur, I have argued that we have an environmental ethical intention. People do, in some cases, have environmental notions of the good life, live with and for others in ecological concerns, and seek to live in environmentally just institutions. We can, further, include animal others. We can (and some people do) acknowledge that animal others, while not aiming at a good life to the depths that humans do, seek better or worse states of affairs and have some level of agency as such (and the greater depth of human aiming does not imply greater value). Ultimately, Ricoeur's ethical intention will vary for different persons based on interpretations, and so can be applied both to other humans in environmental contexts and to at least some nonhuman others. An environmental application of Ricoeur's ethical intention provides fruitful discussion for further thinking on environmental hermeneutics, environmental philosophy, and overall considerations of environmental issues. My argument also shows along the way that environmental concern is shot through with interpretation: At each step of the way, interpretation, of the self and of others, has been a major factor.

The question that arises now is this: How can we argue and genuinely persuade others that interpretations that are more positive toward, or inclusive of, the environment are more valid? How do we argue that people *should* include nonhuman others? How can we argue that a valid interpretation *necessarily* includes environmental aspects in justice? I have only implicitly answered this question in the present work—environmentally positive interpretations seem to follow from a greater openness to the natural world and to others, and a more environmentally considerate identity seems to be a more full identity. In the case of environmental justice particularly, a person with a more inclusive environmental identity is more likely to work against environmental injustices, leading to more people achieving the good life.

The issue still remains—we need a stronger case for why, from a hermeneutic standpoint, environmentally positive interpretations and environmentally inclusive identities are more valid and appropriate than interpretations and identities are otherwise. The present volume as a whole shows that environmental hermeneutics is not a relativistic endeavor, and that there is validity to environmentally positive interpretations. At the same time, however, environmental hermeneutics must continue to strive to make a stronger case for the greater validity of environmentally positive interpretations. While this issue cannot be resolved here, in the present work I hope to have achieved further steps for those who aim at the ecological good life, with and for human and nonhuman others, in environmentally just institutions.

Bodily Moods and Unhomely Environments: The Hermeneutics of Agoraphobia and the Spirit of Place

Dylan Trigg

I have often said that all men's unhappiness is due to the single fact that they cannot stay quietly in a room.[1]

Pascal's Abyss

Shortly after his coach was nearly thrown into the Seine while crossing the Neuilly-sur-*Seine* Bridge in 1654, the French philosopher Blaise Pascal became convinced that an abyss had formed on his left-hand side. Quite apart from the logical improbability that such an abyss was real, this near miss of the Seine had set in a place a reality of Pascal's own, and one that was entirely independent of the objective properties of the world. Such was the extent of his anxiety that for a while Pascal would require a chair beside him to feel reassurance that he was not on the verge of falling. In a letter written by the *Abbé Boileau*, we are told the following:

> His friends, his confessor, and his director tried in vain to tell him that there was nothing to fear, and that his anxiety was only the alarm of an imagination exhausted by abstract and metaphysical studies. He would agree . . . and then, within a quarter of an hour, he would have dug for himself the terrifying precipice all over again.[2]

What is striking about this statement is that in it, we witness a struggle that takes place between Pascal's rationality assenting to the illogical nature of his fears and the irrationality of those fears resuming their powers. Evidently, the incident on the Neuilly-sur-*Seine* Bridge had shaped his bodily experience of space, meaning that no abstract mediation could alter his visceral experience of an abyss beside him, leading Baudelaire to remark that Pascal "sees only the infinite through all windows."[3]

Originally termed "Pascal's disease," the condition would later develop into various spatial phobias, including "*la peur des espaces*" [the fear of spaces], "*horreur du vide*" [horror of the void], *Platzschwindel* [place dizziness], and, finally, agoraphobia."[4] Marked in each case by an intense anxiety brought on by being in particular places, especially those in which the urge to flee would prove difficult, the condition has moved from a simple fear of public places to the more complex anxiety surrounding places that are in some broad sense unfamiliar. In such conditions, the attack can include sensations ranging from heart palpations, shortness of breath, vertigo, trembling of the limbs, tunnel vision, nausea, derealization, and, at its most severe, an abiding sense of impending doom or death. Because of this anxiety, the agoraphobe's experience of the environment tends to be characterized by a series of invisible boundaries and borders that demarcate familiar or homely places from unfamiliar or unhomely places. For this reason, interstitial environments such as elevators, bridges, and airports are especially hazardous for the agoraphobe, given that physical retreat would be comprised once those places had been entered.

As to the underlying anxiety that takes place in the agoraphobe's environmental experience, writing at the close of the nineteenth century, Freud's claim that "the recollection of an anxiety attack" is at the heart of agoraphobia's powers retains its theoretical accuracy.[5] At its core, agoraphobia is an anxiety disorder rather than a fear of places qua places. In this respect, agoraphobia is the necessary outcome of an anticipatory anxiety about experiencing anxiety. In their highly sensitized mode of being in the world, agoraphobic people tend to interpret their bodily response to the environment through a filter of apocalyptic doom. The formula invariably adheres to the following structure: "If I don't get out of here immediately, then X will happen," where "X" might include heart attack, fainting, or vomiting. Because of this catastrophic bodily interpretation, constant supervision of the body's sensations as well as a "superstitious avoidance behavior" accompanies all environmental experience, with an unflinching vigilance directed toward any unfamiliar or uncomfortable sensations.[6] Once detected, those bodily sensations urge the agoraphobe to flee the scene immediately, in the process reinforcing the sense that particular places are marked by the potential of danger.

Where does the agoraphobe flee to? The answer invariably accents the importance of the home in the agoraphobe's worldview. The idea of "home" is inherently ambiguous, ranging from a particular building or room, to a general neighborhood that assumes the appearance of being familiar. Various anecdotal reports testify to this ambiguity. Thus in some cases, the centrality of the physical home, with its borders and boundaries, marks a threshold from agoraphobic embodiment to non-agoraphobic embodiment, as Stewart Sadowsky writes: "Other clients are perfectly comfortable in their own homes despite the fact that they are alone and helpless should they have an 'attack.' The cozy familiarity of their home exudes a physiognomic air of safety and shelter. This permits them to cast off their constant concern about 'attacks.'"[7] In other cases, the locality of the physical home extends to a broader region, and might contain a multiplicity of different homely places. Kirsten Jacobson writes, "It is only when the agoraphobic is in his safe places or somewhere from which there is a reliable avenue to what he considers a home base that he feels comfortable and capable of carrying out his daily activities and interests."[8] As I shall spell out later on, this ambiguity is doubled by the fact that home relies as much on the environment as it does on the bodily way of being in that environment.

The Mood of the Body

In Pascal's brush with the Seine, we witness an incipient agoraphobia being conceived. For him, the "eternal silence of these infinite spaces" took on a concrete form when the terror felt on the Seine became a component of his environmental experience as a whole, implanting anxiety in the materiality of both his body and the world.[9] What is revealed in this confrontation with the abyss is an amplified facet of our everyday environmental experience, namely, *that we carry places with us*. Pascal's vertiginous experience at the Neuilly-sur-Seine Bridge was not insulated to that place. Instead, the experience became incorporated into his bodily schema as a whole, affecting a topophobic relation with place thereafter.

In this historical illustration, there are at least two insights that shed light on environmental hermeneutics. First, the *genius loci* of a given place—be it an abandoned monastery, a forest in the night, or the deck of a ship—is determined as much by the formal properties of a place as it is the "mood" we carry to that place. Only in the materiality of the body is a place given life, even if that life is essentially anxiety inducing. Second, our interpretation of an environment is primarily bodily in structure rather than cognitive. At stake in this interpretation is a reversibility between the world and the body, meaning that if we can speak of an anxious or ecstatic place, then it is only because we can also speak of an anxious or ecstatic body.

In this chapter, I will investigate these two claims by focusing on the phenomenology of agoraphobia. As indicated in the account of Pascal's brush with death, spatial anxiety affords us a clear illustration of how our experience of the world depends as much on the objective features of the world as it does the bodily mood with which we interpret those features. The epistemic advantage of agoraphobia in particular is that it draws our attention to that which is taken for granted in our everyday experience of the world. In doing so, agoraphobia foregrounds themes that are otherwise tacit: the contingency of boundaries, the vulnerability of home, and the unfamiliarity of our experience of the world.

Applying this phobic background to the present context, my central question is as follows: How does our bodily mood affect our experience and interpretation of an environment? The question points to a central concern for environmental hermeneutics, namely, the role of the body in not only structuring and constituting space but how bodily experience defines the very character of an environment. To do justice to the nature of environmental experience, we need to address the relationship between the body's moods and the variations in environmental interpretation that arise from those moods.

The inclusion of the term "mood" points to the Heideggerian notion of *Stimmung*, as it features in *Being and Time*.[10] For Heidegger, mood refers to the manner in which we "attune" ourselves to the world.[11] Attuning ourselves to the world is a fundamental given of being-in-the-world, given that, for Heidegger, being-in-the-world is affective in structure. In a word, *things matter to us*. The world is not a neutral canvas upon which we project our emotions and thoughts. Instead, we are "always already" in a mood with the world, as he writes: "The fact that moods can be spoiled and changed only means that Da-sein is always already in a mood. The often persistent, smooth, and pallid lack of mood, which must not be confused with bad mood, is far from being nothing."[12] Mood gives coherence to the world, disclosing and closing possibilities depending on the mood we are in. That we speak about being *in* a mood rather than approaching mood as an experience that is forced upon us confirms Heidegger's point that mood has an ontological validity to it. At no point are we *not* in a mood with the world.

In this respect, Heidegger's understanding of mood extends beyond the private domain of the individual world. My mood is not, as it were, cut off from a world external to me. Rather, my mood sheds light on an aspect of the (social) world that is already there. Far from the province of the psyche alone, mood thus implicates both the intersubjectivity of the world alongside the world's materiality. "It comes," so he writes, "neither from 'without' nor from 'within,' but rises from being-in-the-world itself as a mode of that being."[13] In this way, mood acts as a bridge between being and world, disclosing ethical and aesthetical values in the act of shaping the world in a particular manner.

Mood is that by which our relationship with the world is possible in the first place. Mood is the context out of which our intentional states are rendered possible. This is clear enough in the notion of a public mood, such as mourning, celebration, or a social convention that has embedded itself in a heterogeneous culture. In each case, mood becomes a phenomenon that is both shared by a collectivity of people and at the same time experienced in a first-person manner. Consider here the different sensibilities marking various cultures, varying from the "stiff upper lip" of a certain class of Englishness to the expressiveness of Italian body language. Such ways of being-in-the-world are sedimented to varying degrees in the sensibility of a public mood.

This mention of mood as manifesting itself in the body alerts us to a shortcoming in Heidegger's account. For although Heidegger has the distinction of elevating mood to a legitimate philosophical category, with its own unique mode of disclosure, what he omits in this study is the role our bodies play in shaping and determining the way in which mood is experienced. Given the pervasiveness of mood, this omission of the body is surprising. For it is not as though when we are bored or afraid that such moods color the world in an abstract fashion. Rather, we experience the tenor of those moods in and through our bodies. When depressed, the state of depression does not end with an abstract state of mind. Instead, the mood of depression takes place in the materiality of the body, in turn affecting a particular bodily experience of being-in-the-world. From this felt materiality, the world itself becomes cloaked in a depressed aura. Things manifest themselves *as* constitutive of depression—the world assumes a particular color, texture, smell, sound, and so forth.

Likewise, if I experience overwhelming lethargy each time I am obligated to meet a friend from my past, then there is nothing mechanical in this response. Lethargy is a physiological experience that carries with it an evaluative framework. Indeed, the accompanying sensations of tiredness and indifference are structured by an intentionality that is directed toward the old friend. If I experience a withdrawing of my body in the company of this person, then this is also a withdrawing of my world from this person. For instance, if this person attempts to rouse me from a state of indifference, then I experience this as a violation of a boundary that I have constructed in my bodily being. My body's refusal to engage with this person is the means by which this relationship works in the first place. Thus, if I am walking or sitting with this person, then I will do so with certain restrictions and limitations, so as to avoid establishing a reciprocal space between us. At all times, the relationship is mediated by an asymmetry in our bodily comportments. In this way, the body is the principle manifestation for values in the intersubjective/intercorporeal world. The body is a sensing and thinking organ: In its flesh, values manifest themselves. Indeed, it is only through the body that the felt experience of value is possible.

At its basis, therefore, the body's moods carry with it an interpretive structure. More than an affective component to lived experience, mood is a mode of knowing the world through the primacy of the body. Although we may remain self-consciously unaware of how our body interprets the material world, the fact that this interpretation persists nevertheless is clear from how we can suddenly find ourselves *in* a mood without even knowing how we got there. Mood is pervasive; it absorbs the world and our experience of the world, such that the very act of interpreting the world is incorporated into our bodily sensibility. It is for this reason that the particular experience of interpreting the world as a singular phenomenon is seldom noticed, except as a vague and inarticulate drone.

Because a mood like depression has such a pervasive power, there can be no reconciliation between a depressed and nondepressed mood, as Heidegger writes of "bad moods": "In bad moods, Da-sein becomes blind to itself, the surrounding world of heedfulness is veiled, the circumspection of taking is led astray."[14] The overarching power of mood to render Da-sein "blind to itself" is only possible because mood embeds itself in the fabric of the body. Indeed, if it is the case that our bodies are the first point of contact with the world, then it is only through the body that the pervasiveness of mood can have a reality. Through the body, the mood of the world assumes a felt reality, conferring a thematic and affective unity upon the world that would be fragmented were it the case that our living bodies were homogeneous blocks of materiality.

The relationship between the bodily nature of mood and the interpretative experience of the environment is implicit in Pascal's abyssal experience. For him, an agoraphobic mood opened up a certain mode of being-in-the-world, from which objects in the world reveal themselves as assuming a phobic or fearful quality. What this illustration provisionally shows, therefore, is how mood, interpretation, and embodiment form a tripart unity, with there being no possibility of one existing without the other, as John Russon says in his eloquent account of mood: "Mood and interpretation are not separate spheres of our existence; rather, we exist interpretively, and mood is the fundamental way in which interpretive existence is experienced by us."[15] In claiming "we exist interpretively," Russon highlights the necessary connection between affectivity and hermeneutics. To understand the materiality of the surrounding world, we need to address the mood of the interpreting body placed within that environment. In order to render this point clear, I would like to connect Heidegger's notion of mood with Merleau-Ponty's account of embodied subjectivity. The result of this dialogue will be the grounds of hermeneutic understanding of mood. Once this connection has been made, we will thus be in a position to see how a bodily mood such as agoraphobia generates the very spirit of a place and how that spirit can also displace us from our bodies.

An Agoraphobic Mood

To begin putting the body in mood, and thus the mood in hermeneutics, I would like to return to our agoraphobic case study. In particular, I would like to consult the example of Allen Shawn, a composer who has written a history of his agoraphobia. Let us join Shawn as he was faced with an empty road "bound on both sides by large open fields."[16]

> [W]hen I got halfway down this empty road, I would freeze in place and balk at continuing, exactly like a dog who freezes at

the door to the veterinarian's office or a horse who refuses to walk over a rotten bridge. I couldn't be convinced that I could continue to walk despite whatever symptoms I felt and that if I did so, I would in fact get to the end of the road and still be the person I was four-tenths of a mile back. The physical reactions included my becoming short of breath and beginning to breathe rapidly (in fact to pant like a dog), feeling my heart beat at twice the normal rate, getting extremely warm and sweaty and feeling like discarding my coat and jacket, finding my vision growing dark and blurred, feeling my face grow cold, and my legs tremulous, weak, and then extraordinarily stiff.[17]

The critical question that must be pursued in light of this passage is as follows: How can an innocuous environment be interpreted in such a way so as to present a threat of total destruction? At stake in this question is not the issue of what lies "beyond" the mood of agoraphobia, as though a psychoanalytic explanation could decode the symbolic ciphers latent in the agoraphobe's experience. Rather, it is a question of how this particular mode of being-in-the-world is opened up for the agoraphobe—a world that is seemingly at odds with a rational assessment of how it is manifest: in other words, as nonthreatening. What is striking about Shawn's account is the intensity of his physical response to an invisible threat, such that the very continuity of his personal identity is called into question on this empty road. As though he could be swallowed up by the emptiness of the road itself, Shawn's experience testifies to the mysterious way in which an environment can become the expressive medium for the body's moods.

How is this relation between the materiality of the world and the lived experience of mood possible? In a critical passage, Merleau-Ponty gives us a clue: "The body is the vehicle of being in the world, and having a body is, for a living creature, to be intervolved with a definite environment, to identify oneself with certain projects and be continually committed to them."[18] As the vehicle of our being-in-the-world, the body is the material reality of our commitments and values. It is only through the lived body that the world can be augmented to a form a coherent union with the I. As Merleau-Ponty argues, if my experience of the world is ruptured by physical disability, then thanks to the intelligibility of my body, a new relationship will be formed that maintains existential unity despite any physical deficiencies. The world will not remain frozen in my able-bodied memory of the past but will instead adapt so that my being-in-the-world has a spatio-temporal continuity to it, despite the objective destruction imposed upon me. In this way, the structure of being-in-the-world attests to a circularity between body and world, with each aspect thinking through the other in a shared dialogue. In its instincts, habits, and desires, the body's being only has a reality in relation to the world. And the same is true in reverse: The materiality of the world—its particular affective tone—is only animated in light of the body. Each aspect gives the other their life, and this life is given context by the body's mood.

As we see in the illustration from Allen Shawn, this dialogical structure takes on a particular clarity in relation to the environment, such that the texture of spatiality corresponds with the felt experience of the body. In his phobia and anxiety, the materiality of the empty road that Shawn stands before becomes imbued with an existential value. More than the mere means to get from one point to another, the road becomes the material manifestation for the specific way that Allen Shawn comports himself in the world, as he writes: "I couldn't seem to get past the point at which I would be closer to the destination than to the point of origin."[19] The struggle inherent in this journey comes alive in the circularity between body and environment, in the process generating the conditions under which value is felt through the flesh.

The manifestation of value in the flesh is clear enough in Shawn's account of walking down an empty road. What is notable is that a particular localized part of the body—in this case, the legs—becomes the focal point of an existential struggle: "I was convinced that when I reached the midpoint, my legs would not move at all and that I would be trapped in place there. I had a vivid picture of myself standing at the centre of emptiness, screaming."[20] This focused intentionality—what Merleau-Ponty calls "pain-infested space"—gains a meaning that reaches beyond the localized sensation itself, affecting the totality not only of the body but also of the environment.[21]

Seen in this way, the mistake in approaching the relation between body and world is to treat specific affective responses such as phobia as a deviation from the self. At stake in this approach would be a relegation of these affective responses to an arbitrary and mechanical response lacking existential value. Indeed, during periods in which he finds himself in "secure surroundings," this is precisely how Allen Shawn reasons his agoraphobia: "I even pretend to myself . . . that my 'personality' . . . is somehow incompatible with agoraphobia.

Sometimes it feels like just an unfortunate fluke that I inherited this trait."[22] As Merleau-Ponty indicates, if the "body is the vehicle of being in the world," then everything within that world shapes itself to the mood of the body. In this respect, to think of environmental experience as having a true or false aspect to it is misleading. True, we can speak of certain experiences of the world being more preferable over others, but this level of preference is not an indication of the normative character of place. Rather, the character of place stands in a reciprocal relation to the body in the place. That Shawn experiences his agoraphobia as somehow "incompatible" with his "normal" personality only reinforces the anchoring and normative role his phobia plays in defining his experience of his self and the world.

Interpreting the Mood of Place

Having considered how mood is bodily in nature, we now need to consider in more detail how that bodily mood affects our interpretation of the environment. At stake in this issue is the question of how a place receives its specific, heterogeneous character in and through time. It is thus a question that points to the genesis of a place's particular *genius loci*. The full scope of this question would incorporate subjective, cultural, political, and economic factors. But within that complexity, we can isolate the role mood plays in giving place its dynamism and meaning. Such a task is only possible within a hermeneutic framework.

A hermeneutic approach to the genesis of place encourages us to take heed of the manifold ways in which environments can be understood as being both existentially meaningfully and materially real, as Clingerman writes:

> Understanding environments ("built" or "natural") requires a middle path to overcome certain intellectual obstacles and limitations of perspective. Structurally, a hermeneutics of nature "reads" nature by navigating through the way that our experience is irreducible to explanations that are naïve realist or social constructivist.[23]

Clingerman's suggestion of a middle path highlights the methodological advantages of a hermeneutic approach to the environment. Able to mediate between the materiality of place and the existential significance of that materiality, by focusing on the role interpretation plays in shaping our experience of the environment, hermeneutics draws the questioner into the question.

We have already seen this circularity between the subjectivity of experience and the correlating objectivity of world in the case of Allen Shawn. There, the bodily mood of agoraphobia set in place—in a literal way—an agoraphobic world. Understanding the agoraphobic body, we thus understand the agoraphobic environment, and vice-versa. Far from sanctioning a dualism, this two-way reversibility demonstrates an ambiguity at the heart of our environmental experience, in which the lived experience of the environment meshes with the materiality of the environment. Clingerman writes incisively, "Discovering the textuality of place is not simply an objective reflection upon nature 'out there,' but equally reflexive thinking of how we are as subjects understanding place through a placial pre-understanding."[24] The mention of a "placial pre-understanding" can be seen, I would argue, in terms of a pre-theoretical mood that is incorporated into a bodily schema.

In this way, mood presents itself as a mode of meaningful intentionality, a pre-reflexive grasping of the environment that is aided at all times by what Merleau-Ponty describes as the "expression of the total life of the subject, the energy with which he tends towards a future through his body and his world."[25] Merleau-Ponty draws our attention to the lived significance of environmental hermeneutics. Far from a mere "state of consciousness," the perception of space is at all times rooted in a "latent significance diffused throughout the landscape or the city."[26] At the heart of this significance are the memories, values, and desires that constitute each and every lived experience of the built and natural environment. In a word, we come to place with that place already within us. This is possible thanks to the hermeneutic circulatory between self and world. As Clingerman states, "The self describes and redescribes itself in response to the confrontation of place; place coheres in its own work of identity and encounter with otherness. In other words, in a reflexive manner place similarly is manifested and reconfigured in response to the selves of its inhabitants."[27] The cohering power of place underscores its importance in defining who we are as selves. This redescription of the self is a task that is constantly being renewed, with each "confrontation of place" being an opportunity to either affirm or evade the self we have become.

Does the centrality of the bodily self entail an anthropomorphized interpretation of the environment? There is no reason to think so, given that the relation between self and the environment instills what David Utsler terms a "creative tension," as he writes:

> Self-understanding comes by way of a reflective, analytical detour and the dialectic of the self and the other-than-self over against the immediate positing of the subject in the cogito. Thus, a hermeneutics of the self as an account of personal identity would not oppose the anthropocentric to the eco-centric, but would actually require a creative tension between both to develop what I call "environmental identity"—i.e., self-understanding in relation to the environment.[28]

As Utsler indicates, there is a self-reflexive aspect to environmental experience, which hermeneutics elicits. The self is not placed in the world in a detached way, before then surveying the contents of the phenomenal world, as though the reality of the self were autonomous from that environment. It is precisely because the self is placed within an environment that self-understanding is possible. In the following section, we explore how this self-understanding helps us understand the importance of the home within environmental hermeneutics.

Agoraphobia and the (Un)homeliness of Home

In the mood of fear and anxiety, the world reveals itself in a particular way, carrying with it a particular set of bodily sensations that are peculiar to the agoraphobic self. The question of how we might hermeneutically understand this relation between the phobic body and the materiality of the world is complicated, given that the sensations experienced do not *prima facie* appear to have a place in the schema of the "total life of the subject."[29] After all, if I am in the midst of a supermarket, feeling myself to be "far from home" (a key expression that haunts the agoraphobe), and thus threatened on all sides by contingency and uncertainty, *despite* the fact that I objectively know little harm could come to me in such an environment, then the relational circle between self and world appears to have come undone. My body appears to be speaking on behalf of a threat that "I" cannot detect. How can we explain this paradox in experience? On the one hand, for the agoraphobe, the body's urge to flee is so strong that no logic or reason could dissuade the subject from the compulsion to escape. On the other hand, as soon as that compulsion to flee is succumb to, then the reality of the escape assumes the appearance of an absurd drama.

At the heart of the phobic interpreting of the environment is thus a dissonance between cognitive and mental phenomena. Such a tension frames agoraphobia as an irrational phenomenon. Yet agoraphobia's supposed irrationality only gains a currency if we grant rational thinking an ontological primacy over the body's experience of the world. This, we recall, is exactly the pose Allen Shawn assumed with his own agoraphobia, as when he writes: "Sometimes it feels like just an unfortunate fluke that I inherited this trait."[30] This sense of having accidently fallen into agoraphobia, as though it were a bad habit and nothing more, overlooks the communicative and hermeneutic layer embedded in the phobic experience of the environment. To begin charting the meaning of this layer, we need only remind ourselves that a central theme for the agoraphobe is the elevation of the *home* to an ontologically privileged position. Not only is the home the place the agoraphobe flees to, it is also the existential and physical center of their reality and well-being. Consider the following quotes:

> There's an emotional quality to interpreting the landscape—on the way (nervous) I thought: "this house looks abandoned" on the way back (more relaxed) I thought: "People are probably home. Maybe they are making dinner.[31]

> When I was home I felt I could control what was happening to me whereas if I went out, there were other people. . . .[32]

> The fear begins as soon as the houses leading to an open area increase their distance from him. . . . A feeling of insecurity appears, as if he were no longer walking secure, and he perceives the cobble stones melting together. . . . The condition improves by merely approaching houses again. [33]

In each of these quotes, the importance of the home is indicated from different perspectives. In the first case from Allen Shawn, the affective experience of journeying through a place is mediated by the figure of the deserted and the inhabited home. When departing, Shawn's anxious experience of the environment leads to an interpretation of the home as unhomely, absent, and altogether uninhabitable. This landscape of anxiety is reversed in his return journey. As his anxiety subsides, so the act of bodily interpretation proceeds to put the home back in place, and thus to once again render it a beacon of existential security. At work here is a hermeneutics of the environment, which is guided at all times by the body's bearing in the world. Departing and returning are experiences that are fundamentally interpretive in structure rather than strictly spatial.

In the second account from an agoraphobic patient named Iris, the home is elevated to a locus of self-control. Given that a large aspect of the agoraphobic's anxiety is losing control in the face of others, the materiality of the home acts not only as a boundary against the outside world but also has a center of reality. Against this reality, the "look" of the other is denied access. Regarding one agoraphobic person, Joyce Davidson writes: "Sufferers' homes are frequently organized to minimize the fear of the look. Brenda arranged for two garden sheds to be positioned immediately outside her patio doors, thus completely blocking the view (both ways) out over her garden, towards the back fence and beyond."[34] This approbation of the home as a mode of diverting the look of the other provides us with another way in which the home acts as the defining environment in the agoraphobic's physical, physiological, and psychological well-being.

The final illustration, given to us by the founder of agoraphobia, Carl Friedrich Otto Westphal, is visceral proof that the agoraphobic's orientation in space is only possible with reference to the home. Outside of this zone, the very fabric of the world becomes insecure, as though "the cobble stones [were] melting together." Already in Westphal's original description of agoraphobia, the critical issue is not the objective features of space—openness or closeness—but the relational distance to home. As Westphal's patient leaves the confines of his zone of safety, so his body opens up to a different way of being. Now, movement is stifled and vertiginous, the very materiality of the world suffering from a lack of reality. Into this abyssal unreality open space becomes problematic, not because of the space itself but because certain aspects of the environment serve to divide the home from the non-home. Crossing the square—the archetypal agoraphobic motif—the danger is not of the square itself, nor even of the public eyes that descend upon the agoraphobe. To be sure, all of these things contribute to the agoraphobe's concern, but the kernel of his anxiety is the question of how he finds his way back in the world. The abysmal quality of the agoraphobic world centers on the pathological need to be orientated at all times, where only disorientation and distance are possible.

This relation between being orientated and being at home in the world points to agoraphobia's power to disrupt our idea(l)s of home. In the previous passage, what we are contending with is literally an *un*-homely environment. Because of the agoraphobic's intense attachment to the home as a site of refuge from the world, anything outside of that circumscribed zone suffers not only from a loss of reality but also from a lack of orientation. Into this disorientated environment, the world assumes an uncanny appearance, as Freud has it: "The better oriented he was in the world around him, the less likely he would be to find the objects and occurrences in it uncanny."[35] No wonder, then, that outside the home, the agoraphobe feels his world turn unreal. In the high vaulted aisles of city streets and across the populated avenues that divide the space into atomized segments, the agoraphobe's body breaks down. That there is such a world outside of his home, for him, is an affront to his ontological reality.

Considered together, at least three things can be said about the agoraphobic's attempt at homemaking. One, for the agoraphobic person, home is a site that is fixed in space in a circumscribed and rigid manner. The home is a center, both figuratively and literally. Only instead of being a center from which life is projected, it has become a place in which life is withdrawn, as Jacobson puts it: "Home is a place of refuge and retreat for the agoraphobic, *not*, as it would be for the healthy person, a supportive base from which projects can be launched."[36] Beyond the walls of the home, the failure to make a home in the world is felt in the agonizing bodily sensations that accompany the agoraphobic experience of being-in-the-world. That these symptoms are experienced as a sense of impending death is thus entirely logical: Outside the home, the agoraphobic subject's world collapses, their reality now dwarfed by total disorientation and unfamiliarity. To this end, the agoraphobe's topophilia is equally topophilic. In other words, the agoraphobe's love and attachment to home is predicated on an equal revulsion and alienation from the non-home, with each realm structurally implicated in each other. In his ambivalent relation to home, what he is ultimately lacking is the resources to find home, not simply in the immediate place beyond his frontier of safety but within the world itself.

It follows from this that the agoraphobe's experience of being oriented in the world is at all times reliant on objective—largely visual—cues. Instead of taking the home for granted as a background presence, an enforced vigilance is constructed so that the agoraphobic person knows where he or she is at any given moment. This insistence on a topographical approbation of space stands in contrast to a more fluid and bodily sense of being at home. Lacking trust in the body's power to get placed in the world, the agoraphobe's experience of being outside of the home remains hesitant and stifled. For the agoraphobe, the expression of being oriented in the world is only possible within the confines of his or her zone of homely safety. Beyond those margins, bodily and spatial constrictions take precedence.

Finally, the agoraphobe's experience of being at home and thus of not being at home reinforces the fact that "home" is ultimately a contingent idea, which at any point could reject our experience of intimacy and familiarity, as Shawn writes: "How can it be—we ask ourselves—that for some people *this* alien place is home, *this* their daily view—one so familiar that they no longer even look at it—*this* their idea of 'normality.'"[37] What Shawn's reflection captures is that the experience of being "far from home" is not a question of spatial distance but a mode of being-in-the-world. The unhomeliness of agoraphobia is located in its resistance to experiencing the world as a familiar phenomenon. In the place of familiarity, the agoraphobic subject continuously faces a world that withdraws from the certainty of being placed. In such a displacement, the circularity between body and world is evident in that both the agoraphobic person and the agoraphobic environment are fundamentally out of place and not at home. In this way, agoraphobia gives us a clear example that the meaning of home is subject to the varying moods of the body. Home is not something "out there," waiting to greet us after a long day at work. It is a relation we hold to being-in-the-world. That such a relation is subject to the moods of our bodies is understandably anxiety inducing: Failing to construct a meaningful relation to the idea of home means that our world—and so our own self—becomes displaced from its axis.

Conclusion

What can the agoraphobic's experience of struggling to be at home in the world tell us about environmental hermeneutics? As I have sought to argue in this chapter, I believe that agoraphobia shows us two things, each of which is tacit in non-agoraphobic experience. First, far from an abstract reflection on the world, mood is the primary

way in which the body interprets the environment. This is especially clear in the illustration of agoraphobia, as the body in question
is heightened in its affective and interpretive sensibility. Because of
this heightened status, the symbolic and existential depth of the
world is given a visceral expression. Our experience of the world is
never neutral, and our bodies are never blank canvases upon which
the materiality of the environment is posited. Instead, our experience
always involves a circularity between body and environment, with
each aspect dependent on the other for its voice. Of the voice of the
agoraphobic's body, Sadowsky writes: "Agoraphobic symptomatology
involves a direct and immediate affective reactivity to certain spatial
physiognomies possessing their own symbolic depth."[38] Even if
those symbolic depths are unregistered by the agoraphobic person,
the body's perception of the environment nevertheless maintains its
principle aim: *to be oriented in a home-world*.[39]

That agoraphobia presents itself as an obstacle in the body's aim
of getting placed means that it reveals the centrality of the home—
not only for the agoraphobe but for all interpreting bodies. Far from
a pathological deviation incommensurable with "normal" environmental experience, agoraphobia amplifies what is common to all
bodily subjects: that home is something that must continuously
be reinterpreted, not only cognitively but corporeally, too. This can be
seen in the fact that home is not only a place in the world but, more
importantly, a relation to that place. Without building a relationship
with the places in which we live—indeed, the world in which we
live—human beings fall prey to ontological insecurity. The insecurity
is not, as normative science would have it, a morbid psychic category
of the agoraphobic personality but a given of what it is to exist, which
each of us must contend with.

In both the bodily mood of the agoraphobe as well as their troubled
relation to home, we gain an understanding on the hermeneutic
structure underlying all bodily experience. The environment is spatial
and bodily at once. It is the living nexus, from which our total
subjectivity is both projected and introjected. The implication of this
is that *without a hermeneutics of the body, a hermeneutics of place
remains unattainable*. Places come to life thanks to the moods we
carry into them. What this means is that the places that hold value
for us are not only spatial in character but developed through our
bodily involvement with them. The environment invites a response
from us. In different ways, the hallway of an auditorium or a tunnel
buried in the jungle urge a response from us. How we respond to

these spatial encounters will depend in large part on the mood of our body, which in turn defines the environment through interpreting it. Equally, our moods are never disembodied or despatialized but always involve a relationship to the living world. To overlook one of these dimensions is to ignore the fact that environmental hermeneutics is a hermeneutics of the bodily self.

Situated in the broader field of environmental philosophy, the contribution that a hermeneutics of the bodily self can make is twofold. First, if environmental philosophy has for the most part concerned itself with specific issues in the world—ranging from pollution, the growth of skyscrapers, the use of cars, the shortage of natural resources, the extinction of species, and so forth—then what has been overlooked in this work is our very (bodily) relation to these issues. By placing the body central, environmental hermeneutics serves to remind us that the environment is not a static, objective world existing "out there." Instead, the environment encompasses our bodies as our bodies encompass the environment.

Second, it follows that if our bodily relation to the environment is foremost in our grappling with environmental issues, then the question of what the environment can ethically teach us is an issue that must be called into a question. The reason for questioning this thought is that it presupposes a normative status to environmental ethics, and a status that is autonomous from the subject. Yet if the character and mood of the environment is provided by the relation human subjects adapt to the environment, then the extent to which the environment can serve as a beacon of ethical insight must be reconsidered. This is not to reduce environmental ethics to the whim of a relativist position. Instead, it is to acknowledge the dependence ethical positions have on the bodily moods that shape our understanding of the environment.

Narrativity and Image

Narrative and Nature: Appreciating and Understanding the Nonhuman World

Brian Treanor

The universe is made up of stories, not atoms.

—*Muriel Rukeyser*

If you want to build a ship, don't drum up the men to gather wood, divide the work and give orders. Instead teach them to yearn for the vast and endless sea.

—*Antoine de Saint-Exupéry*

Common convention, as well as numerous philosophical and scientific accounts, suggests that there are two primary ways of gaining understanding: theory (*theoria*) and practice (*praxis*). In this context, I mean by the former all sorts of abstract ways of coming to know or understand things, with the caveat that in our age pride of place is given to scientific understanding. We tend to think we know things when we can prove them—often without reflection at all on the nature of "proof"—and, consequently, we subject all sorts of inquiry to this quasi-scientific standard. Imagine, for example, a clichéd exchange between two young sweethearts: "But I love you." "*Prove* it!" The same search for demonstrable, indubitable certainty haunts scores of other discourses, including the politicized account of science in which the overwhelming scientific consensus on climate change is seen by some as "uncertain" because, so the claim goes, it has not been "proven." By *praxis* or practice, again in the context of this chapter, I mean direct, visceral, firsthand knowledge as opposed to secondhand, vicarious, or indirect knowledge. After "prove it," the second most common epistemological demand is "show me," that is to say, "let me see *for myself*."

We often assume that we know something either only when it can be proven to us by science or when we experience it directly. However, without discounting either the power or legitimacy of these ways of understanding, I want to consider an underappreciated third way: narrative (*poiesis*).[1]

Appreciating and Valuing Nature: *Praxis* and *Poiesis*

Jack Turner claims it was "Muir's mistake" to think that his experience of the wild could be reproduced and made available through things like Sierra Club outings.[2] Experiencing the wild indirectly through narratives and photographs gives us only a mediated experience, a semblance of the real thing, an abstraction. Dealing with such abstractions blurs the boundaries "between the real and the fake, the wild and the tame, the independent and the dependent, the original and the copy, the healthy and the diminished."[3] This, to some degree, can account for our fascination with the picturesque, odd, or bizarre in art and nature writing (at least since the Romantics). Insofar as art and nature writing influence the way in which we appreciate nature, this tendency to focus on the picturesque or bizarre also shapes our actual interactions with wild nature. Why is there no Tall Grass Prairie National Park?[4] Such a park would preserve the dominant ecosystem of North America prior to European colonization, but tall grass prairies do not lend themselves to drama and fascination in the way that the multicolored geothermal vents and pools of Yellowstone or the massive granite monoliths of Yosemite do.

To appreciate wilderness and wildness, Turner asserts, we need "gross contact," intimate personal experience with wildness. Such contact is essential both for grasping the truth of a thing or phenomenon (to which we will return in the next section) and for appreciating and valuing it: "[W]e only value what we know and love" and, lacking precisely the visceral experience of gross contact, most people "no longer know or love the wild."[5] Experiencing wildness secondhand, as through narrative, leaves us with only the "abstract wild."[6] It is, perhaps, in this vein that Thoreau placed such a strident emphasis on the importance of personal experience. In a pithy *apologia* for speaking so much about himself, he claims he is "confined to this theme by the narrowness of [his] experience."[7] He is in favor of "practical education," choosing to grow his own food and build his own shelter so as not to be "defrauded" of the experience by "resigning the pleasure of construction" to the carpenter.[8]

However, such an uncompromising commitment to personal experience has at least two drawbacks for environmental philosophers. First, there are many parts of nature—animals, plants, ecosystems, or natural phenomena—that most people will never see or experience firsthand. If truly valuing the Arctic National Wildlife Refuge depends on loving it, and loving it depends on gross contact with it, then ANWR will be lost. Not everyone has the luxury to take off weeks, months, or years to immerse themselves in nature. Even if they did have the time, many would not have the inclination to do so; and those with the inclination might not have the means. Moreover, even if they did have the time, inclination, *and* means to do so, so many people seeking gross contact with a wild ecosystem would surely overwhelm and destroy it. The desire for gross contact with certain manifestations of wilderness or wildness can be destructive—a place becomes "loved to death"—as has, arguably, happened to Yosemite Valley.

What, then, are we to do? Fortunately, while Turner's "rant" (his term) in favor of gross contact makes several critical points, there is, in fact, reason to think that he overreaches in his condemnation of mediated experience.

First, we might point out that, speaking absolutely, no experience is unmediated.[9] All our experience is filtered. However, while this is an essential point, I think it is more worthwhile to engage in a charitable—and, I think, more accurate—reading of Turner's claims. His preference for gross contact need not suggest that we can actually have a *completely* raw experience, unfiltered by any concept or narrative. Rather, we can frame the preference for gross contact in terms of privileging direct experience as undiluted *as possible* by concepts, categories, and narrative structures, especially those that are rich, complex, or highly developed or those that are foreign to the environment of the experience (e.g., finding pictographs in the desert and thinking of them in terms of the revenue they could generate as a tourist attraction).

Second, there is good evidence that secondhand narrative accounts *can* bring people to value things. It was precisely the genius of Muir—as well as his successor, David Brower—that he successfully induced many people to value and, after a fashion, to love places that they themselves would in all likelihood never see or experience in person. Bringing them to value nature in the right way, or one of several possible right ways, is something to which we must return in the next section. For now, let's concentrate on how and to what extent mediated experience, of which *narrative* is the archetype, can stand in for actual experience.

On some level, people have an intuitive sense of the power of narrative to elicit appreciation, concern, understanding, and even personal transformation or conversion. Take, for example, the way in which noted philosopher and virtue ethicist Rosalind Hursthouse came to her vegetarianism. This took place not due to the philosophical arguments of Peter Singer or Tom Regan, two of the most celebrated exponents of animal rights, with which Hursthouse was very familiar. Rather, it took place over a period of time spent reading about the virtues and a slow conversion in which she began to see a deep dissonance between her "interest and delight in nature programmes about the lives of animals on television and [her] enjoyment of meat."[10] These narratives led to a sort of epiphany. "Without thinking that animals had rights, I began to see both the wild ones and the ones we generally eat as having lives of their own, which they should be left to enjoy. And so I changed."[11]

Indeed, Turner's own insistence on the ultimate necessity and irreplaceability of intimate, visceral gross contact is balanced by a deep appreciation for the power of narrative.

> Most of us, when we think about it, realize that after our own direct experience of nature, what has contributed most to our love of wild places, animals, and plants—and even, perhaps, to our love of wild nature, our sense of citizenship—is the art, literature, myth, and lore of nature. For here is the language we so desperately lack, the medium necessary for vision.[12]

It is in this vein that Turner concedes that "old ways of seeing do not change because of evidence, they change because a new language captures the imagination."[13] Richard Kearney makes a similar point regarding narrative, imagination, and ethical sensitivity:

> The power of empathy with living things other than ourselves— the stranger the better—is a major test not just of poetic imagination but of ethical sensitivity. And in this regard we might go so far as to say that genocides and atrocities presuppose a radical failure of narrative imagination.[14]

Turner is correct in pointing out that narrative gives us a more overtly mediated experience and, therefore, that narrative cannot entirely replace actual experience. However, the power of narratives to help us "see things anew" can hardly be overstated. Indeed, while it is true that narrative accomplishes "less" than actual experience in many ways, it has numerous other salutary features, at least one of which arguably accomplishes "more" than actual experience. In order to see how this is so, we need to understand the extent to which we are, fundamentally, narrative beings, the degree to which narratives shape our experiences at the most basic level. Here I can offer only the briefest gloss on the structure of narrative identity as developed by Paul Ricoeur—alongside whom I might mention, in varying degrees, Richard Kearney, Martha Nussbaum, and Alasdair MacIntyre— before moving on to focus on the question of how narrative can elicit understanding of and concern for the natural world.

Ricoeur distinguishes between two sorts of identity: *idem* (the self-identical) and *ipse* (the self-same). The former is fixed and corresponds to the "what" of identity, but the latter admits of change over time and is best understood as a narrative identity, corresponding to the "who" of identity. Taken as a practical category, identity is the answer to the question of "who": Who did this? Who is responsible? But to answer these questions, questions of "who," we tell the story of a life. Ricoeur develops his understanding of narrative identity in terms of *mimesis*—taken in the Aristotelian sense of the imitation of action—which he articulates as having three aspects: $mimesis_1$ (prefiguration), $mimesis_2$ (configuration or emplotment), and $mimesis_3$ (refiguration). As narrative beings, we have a certain pre-figurative understanding—a grasp of the structural, symbolic, and temporal aspects of narrative that give it meaning—which allows us to make sense of narratives we encounter. Emplotment desig-nates the way in which we actively link together—an activity that can be more or less conscious and intentional—isolated incidents into narrative events. It includes the "organization of events"[15] but also includes relating them to each other and to a larger picture, including a conclusion or other orientation, in which they make sense. Emplotment mediates: (1) between individual events and the plot as a whole; (2) between heterogeneous elements in the plot (agents, goals, unexpected results, and so forth); and between temporal characteristics of the plot—the sequencing of the episodes, the "grasping together" of the incidents into the events of a story, and leading toward a conclusion.[16] To emplot or to understand a narra-tive is to understand how and why the episodes of the plot lead to a certain conclusion. Thus, there are thick and thin narratives, the former including literary works with complex storylines (e.g., *The Brothers Karamazov*) and the latter extending as far as, for example, relatively brief illustrative examples that ask us to "picture this. . . ." The process of emplotment is never completely impartial; it is selective and based on our perspective, beliefs, knowledge, history, and so forth. The final step in this mimetic process is the refiguration of a life. After reading a story, we return to living a life, but not in the same way we did prior to reading the story. Every story we hear affects us in some way, sometimes trivially but sometimes profoundly. As Thoreau asks rhetorically in *Walden*, "How many a man has dated a new era in his life from the reading of a book?"[17] It is because of the ability of narrative to facilitate the refiguration of our lives that a text can genuinely teach us something *beyond* facts of the persons, places, and events of the story.

The narratives I read or hear change the narrative that I am (the *who* that I am as opposed to the *what* that I am). They do this, says Ricoeur, by offering a sort of "as if" experience in which we can try out, as it were, different possibilities. For this reason, Ricoeur calls narrative an "ethical laboratory."[18] We use narratives to experiment with possibilities, exploring different situations and different ethical responses, projecting ourselves into stories and making judgments about the actions of characters; it is for this reason that there are no ethically neutral narratives.[19] "To understand what courage means, we tell the story of Achilles; to understand what wisdom means, we tell the story of Socrates; to understand what *caritas* means, we tell the story of St. Francis of Assisi; and, as I've said elsewhere, to understand simplicity, we tell the story of Henry David Thoreau; to understand attention and observation, we tell the story of Aldo Leopold; to understand love of wildness, we tell the story of John Muir.[20] Kearney points out that "the validity of Ricoeur's observation can be seen in the simple fact that while ethics often speaks generally of the relation between virtue and the pursuit of happiness, fiction fleshes it out with experiential images and examples—that is, with particular stories."[21]

Ricoeur argues that narrative facilitates our self-understanding and our understanding of others. But to what degree can stories replace the visceral gross contact on which Turner insists? Can narrative elicit intimacy, love, and concern? At least some theorists seem to suggest that it can. Kearney goes so far as to claim that, at least in certain cases, while narrative experience is vicarious—that is, unreal on the face of it—"it is experience nonetheless; and one more real sometimes than that permitted in so-called reality."[22] Kearney is not alone in his belief in the power of narrative. Authors make similar claims. Dermot Healy writes, "I can still remember the liquid feel of . . . words for rain. How the beads were blown against the windowpane, and glistened there and ran. The words for rain were better than the rain itself."[23] Ernest Hemmingway wrote, in a letter to his father, "I'm trying in all my stories to get the feeling of the actual life across—not to just depict life—or criticize it—but to actually make it alive. So that when you have read something by me you actually experience the thing."[24]

Nevertheless, the idea that narrative actually re-presents reality seems like a stretch. Isn't any story detached, by its very nature, from the reality it is trying to represent, victim to its own kind of abstraction? Implying that narrative can connect us to reality as a substitute for gross contact smacks of naïve posturing. It may be true that, to understand something about courage, we tell the story of Achilles or perhaps some other modern virtuoso of courage; but reading about the siege of Troy or the slaughter on Omaha Beach and *Pointe du Hoc* is a far cry from actually experiencing either battle. Narrative reality differs from lived reality in important ways, and narrative experience remains "as if" experience. This being the case, it's hard to see how narrative can overcome charges of detachment and abstraction and, therefore, hard to see how (virtual) narratives can be substituted for gross contact, inducing us to really love and value (actual) wildness.

What, then, are we to make of this situation? On the one hand, both intuition and reason suggest that narrative plays an important role in understanding our selves and the world. On the other hand, it seems commonsense that there are significant differences between narrative experience and "real," firsthand experience. How could the words for rain be "better," as Healy claimed previously, than real rain? Can Hemmingway's stories give us the "actual" experience he seeks to communicate? It's true that narrative experience *cannot* replace actual experience insofar as it cannot, for example, ulti- mately habituate us to the virtue of courage (following the reference to Achilles discussed earlier). Nevertheless, narrative *can* play an important role in self-cultivation and, therefore, the cultivation of virtue.

Ricoeur correctly maintains that we end up refiguring our lives as the result of the narratives we emplot; the stories we read, hear, and tell shape the way we are in the world.[25] The fact is, if we are to believe what people actually report about their experiences, stories can and do change people's lives. Epiphanies are no doubt often the result of "real" experience; however, narratives can also bring about these radical transformations, and indeed have a particular facility for doing so. The imaginative play of narrative and the "synthesis of the heterogeneous" that is part of emplotment are capable of highlighting paradoxes, emphasizing essential truths, and revealing hidden mysteries that can be overlooked in the hustle and bustle of actual experience. Emily Brady notes that imagination can intensify experience and in so doing play an exploratory, projective, ampliative or revelatory role.[26] Returning to Dermot Healy's claim, we might say that the words for rain are "better" than the rain itself insofar as the words, the story or narrative, can be, at least in some cases, *more* effective than actual experience at bringing about the life-altering insights that amplify our concern, increase our understanding, or drive personal transformation. The words for rain can, through this narrative power, transform the "simple contingencies of everyday life . . . into narrative epiphanies."[27] In the chaos and rush of certain experiences, or in the busyness of everyday life, or in the idleness of reverie, we can miss essential truths that only come to light later, in the retelling. We miss in the moment of experience (i.e., in actual experience) what we find, in retrospect, to be essential about the experience (i.e., as we retell the story of what we experience to others or to ourselves). This fact demonstrates the power of narrative, and it is a short step from that conclusion to the realization that other narratives, narratives of places and people we have not experienced and which may not even exist in reality, can also affect us in this way.

Understanding Nature: *Theoria* and *Poiesis*

So, narrative can play a useful role in terms of stimulating our love for nature and, therefore, our valuing of it. Not, to be sure, the precise role played by gross contact but an essential role nonetheless. But doesn't narrative also *distort* the reality to which it exposes us? If so, this narrative familiarity and concern may come at the cost not only of misunderstanding nonhuman nature but, as a consequence, also of harming it. Perhaps this belief helps to explain the substantial number of people for whom science provides the gold standard—indeed, for some the only *real* standard—for knowledge and truth.[28] And this number increases as one confines one's sample to the "scientifically literate," which tends to have a high degree of overlap with "those who value ecology."

Holmes Rolston argues that scientific knowledge is critical not only to fields like ecology but also to the *aesthetic appreciation* of nature—of which narrative is, presumably, one type. On the face of it, aesthetic appreciation is not, for most people, dependent on scientific understanding. We tend to think that the aesthetic appreciation of Caravaggio's masterpiece *The Calling of Saint Matthew* is independent of detailed knowledge about the chemical properties of the paint applied to the canvas. Nevertheless, with respect to the appreciation of nature, Rolston argues for the "superiority" of science, which provides the "definitive interpretation" of natural phenomena: "science corrects for truth."[29] Rolston has some good company here. Allen Carlson also argues that "aesthetic appreciation of nature must be directed by knowledge about it. The kind of knowledge necessary is that provided by ecology, namely understanding of different environmental systems and their interactions."[30] This "cognitive model" for appreciating nature is opposed, frequently, to the "imaginative model" evident in the work of thinkers like Emily Brady.[31] To be fair—I don't want to overstate the case—Rolston does argue that it is important to both understand and to "stand-under," that is to grasp scientifically and to experience personally. However, it clear that science remains *primus inter pares*. And, in any case, the acknowledgment of "standing under" is a concession to lived experience, not narrative "as-if" experience.[32]

Like Rolston, Marcia Eaton attempts to offer an approach that values both the scientific and the imaginative approach, but which clearly and firmly claims pride of place for the scientific approach. She agrees that "designers, managers, and theorists must give due attention to ways in which fiction and other art forms shape thought in our efforts to establish successful and sustainable practices,"[33] but she is very wary of what she takes to be imagination's tendency to take flight, unfettered by scientific reality. Responding in part to Brady's contention that the free play of imagination allows us to imagine a tree as "a person or an animal or a tower or a mountain or whatever,"[34] Eaton takes issue with undue reliance on imagination. Although Brady suggests such imaginative fancies can give us insight into "aesthetic truths," Eaton counters, quite convincingly, that they result at least as often in aesthetic falsehoods.

Looking at Felix Salten's enormously influential *Bambi* and its subsequent interpretations in film and literature—"enormously influential" in the sense that they have shaped our cultural perception of deer—Eaton points out that an apparently innocent imaginative story has resulted in a "totally false" view of deer and their relationship to the ecosystems in which they live. "Fiction," she writes, often has a "tendency to sentimentalize . . . or to demonize," and both result in misconceptions.[35] While *Bambi* may well be useful if one wants to teach children certain moral lessons, it can cause epistemic distortion and practical disaster when we allow it to shape our understanding of actual deer in the actual world, how they interact with and function within their ecosystems, how we relate to them, and how we should relate to them. The antidote, argue Rolston and Eaton, is sound scientific and ecological knowledge. Although imagination does serve a purpose according to Eaton, we must cultivate "informed imagination."[36] Only then will imagination's benefits outweigh its tendency to distort. "I do not want to claim that there is no positive role for fiction—for imagination in general—in developing a sound nature aesthetic. I do insist that it must be based upon, tempered by, directed and enriched by solid ecological knowledge."[37]

Neither Rolston nor Eaton suggest that we should do away with or completely discount personal experience or imaginative play. Each thinker attempts to carve out a role for what he or she takes to be the legitimate use of imagination. However, in both cases this role is very clearly subordinate to scientific and cognitive engagement. Rolston calls science the *definitive* interpretation of nature; Carlson suggests that science should *direct* the imagination. Eaton concurs with Carlson and adds that we must begin with sound ecological knowledge before we let the free flight of fancy run amok, that the cognitive model must supply the "foundation" on which our imaginative projections are built. "I see no choice but to insist that fancy *take off from* a solid knowledge base."[38] In addition to implicitly affirming Rolston's belief that "science corrects for truth," Eaton's description suggests a linear model in which we must begin with science and ecology, which should form the foundation on which any and all imaginative and narrative structures are built. Our nature writers, artists, photographers, and documentary filmmakers should, presumably, be trained first and foremost as ecologists. Only then can we trust them to tell "true" stories about the subjects they love.

But isn't the actual relationship between imagination and knowledge much less linear, much more mutually enriching? The supposed unbiased and objective detachment of scientists has been critiqued from many angles, perhaps most famously by Thomas Kuhn in *The Structure of Scientific Revolutions.*[39] We tend to think that science studies objective reality, unsullied by opinion or bias, and that scientific progress is progress toward the *really* real and the *truly* true. However, according to Kuhn, science is not engaged in inevitable, teleological progress toward objective, ultimate truth.[40] "An apparently arbitrary element, compounded of personal and historical accident, is always a formative ingredient of the beliefs espoused by a given scientific community at a given time."[41] "Paradigm shifts"—a phrase made famous by Kuhn, which has now entered our collective lexicon—do not happen because someone sees the truth more clearly. Often "the new paradigm . . . emerges all at once, sometimes in the middle of the night, in the mind of a man deeply immersed in crisis. What the nature of that stage is—how an individual *invents* (or *finds he has invented*) a new way of giving order to data now all assembled—must here remain inscrutable and may remain so."[42] In any case, the "transfer from paradigm to paradigm is a *conversion* that cannot be forced"[43] and cannot be justified by "proof."[44] There are, however, aesthetic appeals to simplicity, suitability, neatness, elegance, or similar traits and, critically, "*the importance of aesthetic considerations can sometimes be decisive.*"[45] There is no "raw data" or "pure given" unsullied by interpretation[46] and there is no objective decision procedure to guide us in choosing a new paradigm over the old one.[47] Seen in this light, science itself, at any point in its history, is merely one interpretation of the world.

Rolston seems to acknowledge this when he argues that science provides the "definitive interpretation" of nature, which implicitly affirms that there are other, less definitive interpretations. But in the face of multiple interpretations, it seems worthwhile to carefully consider whether science provides either our most basic or our principal interpretation.

First, it seems clear that science is not our most basic or fundamental interpretation of nature in terms of eliciting concern for interest in it. I suspect that few of our nature writers, artists, photographers, and documentary filmmakers, including those of whom Eaton and Rolston would approve, were drawn to their subjects through soil microbiology (though perhaps some may have been). Most, I'd wager, were drawn to their subjects by a *non*scientific experience of some sort, whether narrative or empirical. In fact, given that few, if any, children receive their introduction to nature via a Petri dish, it's safe to say that even ecologists and conservation biologists began not with science but with a narrative or empirical experience of nature (it amounts to the same thing insofar as we understand our empirical experience retrospectively in narrative). Eaton herself reminds us that "many ecologists describe the aesthetic experiences [including the imaginative] that drew them to their work in the first place."[48] This being the case, it would be misleading to imply that we must *begin* with sound scientific and ecological knowledge from which, and only *subsequent* to insuring the scientific merit of the platform, we ultimately risk allowing our imagination to "take off" and engage the natural world.

Second, though perhaps more controversially, there is no reason to suppose that science alone provides "the definitive" interpretation of nature or that it alone "corrects for truth." In reality, sound appreciation and understanding of nature involves both ecological knowledge and imaginative engagement (of which narrative emplotment is the primary manifestation), and these two ways of relating to nature mutually inform, support, and correct each other in a complex and symbiotic relationship. It is true that ecological knowledge should rein in certain imaginative excesses and that, in certain cases, science corrects for truth. However, it is also true that the insights of science are often sterile and without meaning or value—after all, ecology can tell us what role deer play in their respective ecosystems, but it cannot tell us whether we should value deer in anything other than instrumental terms—and that, in this way, imaginative and narrative engagement also correct for truth.

While the scientific interpretation arguably reigns supreme in modern discourse, it is not without its own problems and limitations. The very detachment and disinterestedness that gives science its power opens it to charges of being removed from the lived reality of the world. Indeed, this tendency is part of what ecology, as a scientific discipline, attempts to correct. Although abstraction is not bad or distortive *per se*, it can metamorphose, eventually succumbing to what Gabriel Marcel calls the "spirit of abstraction."[49]

[T]he process of abstraction can quickly overcome the concrete, embodied existence from which it is abstracting, detaching itself and becoming an independent system. As Marcel describes it: "it can happen that the mind, yielding to a sort of fascination, ceases to be aware of these prior conditions that justify abstraction and deceives itself about the nature of what is, in itself, nothing more than a method." It is necessary to abstract, but equally necessary to acknowledge both that one is abstracting and that the abstraction is not the point of the exercise. To confuse the two is to initiate "a violent attack directed against a sort of integrity of the real."[50]

The spirit of abstraction should not be confused with science, nor with theoretical thinking, nor even with abstraction itself (for it is possible to abstract without succumbing to the spirit of abstraction), each of which, used properly, can be a useful tool in our quest for greater knowledge and understanding of the world and our place in it. The spirit of abstraction is, rather, something along the line of losing the forest for the trees, seeing the part as the whole, perhaps even going so far as the reification of things or idolatry.

The power of abstraction lies in its ability to help us to understand some things better—more precisely, with greater detail, and so forth—due to the *focus* that it allows. However, because abstraction can succumb to the spirit of abstraction, abstractions can lead to different *mis*understandings. These misunderstandings are often tied to the abstraction itself, from looking at a thing in isolation, in a vacuum, rather than in context, where it interacts with and depends on many other things in a complex web of relationships. Instances of scientific misadventures stemming from abstraction abound. Take, for example, the deleterious effects of abstraction that results in the kind of "nutritionism" or "nutrition science" critiqued in Michael Pollan's *In Defense of Food*.[51] Nutritionism encourages us to "understand and engage with food and our bodies in terms of their nutritional and chemical constituents and requirements—*the assumption being that this is all we need to understand*."[52] Nutritionism is a reductionist approach to food that is born of an abstraction run amok, an abstraction that looks at parts (nutrients) and assumes that they collectively equal the whole (food). What's remarkable about this approach is not the obvious fact that it completely misses a number of significant things about production, preparation, and consumption of food—the social and cultural aspects, for example—but that it *also* misses the one thing at which it does aim: more healthful eating. It is nutrition science that brought us breakfast "cereals" that are nothing but corn soaked in sugar and "fortified" with "vitamins and minerals," and which brought us the obesity, heart disease, cancer, and diabetes that go along with such nutrition-endorsed "food-like substances."

Just the Facts?

The Limits of Simple Facts

Recent research in psychology and political science illustrates well the limitations of a purely cognitive approach to truth, based on arguments regarding facts.

> In a series of studies in 2005 and 2006, researchers at the University of Michigan found that when misinformed people, particularly political partisans, were exposed to corrected facts in news stories, they rarely changed their minds. In fact, they often became even more strongly set in their beliefs. Facts, they found, were not curing misinformation. Like an under-powered antibiotic, facts could actually make misinformation even *stronger*.[53]

It would be tempting to conclude that the fault lies in a lack of education, that educated people, even when misinformed, possess the critical tools to analyze their beliefs in light of new information. Indeed, this is one of the promises of education, especially in the liberal tradition. That conclusion, however, would be incorrect: "A 2006 study by Charles Taber and Milton Lodge at Stony Brook University showed that politically sophisticated thinkers were even less open to new information than less sophisticated types."[54] Although the well-educated, politically sophisticated people may be correct in a large number of their beliefs about states of affairs, their justified confidence in this body of knowledge makes it almost impossible to correct them on the things about which they are completely wrong or misinformed. Their education and the fact that they are generally well informed make them overconfident in even their mistaken beliefs.

This phenomenon, which researchers have dubbed "backfire," presents a real problem for those who believe that more accurate facts, more information in our already information-saturated society, are the panacea to our woes, political or environmental. The majority of our opinions are based on unverified beliefs. Indeed, this is true of necessity, since it would be impossible to attempt to personally empirically verify all the things I believe to be true. Much of what we believe we believe, with entirely reasonable justification, based on what we glean from others—others' experience, others' empirical verification, others' expertise, the wisdom of tradition, and so forth.

The problem for the "cognitive approach" is that these *beliefs* seem to dictate what we accept as *fact*. This phenomenon was recognized long ago by William James: "Our minds [and knowledge] thus grow in spots; and like grease-spots, the spots spread. But we let them spread as little as possible: we keep unaltered as much of our old knowledge, as many of our old prejudices and beliefs, as we can. We patch and tinker more than we renew."[55]

Suppose a person spent a long time, a very long time, attempting to assemble a jigsaw puzzle—call it "The Cloud." The puzzle has many thousands of pieces and is uniformly white, without any picture to distinguish the pieces. Moreover, the jigsaw lacks any traditional shape that would give it discernible edges to facilitate solving the puzzle. Having sacrificed many months, perhaps years, our hypothetical puzzler raises the final piece to snap into the completed canvass, only to find it does not fit. First perplexed, then frustrated, then quite angry, she tries various orientations of the piece in a vain attempt to get it to fit the rest of the assembled puzzle. However, in each and every orientation, there is an extra bulge of material that prevents the last piece from fitting into the final gap. Clearly, reasons our puzzler, there has been an error at the factory and this puzzle is defective. After so much time invested, the puzzler is unlikely to scrap the entire effort and start over in order to figure out how and why this last troubling piece connects to the existing framework. After all, it is clearly the final piece that is defective (i.e., "wrong") since the rest of the pieces already fit together in what appears to be—or appeared to be until the discovery of the ill-fitting piece—a well-organized unit. It's not hard to imagine that, in such a situation, the next step in the process would be to go to the kitchen for a pair of scissors to "fix" the offending puzzle piece in order to insure that it fits in the way it was, obviously, supposed to.

James was writing in 1907, but the contemporary research confirms his insights regarding the way in which our knowledge "grows."

Generally, people tend to seek consistency. There is a substantial body of psychological research showing that people tend to interpret information with an eye toward reinforcing their preexisting views. If we believe something about the world, we are more likely to passively accept as truth any information that confirms our beliefs, and actively dismiss information that doesn't. This is known as "motivated reasoning." Whether or not the consistent information is accurate, we might accept

it as fact, as confirmation of our beliefs. This makes us more
confident in said beliefs, and even less likely to entertain facts
that contradict them.[56]

If we want people to understand and accept facts, we have to address
their underlying worldview, for it influences what they will even
accept as a fact. But a person's worldview, like her personal identity,
is fundamentally narrative. To get people to change their minds
about, for example, climate change, it is necessary to address the
underlying narrative into which facts about climate change fit or do
not fit. Simply bombarding people with more information, more
facts and arguments, is unlikely to have any effect.

The Power of Narrative

Building on his claim that we learn about courage from reading
about courageous exemplars such as Achilles, Kearney writes:

> If we read *Anna Karenina*, we experience the tragic fate of a
> passionate woman in nineteenth-century Russia. If we read
> *Scarlet and Black*, we relive the life of an erratic, willful youth
> in Napoleonic France. And if we read *The Jaguar* by Ted
> Hughes, we can even transport ourselves into the skin of a
> 'non-rational' animal. What is impossible in reality is made
> possible in fiction.[57]

I should point out, lest we think these are the romantic musings of
an urban philosopher inexperienced with the wild, that Jack Turner
engages in the very sort of narrative projection for which Kearney
argues. Dismissing scientific study of pelicans in favor of gross
contact, Turner then goes on to say, "what interests me is not that
pelicans can soar, that soaring is useful, or that they soar here. What
interests me is the question of whether pelicans *love* to soar."[58]
Considering the inner lives of the pelicans he admires, he admits,
"I don't think we can say why [certain animals do what they do] without using analogies and metaphors from human emotional life."[59]

But doesn't this amount to crass anthropocentrism of the very sort
that justifiably concerns Marcia Eaton? Possibly, but not necessarily.
It is true that our narratives are hopelessly anthropocentric; we are
essentially narrative beings and, as far as we know, the only narrative
beings. Our experience and our perspective shape both our selves

and our world(s) at the most fundamental level. However, our narrative perspective no more condemns us to a vicious sort of anthropocentrism than it does to a vicious cultural bias. Just as narratives can open us to genuine, if complicated and partial, understanding of other human experiences and worlds (e.g., the Holocaust, or antebellum slavery in the United States), narratives can open us to other nonhuman experiences and worlds.[60] Turner and Kearney both recognize this, affirming the ability of narrative to open us to communion with others, human and nonhuman.

> It is not popular now to attribute human characteristics and processes to wild animals, since it projects onto the Other our biases and perceptions and limits our view of their difference. But all description is merely analogy and metaphor, and as such is forever imperfect and respectful of mystery. . . . We also fail to appreciate that many of our descriptions of human behavior are appropriations from wild animals: the lion-hearted hero, the wolfish cad, the foxy lady. And this suggests that life is a spectrum where unity is more pervasive than difference. . . . It is no more odd to say that pelicans love to soar and do so in ecstasy than it is to say what we commonly say of human love and ecstasy: that our heart soars.[61]

Of course, we must remain ever vigilant against allowing our inescapably anthropocentric perspective from sliding into a naïve and vicious anthropocentrism. Nevertheless, this should not dissuade us from embracing narrative in a manner that is (1) humbly aware of its inevitable bias and, at the same time, (2) excited about its genuine ability to open us up to other worlds.[62]

Conclusion

Despite Turner's protestations, John Muir's narratives about Yosemite and the High Sierra *did* induce people to love, and consequently preserve, these natural treasures. And while those narratives are notable for their romantic and religious tone, to which Marcia Eaton might object, they *also* gave those willing to read carefully an accurate account of many natural features, including the then-controversial theory that the valleys of Yosemite and Hetch Hetchy were formed by glacial erosion rather than by rivers. I don't claim that Muir's narrative account is better than either personal experience or scientific study of the Sierra Nevada in an absolute sense, or that narrative can simply replace either one. However, if part of the goal for both Turner and Eaton is the appreciation of and consequent care for the natural environment, it is worth pointing out that Muir's narratives were *more effective* than either scientific treatises (e.g., Whitney's geological reports) or personal experience (given that most people never go to Yosemite). Muir's narratives about the High Sierra, and what it did for and to him, moved his readers to value mountains they would never climb and for whose scientific history they cared not a whit.

So while both science (*theoria*) and practice (*praxis*) are useful, neither one is without problems and limitations. Science, as noted previously, frequently falls prey to the spirit of abstraction. Personal experience, in turn, is not always a possible or practical source of understanding. My purpose here is not to discount the understanding we get from scientific study or personal experience; both remain essential sources of understanding and illustrating their limitations in no way suggests that either one should be abandoned. Narrative is neither inherently superior to nor—at least in the case of narratives of nature—independent of gross contact and scientific understanding. The fact is that all three ways of understanding can, and ideally should, be mutually illuminating.

Scientific understanding, as Rolston points out, should be open to the wisdom of tradition and personal experience ("standing-under").[63] The value of such personal experience lies, I take it, in the way it tends to check the excesses of scientific abstraction. Personal experience grounds scientific insights in concrete reality and, as Rolston notes, scientific understanding can, in some cases, enrich personal experience of nature.[64] Eaton acknowledges that ecologists are most often inspired, both initially and as their careers proceed, by personal experiences. But we remember, recount, and understand those personal experiences, to ourselves and others, in a narrative form. We cannot understand, express, or even *think* of our experience—especially the powerful experiences with which Turner is concerned—outside of a narrative that gives it structure. Moreover, these interpretive acts are *not essentially distortive*, at least not in the pejorative sense of being delusional, deceptive, or fraudulent. That's important because such interpretation is also inescapable. Interpretation, experiencing life from a certain perspective and relating it in narrative, goes all the way down. There's no escaping the hermeneutic circle.

So scientific understanding needs personal experience and narrative, and personal experience needs narrative and scientific insight. Finally, we should add that narratives of nature rely on personal experience and science. Mick Smith claims, "[T]he stories we tell . . . depend upon more (or less) skillful and more (or less) conscious admixtures of nature and history."[65] Playing on this insight, we might suggest that the best nature narratives tend to rely upon skillful admixtures of some level of scientific understanding and some degree of personal experience.

Narrative is on some level dependent upon personal experience and gross contact for its inspiration and content. Our narratives must be about *something* and—as was the case with Thoreau, Muir, and Brower—it is often the case that powerful personal experiences make the most moving narratives. Narratives about nature are also dependent on parameters set by good science. In this sense, science *does* "correct for truth," but it is neither the sole nor final arbiter of it. Ricoeur claims that historical narratives are distinguished by a particular sort of truth claim that fiction does not make. This fidelity to historical accuracy means that, with respect to the Holocaust, for example, it is important to *both* tell the stories *and* count and document the bodies of the victims. Historical narratives require both narration and fact checking. Analogously, narratives about nature are constrained by certain limits set by science. Science does not dictate *the* narrative. There are an infinite number of ways to account for and relate both our experiences of nature (gross contact) and the complex web of relationships that make up ecosystems (scientific understanding), just as there are an infinite number of points between three and four on a number line; however, science can eliminate certain narrative excesses as unjustified, just as the numbers two and five are definitively out of the specified range on that line.

This analysis does not deny the fact that there is a physical reality out there that is independent of our interpretation of it—a "thing" that is part of our concern when we talk about "nature," and which science tries to address. However, the scientific way of engaging nature is itself only one layer of the palimpsest, to borrow an exceptionally fitting metaphor used by Martin Drenthen, of the "legible landscape."[66] Nor need these arguments reject the reality of sublime or "saturated" experiences that cannot be fully captured in narrative, or indeed in any sort of language, although there would be disagreement with certain phenomenologies regarding whether interpretation goes "all the way down" or whether it is tacked onto a preexisting experiential foundation after the fact.[67]

The emphasis on narrative in this chapter stems not from a belief that it can or should supplant the other two modes of understanding nature—both of which I consider essential in their own right—but from the fact that narrative has, at least in certain quarters, been misunderstood and, at times, dismissed as less significant than personal experience or rigorous science. Turner, while he admires the ability of narrative to inspire, claims that personal experience, or gross contact, gives us the "real thing." Eaton and Rolston, though they appreciate fiction and personal experience, suggest it is science that gives us the "real thing." However, both personal experience and scientific understanding are themselves shaped by a narrative or quasi-narrative structure, and narratives in the traditional sense can be essential aids in coming to appreciate, value, and understand nature. If this account is correct, we must come to terms with the fact that *all* our understanding, including the way in which we process gross contact and scientific knowledge, is to some degree narrative.[68] Therefore, far from being trivial or inessential, narrative is fundamental to both our understanding and our appreciation of nature.

10

The Question Concerning Nature

Sean McGrath

In this chapter, I situate Timothy Morton's and Slavoj Žižek's "ecology without nature" (hereafter EWN) within the broader history of transcendental-structuralist ontology.[1] I will argue that, notwithstanding Morton's recent turn to object-oriented ontology, his deconstruction of a certain notion of nature, which we provisionally describe as the extra-lingual intelligible order, does not deviate from the a-cosmic trajectory of late modern thinking, from nineteenth-century transcendental philosophy, through hermeneutics and semiotics to Lacanian psychoanalysis. Even if something like nature in fact existed, the argument goes, we would have little to say about it, locked as we are into a self-referential meaning system, whether this be understood as the synthetic whole produced by the transcendental ego as its proper arena of freedom (Kant, Fichte), the tradition-constituted and meaning-saturated order of being which can be understood (Gadamer), or the necessary illusion of stable extra-mental structure (the symbolic) which defends the ego from the truth of its own vacuity and the horror of life (Lacan, Žižek).[2] The popularity of EWN is largely due to its trendiness: Breaking with the folksy and somewhat frumpy environmental holism of the '70s and '80s, EWN is an environmentalism wise to contemporary theory, confirming the growing conviction in continental philosophical circles of a necessary movement beyond phenomenological critiques of calculative science to hypermodern reinscriptions of technological thinking (e.g., Badiou, Meillassoux, Gabriel, and speculative realism). Ultimately, what is discarded by EWN is not the quantifiable and manipulable material "order" (however contingent), which has been the guiding construct of natural science since the seventeenth century, but any account of matter as cohering in an extra-lingual intelligible whole, an organic totality or *kosmos*.

The force of EWN's critique of environmental philosophy derives from its exposure to the political impotence of eco-phenomenology, Heideggerian ecology, deep ecology, and eco-feminism. We have been reading Heidegger for the better part of a century, Timothy Morton reminds us; we have generated countless pages of phenomenological descriptions of embodiment, emplacement, and so on; we have had two centuries of romantic gushing over the ennobling experience of wilderness, and yet we have no political will to change the course of our economic and technological development, which is almost certain to end in total ecological and economic collapse. Witness how quickly global warming fell from the agendas of the major players of world politics with the 2008 financial crisis. The message was clear: In a forced choice between clean industry and economic growth, the latter wins out every time. Environmentalism has found a provisional place in our political rhetoric, but our political unconscious remains driven by capital, or more specifically, consumption. How are we to change this?

No doubt, the failure of environmental concerns to transform the political cannot be denied. When the water from the melted Greenland ice cap submerges most of London, New York, and Dubai, we may see the mobilization of a planetary will to reform our unsustainable way of living, but by then it will certainly be too late, if it is not all ready. However unpopular the question, we must ask ourselves if, from the perspective of environmental hermeneutics, this desperate practical exigency is not beside the point: What is at stake in the question concerning nature is a philosophical problem, perhaps *the* philosophical problem of the West, and its intrinsic interest and contemporary relevance is not reducible to any particular practical concerns. Even when the world is ending, philosophy remains philosophy. Perhaps the end of the world is the perfect time to philosophize without any clear sense of where it will lead, to speculate about what is ending and what is beginning in as open a way as the times allow.

I will make a case for questioning EWN's rejection of "premodern" cosmology and argue for a reconsideration of alternative models of material interdependence, specifically, one central to the Renaissance philosophy of nature. Hermetic holism, which is neither premodern nor modern in the accepted sense of the term (I will argue that it is *alter-modern*, a way of being modern that has never been tried), maintained the crucial primacy of contemplation over critique, calculation, and control (the ideological stance of modernity) without, however, aestheticizing nature. It offers us a strong theory of interdependence but does not implicate us, as EWN does, in a posthuman metanarrative. In the end, I aim to rehabilitate Heideggerian *Gelassenheit* as the most extreme challenge to the reigning *Zeitgeist* and still a central theme for the hermeneutics of nature, for it disturbs our age in exactly the place where a constitutive ontological attitude makes it all but impossible for us to the think the being of nature.

Ecology without Nature

A statement of the scientific credentials of EWN is found in Žižek's *In Defence of Lost Causes*:

Today, with the latest biogenetic developments, we are entering a new phase in which it is simply nature itself that melts into air: the main consequence of the scientific breakthroughs in biogenetics is the end of nature. Once we know the rules of their construction, natural organisms are transformed into objects amenable to manipulation. Nature–human and inhuman–is, thus, 'desubstantialized,' deprived of its impenetrable density, of what Heidegger called "earth." This development compels us to give a new twist to Freud's title *Unbehagen in der Kultur*– discontent, uneasiness in culture. With the latest developments, the discontent shifts from culture to nature itself: nature is no longer 'natural,' the reliable 'dense' background of our lives; it now appears as a fragile mechanism, which, at any point, can explode in a catastrophic direction. Biogenetics, with its reduction of the human psyche itself to an object of technological manipulation is, therefore, effectively, a kind of empirical instantiation of what Heidegger perceived as the 'danger' inherent to modern technology. What is crucial here is the interdependence of man [*sic*] and nature: by reducing man [*sic*] to just another natural object whose properties can be manipulated, what we lose is not (only) humanity but nature itself. In this sense, Francis Fukuyama is right: humanity itself relies on some notion of 'human nature' as what we simply inherited, namely, the impenetrable dimension in/of ourselves into which we are born/thrown. The paradox is, thus, that there is man [*sic*] only insofar as there is impenetrable inhuman nature. With the prospect, however, of biogenetic interventions opened up by the access to the genome, the species is able to freely change/redefine itself, its own coordinates; this prospect effectively emancipates humankind from the constraints of a finite species, from its enslavement to the 'selfish genes.' However, there is a price for this emancipation. . . . Should we not apply here the fundamental lesson of Kant's transcendental idealism: the world as a Whole is not a Thing-in-itself, it is merely a regulative Idea of our mind, something our mind imposes onto the raw multitude of sensations in order to be able to experience it as a well-ordered meaningful Whole? The paradox is that the very 'In-itself' of Nature, as a Whole independent of us, is the result of our (subjective) "synthetic activity."[3]

Here is EWN in its starkest outlines: Biogenetic developments divest us of our illusions of possessing a depth dimension, a human nature, and simultaneously demolish time-honored distinctions between culture and nature, artificial and natural, organic and mechanical. From the perspective of what we actually believe when we do science (always for Žižek the final court of appeal), there is no natural order. Žižek adds that there never was: Nature as the whole that contained humanity was a necessary illusion, a quasi-Kantian ideal, a Lacanian "fundamental fantasy" to be traversed, if never totally abolished. Breakthroughs in genetics demonstrate that no "nature in itself" prevents us from altering the course of our evolution or artificially creating life. When these things happen, as they inevitably will (if they have not already), the comforting fiction of a natural order grounding the intelligibility of our language and the morality of our actions is shown for what it is: a projection, however inevitable or even necessary on a day-to-day level.

According to Morton, the nature that appears to have recently died is not to be mourned at all; it was, in fact, never alive, for "nature" as such is nothing more than an environmentally dubious construct, a romantic leftover, even if it still inspires eco-phenomenologists, eco-feminists, and Walt Disney alike, something over and against us, to be aesthetically enjoyed as the beautiful and sublime backdrop of our lives.[4] The distance assumed between nature as an aesthetic object and the natural aesthete, is, according to Morton, the nub of the ecological problem itself. By holding ourselves apart from an idealized nature, we indulge a number of ecologically violent beliefs: that the structure of the visible is an accurate measure of the structure of the invisible; that things will go right if we simply let nature be; that we are at the center of universe of meaning that subsists apart from us; that we have a "natural right" to the ownership of matter. In opposition, Morton offers us a neo-Darwinian "mesh of interdependence," which, not accidentally, has a Buddhist ring to it (Morton is himself a Buddhist), a material ontology in which "everything is related to everything else" to such an extent that any effort to distinguish things or species of things must be ultimately seen as empty abstraction. Morton combines claims from contemporary biology and physics with deconstruction (the somewhat tired critique of "essentialism") to mount a multi-fronted attack on the self-identity of "nature" and "natural" beings: Things do not stand apart from other things but are assemblages of them. If we look beneath the skin of our most cherished natural being—say the drowning polar bear desperately in search of an ice floe—we do not find a definable essence which justifies our distinguishing it from another, but rather an interweaving net of chemical and physical properties, which are not themselves stable but in a state of continual transformation.

Morton draws a neo-Marxist conclusion to his monistic metaphysics: At bottom, there are no natural substances, only collective (political) relations. For Morton, this natural scientific-deconstructive version of the Buddhist doctrine of "no self" becomes the metaphysical departure point for a more muscular eco-politics, ostensibly an ecology with fangs, by distinction from failed deep ecology with its crypto-essentialism, insipid eco-phenomenology, with its self-indulgent romanticization of the wilderness, or dangerously a-political, quietistic Heideggerian holism. These passé environmental philosophies share a common assumption which is lethal to the environmentalist cause: the assumption that nature is a coherent and intelligible whole to which the human being primordially belongs. This notion is not just innocent fantasy; it is at the root of the eco-crisis, for it is, in Morton's peculiar genealogy, inextricably bound up with foundational assumptions of liberal capitalism and consumerism—the proximate causes of global warming, resource depletion, and the poisoning of land, water, and air. As the crowning achievement of evolution, the liberal self was licensed to lord it over "inferior" material entities, even if benignly, as "the shepherd of being," or innocuously as the enraptured aesthete who cannot get enough fresh air and enthuses, with Thoreau, "in wilderness is the preservation of the world." Morton's notion of interdependence means there is no environment ontologically distinct from those it environs just as there is no essential difference between human and animal or no real distinction between artificial and organic. The eco-freaks who fetishize pristine wilderness are the heirs of Hegel's "beautiful soul": neo-romantics longing for union with that which they themselves have alienated for the sake of constituting their own spiritual identities. EWN will do without the fantasy of a subsistent natural order and its attendant organicism, Heideggerian holism, vitalist holism, ecological holism. Every version of organicism is guilty of arbitrarily excluding some part of the material real which it defines as *not belonging* to the whole, whether it is the machine, the inanimate thing, or people whose skin color, ethnic origin, or sexual orientation we happen not to like. By distinction, Morton offers us hyper-collectivism: Matter belongs to all those who happen to be implicated in it, which means, no part of it belongs to any other part of it, distinctions between parts being what they are, abstractions from underlying levels of interdependence (the human DNA is 25% daffodil, as Morton points out). In Morton's most scathing critique, the nature lovers who drive much of the ecological movement are likened to the decadent *flaneurs* of late romanticism, ostensibly the first consumers, who consumed without consuming, enjoying the show of that which was on display in the shop windows of nineteenth-century Paris arcades.

This kind of hermeneutic-deconstructive conceptual genealogy no doubt advances the environmental philosophical discussion, which, too, often becomes nostalgic and philosophically stale, recycling hackneyed images of a wholesome, refreshing, and invigorating "warm green" nature (to use Bruno Latour's phrase) we remember from summer camp. EWN has raised the question concerning nature and, although Morton and Žižek have no intention of leaving it open, once raised, the question cannot be easily closed again. Morton, in particular, shows us, as much by his literary bombast as by his historical insight, that the concept arguably most central to the environmental philosophies of the last century, the concept of nature itself, remains entirely obscure and uninvestigated. It is indeed time that we asked the question concerning nature. But how to ask it?

Ontology without Nature

Part of the reason for the success of Morton's *Ecology without Nature* is that, contrary to its rhetoric of radicality and novelty, it does not, in fact, challenge the dominant twentieth-century epistemologico-ontological discourse, namely structuralism (the rightful heir of transcendental philosophy). Morton's ecology without nature is preceded by structuralism's ontology without nature; the affinity between them is demonstrated by the facility with which Žižek, the grand master of neo-structuralism, has appropriated Morton's critique of environmentalism into his own tireless Lacanian analysis of Western decline. Žižek celebrates Morton's *Ecology without Nature* for liberating us from "the fundamental fantasy" (the constitutive lie essential, according to Lacan, to subjective identity) that the human being with its values, decisions, and cultures has a home in the universe. By means of countless signs of the times, from the *Eyjafjallajökill* volcano to the BP oil spill, we are agonizingly being divested of the illusion of the possibility of a return of the human community to the bosom of mother earth—a state of balance which never existed in the first place. The romantic nostalgia for a home in the cosmos has become in the present circumstances life-threatening. We must use our instrumental reason, Žižek argues, not renounce it, so as to better organize and distribute our finite resources and postpone the inevitable failure of civilization for as long as possible.

For Žižek, the significance of EWN is that it unmasks the Big Other as a Big Lie: Nature as the aesthetic and moral background of our lives is the fiction constitutive of our vacuous subjectivity. The construct is inevitable for our material situation is unthinkably bleak: We are the arbitrary products of an accidental series of events which made matter favorable to life, a series which can and likely will be reversed without warning. Environmentalism is "eco-ideology," a new opium for the people, replacing the moribund Western religions with an equally self-serving myth that functions as an external authority in human affairs. "There is no Evolution: catastrophes, broken equilibriums, are all part of natural history; at numerous points in the past, life could have turned towards an entirely different direction. Even the main source of our energy (oil) is the result of a past catastrophe of unimaginable dimensions."[5] To traverse the fantasy and mobilize genuine political will for environmental policy (by distinction from the empty rhetoric of our elected representatives), we need to practice that curious Lacanian two-step of acknowledging the inevitability of the Big Lie which constitutes our subjectivity and becoming more at ease with the necessity of lying (i.e., "enjoying our symptom"). "Along these lines, 'terror' means accepting the fact of the utter groundlessness of our existence: there is no firm foundation, a place of retreat, on which one can safely count. It means fully accepting that 'nature' does not exist."[6]

It is no doubt odd that environmentalism has found an ally in structuralism, for nothing seems to be less ecological than Saussure's semiotics. But EWN is primarily about eco-politics, not the "intrinsic value" of clean air, verdant forests, and teeming seas. The upshot of EWN is that there is nothing outside the political and this is clearly familiar structuralist terrain. The structuralist revelation of the arbitrariness of signification is already for the early Lacan sufficient grounds for denying an extra-lingual order of intelligibility. If signification is mere convention, the signified, tethered as it is to the signifier, has no more ontological subsistence than the signifier. In short, we are securely on medieval nominalist terrain: Talk is always only talk about talk. The key to coherent speech (spoken or written) is neither the authentic expression of transcendental concepts nor reference to an extra-lingual state of affairs, but the successful application of a sign within a system of differences. A natural referent (the "transcendental signified") or a pre-lingual order of things has no role to play in language; on the contrary, language is only possible in the absence of the pre-lingual thing. Where we can speak, there nature is not, and where we cannot speak, there we are not.

In the proto-structuralism of medieval nominalism, the problem of universals was ostensibly solved by making thought a self-generative system, independent of an extra-symbolic order. Transcendental philosophy mediates medieval nominalism and late-modern structuralism, for the effect of Kant's making intelligible structure dependent on modes of cognition and perception was the reinscription of nominalism's severance of thought from thing. While Kant was still alive, there were protests against this expansion of the cultural to the compass of being: Schellingian nature-philosophy refused the dichotomy of thought and thing, mind and matter, subject and object: The thing *thinks itself* in me, according to the young Schelling, which means equally that the thought *things itself* outside of me (thought is not an intrasubjective affair).[7] "The whole of modern European philosophy since its inception (through Descartes) has this common deficiency—that nature does not exist for it."[8] But outside a relatively small circle of nineteenth-century speculative philosophers, the Schellingian objection was not heeded, and the human sciences developed in the direction of nominalism–structuralism. Early philosophical hermeneutics found a fertile ground for its history/culture distinction in transcendental philosophy. For Dilthey, the hermeneutical sciences (the *Geisteswissenschaften*) concern the self-generative world of thought: The material thing, if it could still be affirmed to exist, was the subject of natural science. The early Heidegger and his disciple Gadamer, rejected this ontological reduction of hermeneutics to the cultural, but the effect of their collaborative widening of the hermeneutical circle to include ontology was not a return to realism, quite the reverse: Now reality itself became merely a manner of speaking about beings whose existence in every way depended upon our speaking about them.[9]

Freud's psychoanalysis initially appeared to be founded upon a naturalism foreign to transcendental-structuralist ontology. Trained as biologist, Freud believed he was interpreting psyche as a natural phenomenon, if not a material thing, at least a thing obeying something analogous to the laws of matter and explicable, like matter, in a purely non-teleological and mechanistic way. The mind-matter dichotomy of early modernity, like the related culture-nature dichotomy of transcendental philosophy, was to be overcome not by denying the mind-independent existence of nature but by proving that mind and culture were not unnatural; the ego, along with other precious products of human civilization, was a defensive formation of a highly evolved material organism ultimately driven by the same dual drives of life and death that one can observe in an amoeba.[10]

At the heart of Lacan's "turning Freud inside out" like a glove, as Lacan puts it, is the French psychoanalyst's break with this psychoanalytical naturalism.[11] In Lacan, Freud's lifelong effort to make a natural science of psychoanalysis is laid to rest; a natural history of the psyche is no longer possible; psyche is an order of linguistically structured experience to be explicated *sui generis*. In the mirror stage, which gives rise to the ego, the presubjective mind of the infant seeks release from the chaos of its uncoordinated body by disidentifying with its lived experience and learning to identify with the unified and defined body revealed to it in the mirror. Mommy and Daddy help it along by pointing at the reflection and chirping brightly, "It's you!" This misidentification marks the child's entrance into consciousness, first into "the imaginal," which is for Lacan, not the immediately sensible, but the base stratum of false consciousness upon which the full network of the symbolic order rests, and later into "the symbolic," the world of meaning substituted for "the real." The child's identification with the coherent image of itself in the mirror is a repression of the lived experience of its own body. Repression is thus the condition of the possibility of subjectivity—repression not only of basic sensations and drives but of anything that challenges or threatens the ideality around which the psyche consolidates its virtual identity. The child, liberated by the mirror from the coil of appetites and drives that compose its awkward and disobedient body, is henceforth an "I," an "immaterial subject" able to author its actions and enjoy an interior life about which it can then speak with other subjects. But this liberation is bought at a price. "The real" withdraws in the repression only to return in the inevitable slips and breaks, crises and pathologies, which characterize human existence.

Lacan's structuralist theory of subjectivity simultaneously evacuates interiority of authentic meaning and relegates exteriority to the absurd, the remainder which thought cannot think, not because it exceeds comprehension but because it gives no intelligibility to be thought. Subjectivity is a virtuality publicly constituted "in" the psyche of every speaker by the language she speaks. Language, without which it could not exist, forces the subject to expel itself from life and destines it to desire this expelled life as the only thing that can complete it. The life from which consciousness withdraws haunts it as the unimaginable "real." Because regaining "what was lost" can only mean the extinction of the subject, the psyche is trapped in a no-win situation of fantasy, desire, and denial, condemned to perpetually desire an integrity, wholeness, and unity with matter that must always elude it (hence the inevitability of "nature" as a phantasmic aesthetic object: The subject must on some functional everyday level experience itself as in exile from a primordial state of belonging). Žižek again, "There is no subject without some external 'prosthetic' supplement which provides the minimum of his phantasmic identity—that is to say, the subject emerges via the 'externalization' of the most intimate kernel of his being (his 'fundamental fantasy'); the moment he gets too close to this traumatic content and 'internalizes' it, his very self-identity dissolves."[12] The "prosthetic supplement" of eco-ideology is nature, the cosmological whole, the universe of meaning from which modernity has putatively alienated us. The eco-ideologue simultaneously longs for union with the cosmological whole and holds himself at a distance from it by idealizing it as the lost home of humanity. Were he to achieve the object of his desire, union with the cosmological-whole, his identity would dissolve.

Suffering the insubstantiality of subjectivity is at the heart of Lacanian neurosis, but such suffering cannot be avoided; it can only be accepted as inevitable. Lacanian subjectivity is nothing (here is the place of interface with Morton's Buddhism), an absence of reality, a zero constituted by the negation of immediacy. Language, a system of self-contained and self-generative significations which is always incomplete, exposes the nothingness of the subject by denying it the possibility of shaping a "true" self: Every time the subject utters a word his utterance is sabotaged by unintended significations. The Lacanian unconscious—that which is expelled, denied, forever misplaced—is not a source of meaning, creativity, or life; it is the wound left in us by language, by the excision of the infant from presymbolic life, not a reservoir of meanings and intuitions that have slipped beneath the threshold of consciousness but an idiot that apes the logos, playing with language without understanding what it says, mechanically proliferating—without intention or meaning—the differences and ambivalences that are as integral to language as is grammar.

The Lacanian "subject" is the Cartesian subject living in a disenchanted world, the subject of modern science, and, we might add, the Hegelian subject, a subject deprived of "roots" in "nature" by the structure of consciousness itself. Subjectivization is not a reversible procedure: There is no way back into presubjective "nature." This is the significance of Lacan's *Vel*, the mathematical sign for an either/or choice. The human being is faced with an impossible choice, one akin to the pseudo-choice offered by the mugger: "Your money or your life." The choice is between subjectivity (rationality, symbolic life, etc.) and being (natural life, immediacy). Like the mugger's victim who cannot choose to save his money without losing his life (and therefore losing his money as well), so the modern subject cannot choose nature over subjectivity without losing nature (for no one would be left to enjoy what is chosen).[13] Subjectivity is only possible on the grounds of a severance of consciousness from "the real" (the cut effected by "the symbolic"). The unconscious is the trace of this scission: It is not "the remains of nature" (as though some dimension of man's natural origins exists on a subterranean level of the psyche) but rather the other whose exclusion is necessary to maintaining the bubble of the symbolic—a sign of an absence of totality, a gap or lack.

Lacan coins the term "extimacy" ("something strange to me, although it is at the heart of me"; an "intimate exteriority"[14]) to articulate how subjectivity is constitutionally exteriorized by the symbols without which it would not exist. The most interior is the most exterior, and vice versa: The unconscious is noisy with cultural chatter, and "the great outdoors" is filled with subjective constructs. The community of speakers who give me my identity and position in the symbolic are not outside me; they occupy the innermost sanctum of my subjectivity. Conversely, outside me is no natural order but the social political collective that grants me an identity, in other words, a position in the symbolic. At the core my existence is not nature but the marketplace of collective meaning which has engendered my identity. Žižek likens "extimacy" to the *Kinder Surprise* one can find in every German grocery store. The sumptuous chocolate egg promises a rich gooey interior. When one breaks it open, however, one finds that it is hollow. Inside, wrapped in a plastic capsule, is a toy made in China, a piece of indigestible junk.

The Lacanian symbolic (accurately depicted in the previous image) is a reaction to an unthinkable real far more disturbing than Kant's *Ding an sich*, the directionless material origin of our existence. Nonetheless, we must not miss the transcendental echo in EWN: For Žižek and Morton as much as for Kant and Lacan, any experience of nature as organic whole, the universe of meaning, can only be the result of a substitution of a psycho-genetically structured totality for the material chaos of the universe. However necessary

Figure 1   Kinder Surprise Egg. (Photo by Martin Drenthen.)

Figure 2   Kinder Surprise Egg. (Photo by Martin Drenthen.)

our fantasy of interior depth and infinite meaning, EWN demolishes both senses of Aristotelian *physis*: There is no essence within that could render the unconscious a richly mysterious source of meaning, and no cosmological whole without that could ground culture in something other than convention.

Morton and Žižek are not claiming that cosmos has ever been an important scientific ideal for the West: Rather, as a fantasy with little purchase on the real, it has allowed environmentalists to dream about a nature that only truly exists in romantic poetry and Tolkien novels, while capitalist-driven science proceeds to ravage the planet. What remains for EWN is modern techno-science, with its seemingly irresistible will to mastery, a machine that EWN hopes is separable from the capitalist-consumerist ideology which currently holds it hostage. With the collapse of romantic and post-romantic organistic models of nature, there is nothing left for us to do but recognize technology as the only reasonable response to the senseless contingency of matter: At least we might try to hold the monster at bay for a while. Disburdened of our romantic ideal of a universe of meaning, we must, Morton and Žižek agree, begin to think much more realistically and sustainably about how we use the fraction of inhabitable matter the universe has arbitrarily bequeathed to us. What is required in such a hopeless, indeed, Gnostic *Geworfenheit* is not less instrumental reductionist thinking but more. Let us stop waiting for a God; if technology cannot save us, nothing will.

The Gnostic overtones of EWN require no great hermeneutical work to expose. Notice the not-so-subtle nihilism in the following quote from Morton: "Dark ecology is no solution to the problem of nature, which has more in common with the undead than with life. Nature is what keeps on coming back, an inert horrifying presence and a mechanical repetition."[15] One thinks of Lars von Trier's *Antichrist* in which the doomed couple simultaneously come to terms with the disavowed violence at the root of their sexual life, visualized in scenes of genital mutilation, and the repressed monstrosity of "nature in itself" epitomized in the deformed fox that pauses from eating its own entrails to declare, "Chaos reigns" (two sides of the same coin). Von Trier's ludicrous CGI sequence is the new version of Disney's talking animal, it seems, what beasts would really say if they could speak. As in ancient Gnosticism, there are those who can accept the bleak truth that "mother nature" is a monster and those that cannot. The former stand to the latter as the initiated, the elite (those who have experienced *gnosis*) to the uninitiated and ignorant. The difference between ancient and modern Gnosticism is nonetheless dramatic: For modern Gnostics like Lacan, Žižek, and Morton, there is no salvation in *gnosis* save the smug satisfaction of knowing better than others and whatever joy the perfunctory disposal of the history of religion and metaphysics affords them.

The Road Not Taken

It is important to note the difference between EWN and the political ecology of Isabelle Stengers, Bruno Latour, and Michael Serres.[16] Stengers, Latour, and Serres use deconstructive antifoundationalist claims to challenge the entrenched de-politicization of techno-science. The questionability of modernity's cherished fact-value distinction and the related distinction between primary and secondary qualities destabilizes techno-science by removing any last objections to including the fate of the nonhuman in political discussion.

The relativization of "matter in motion" and the demotion of natural science from its place of a-political prestige to one discourse among many, as such, an epistemology founded in certain ethico-political decisions and meaning-generative practices, inclines Latour and company to reconsider older models of nature hitherto deemed unscientific or outmoded. EWN, by contrast, at least in its Žižekian form, is not, in fact, antifoundationalist at all. On the contrary, the undisputed authority on posthuman material interdependence for EWA remains the natural scientist.

As good an example as any of the kind of environmentalism that EWN polemicizes is ecofeminism. In *The Death of Nature*, eco-feminist Carolyn Merchant contrasts androcentric Enlightenment atomization, objectification, and dissection of nature, with Renaissance holism, which still retained premodern ideas of cosmos. "The female earth was central to the organic cosmology that was undermined by the scientific revolution and the rise of a market-oriented culture. . . . For sixteenth-century Europeans the root metaphor binding together the self, society and the cosmos was that of an organism."[17] Nature conceived as an organic whole simultaneously knitted together individuals in concentric circles of belonging—family, community, state, created order—and justified the subordination of the good of the individual to the common good. By contrast, the "dominion" model of techno-science sundered humanity into atomistic, self-interested individuals and reduced the earth to exploitable resource.

EWN does not deny this cultural history; it rejects the assumption that it could be reversed, the ostensibly environmentally fatal delusion that there is a way back into premodern ways of thinking about our relationship to matter. Thus a question, central to the Heideggerian–eco-phenomenological approach to environmentalism, remains unasked by Morton and Žižek: Why did our thinking about nature change in the seventeenth century? Was it the inescapable adjustment of culture to the horrible truth of matter, or the victory of one socially constructed model of nature (scientifically objectifiable, meaningless matter) over another (cosmos)?

It is worth dwelling for a moment on this vanquished "premodern" cosmology, which still captured the imagination of the West, at least until the seventeenth century, if not later (it remained an inspiration for Schellingian nature-philosophy). In fact, it is not premodern at all, for if "modern" is to be a useful adjective, it must include the Renaissance—Leonardo, Martin Luther, Galileo, are not medieval thinkers but the architects of modernity itself. Among the ancient ideas which were retrieved by Renaissance humanists, the Greek notion of *kosmos* was preeminent. It is well-known that the translation of the *Corpus Hermeticum* by Ficino played a key role in the development of Renaissance natural science: In the hermetic idea of kosmos as a living being created by a personal God as medium of his self-revelation, Renaissance humanism found a theologico-metaphysical ground for the rapidly developing experimental sciences.[18] The orthodox Christian concept of creation was not so intimately tied to the life of the creator; although a revelation of his mind, material nature was not held to be essential to God's own experience of himself. In a theology of creation such as Augustine's, matter is to some degree arbitrary, and medieval ascetical-mystical theology habitually directed the soul beyond and above it. Matter was no doubt good, but not that good, and in its tempting quality it posed a grave threat to the soul: best to have as little to do with it as possible.

The hermetic model of nature, by contrast, required that the philosopher/theologian/natural scientist (ideally the same person) contemplate the material structure of nature, not in an objectifying theoretical state of detachment but immersively, for matter was our primary mediator of the divine mind. The thought, commonly associated with nineteenth-century idealism, that God knows himself, through our knowledge of his creation, is a product of Renaissance hermetic cosmology.[19] What we speak of here is not "premodern," as though every hermeneutics of nature prior to seventeenth-century calculative science can be lumped together under one category, but, to coin a term, *alter-modern*, another way of being modern, the road not taken by the West. This alternative cosmology supports the almost universally accepted environmental thesis of ontological interdependence (common to EWN, eco-phenomenology, and depth ecology), but without Morton's antihumanism or Žižek's nihilism. The alter-modern cosmology of hermeticism retained a place for contemplation (the neo-Platonic lineage, which returns in Heideggerian *Gelassenheit*), but without aestheticizing nature in a decadent ("romantic") way. Nature, for the hermeticist, was no sublime show to be observed from a distance but a cosmological process to be participated in, and not in an objectifying fashion, for the first material entity to be transmuted (decomposed down to its base elements or *prima materia* and reconstituted on a higher level) was the alchemist himself. The hermetic kosmos was neither a collection of extended objects nor an aesthetic consumable; it was, to use Morton's preferred expression, *intimate* with man. When we contemplated nature, we contemplated ourselves, and vice versa. When we operated on it, we operated on ourselves. When we profaned it, abused it, or denigrated it to a means to some utilitarian end, we profaned, abused, and denigrated ourselves.

Foucault describes the abrupt turn from hermetic cosmology to calculative science in the seventeenth century not as an *evolution* (the myth of progress propagated in history of science textbooks) but as a cultural *revolution*, the outright rejection of one paradigm of nature in favor of another, for predominantly political purposes.[20] In the culture wars of the late Renaissance, the Cartesian subject won out over the hermetic Magus. Where the latter contemplated an infinite order not of his making and endeavored to understand it as far as possible so as to be able to productively participate in it, the former does not find itself in an order, it rather orders that which is inherently disordered. The implication of this world-generating responsibility is that the subject itself has no place in the order which depends upon it to exist. The power of the Cartesian subject is conditional upon its homelessness in the universe: Only that which wholly transcends extended being, which is in no way part of what it orders, could ultimately control it. At issue was an evaluation of the highest task of reason: Was it contemplation, as the hermeticists, following Aristotle, Plato, the Stoics, and the legendary Hermes Trismegistus believed, or control? For the hermeticist, total control was neither achievable nor desirable. The kosmos was wondrously incalculable because all things mirrored each other in an infinity of correspondences. "The universe was folded in upon itself: the earth echoing the sky, faces seeing themselves reflected in the stars, and plants holding within their stems the secrets that were of use to man."[21]

The hermetic kosmos is depicted in the accompanying illustration from Mylius's 1617 *Opus medico chymicum*.

Everything is connected to everything else, but not in a posthuman mesh of mindless interdependence: Rather, the human, the microcosm is the center of a whole which is permeated by mind, the somnambulistic intelligence of "the world-soul." Nature is depicted here as an absolute totality, including even the Trinitarian God, it has no outside. The hermetic kosmos in Mylius's illustration is divided into four parts: an upper (divine), a lower (earthly), a left (of the central figure) (Holy Spirit/dark/feminine), and a right (Divine Son/light/masculine). The divisions are not separations, for everything occurring on the left is mirrored on the right, and everything

Figure 3 Johann Daniel Mylius, *Basilicae Philosophicae*, Frankfurt, 1617, British Library.

occurring above is reflected below. "That which is above is like to that which is below, and that which is below is like to that which is above, to accomplish the miracle of the one thing," runs the first line of the *Tabula Smaragdina*.[22] These correspondences are not simple repetitions but modifications and developments, sometimes reversals of structure. The divine Trinity, composed of the *Tetragrammaton* (top center), the lamb, and the dove, is reflected in the three figures below, the Magus, the woman and the man. Assuming that the wheel turns clockwise, the Holy Spirit is the luminal being who initiates the theophany of matter. In hermetic systems such as Boehme's, the mundane corporeal world is nothing other than the world of the Spirit. Every natural being therefore images the divine.

Alchemical work is a fundamental theme of Mylius's illustration (the image has often enough been interpreted as a cryptogram of alchemy). The central disc consists of a number of concentric circles that describe symbolic recipes for integrating the worlds above and below. In the middle of the wheel is a triangle containing the symbol of Mercurius, the chthonic spirit of alchemy, who tends to converge with the Christian Holy Spirit in alchemical imagery, but with a set of associations troubling for Christian orthodoxy. He is also called the world-soul (*anima mundi*), the spirit in nature, and the trickster, Hermes, the duplicitous divine messenger. The wheel of three divine persons gives rise to a wheel of five animals below, each symbolic of a stage in alchemical transformation: a crow, a swan, a winged dragon, or a cockerel, a pelican, and a phoenix. He holds the sun, she the moon. He is assisted by a lion, she by a stag. Beneath his feet are the phoenix and the symbols for the masculine elements, fire and air. Beneath her feet are the eagle and the symbols for the feminine elements, water and earth. The Magus in the center is the androgynous Mercurius incarnate, neither male nor female, holding the two axes that can sever the bonds that tie the masculine and the feminine to either side of the kosmos, thus making the opposites mobile and free for recombination in new forms. The Magus stands on a lion with two bodies and one head. Out of the lion's mouth gushes water, the *aqua permanens*, the eternal water, the *aqua vitae*, the water of life. One side of his robe is dark, the other light, for he is himself the *coincidentia oppositorum*.

Most fascinating for our purposes is how hermetic contemplation was not separated from practical and technical work. The alchemical assumption was that nature was not yet finished: God left it incomplete so that the human being might become a participant in its development. Thus science and technology were divinely commissioned, in no way non-natural or transcendental. Technology was as natural as fungi or sexual reproduction. The alchemist did not stand over and against nature like a Cartesian subject or romantic aesthete, his contemplation and works were themselves products of nature and would be all the more successful the more in accord they remained with the deep spiritual patterns underlying matter. Read Paracelsus on the sovereignty of nature over art:

Q. *What is the chief study of a philosopher?*
A. It is the investigations of the operations of nature.
Q. *What is the end of nature?*
A. God, who is also its beginning.
Q. *Whence are all things derived?*
A. From one indivisible nature.

Q. *Give a concise definition of nature.*
A. It is not visible, though it operates visibly; for it is simply a
   volatile spirit, fulfilling its office in bodies, and animated
   by the universal spirit—the divine breath, the central and
   universal fire, which vivifies things that exits.
Q. *What should be the qualities possessed by the examiners
   of nature?*
A. They should be like unto nature itself. . . .
Q. *What matters should subsequently engross their attention?*
A. The philosophers should most carefully ascertain whether
   their designs are in harmony with nature. . . . If they would
   accomplish by their own power anything that is usually
   performed by the power of nature, they must imitate her in
   every detail. . . .
Q. *What is the object of research among philosophers?*
A. Proficiency in the art of perfecting what nature has left
   imperfect.[23]

Alchemical *techne* was not conceived as other than *physis* but
rather as *physis* by other means. The alchemist had to combine
spiritual and moral practice with scientific know-how for what he
was called to perform was nothing less than a sacramental act of
co-creation. The nature which was his guide in everything he did
was not an object, nor the totality of objects, it was not even primarily
material, it was being itself. Let us stop misinterpreting alchemy as
an ancestor of modern chemistry. The aim of "the work" was not
control but divinization of the Magus. No question, the Magus's
participatory transformation of the base into the sublime promised
him immense power over natural processes, but the assumption
that such power was his to command at will was a temptation to be
resisted. On the basis of this ethical imperative to resist the will to
mastery arises the curious blend of traditional Christian piety with
scientific discipline in alchemical engravings: The alchemist was
dependent upon power which, in the end he could not control, for it
was the privilege of Mercurius to initiate the alchemist into infinite
knowledge or leave him bereft with his failed experiment.[24] *Aurum
nostrum non est aurum vulgi,* "our gold is not the ordinary gold,"
the alchemists claimed, but the kings and queens upon whose
patronage they depended paid no attention, and so alchemy became
known for its more utilitarian possibilities. In the accompanying
famous engraving from Heinrich Khunrath, the alchemist kneels in

Figure 4   Heinrich Khunrath, *Amphitheatrum sapientiae aeternae solius verae*, Frankfurt, 1609.

prayer before beginning his work. In the smoke from the censor in Latin text are the words: "ascending smoke, sacrificial speech acceptable to God."

It is worth reflecting on who the Renaissance Magus is. His identity is disclosed in his position in Mylius's map of the kosmos: not at the top, or the center, but in the bottom middle, that is, at the center of the earthly realm. He is not a fallen angel, who really belongs somewhere else, nor is he the sovereign subject of idealism, concealing in his a priori nothingness the mechanism by which the order of things is generated. He is the node in the network of spiritual-material relations, the connecting point in the currents of desire that animate the macrocosm. Deleuze and Guattari's schizoid nomad comes close to the ideal: "Not man as the King of Creation, but rather as the being who is in intimate contact with the profound life of all forms or all types of beings, who is responsible for even the stars and animal life, and who ceaselessly plugs an organ machine into an energy-machine, a tree into his body, a breast into his mouth, the sun into his asshole: the eternal custodian of the machines of the universe."[25]

Renaissance alchemy distinguished itself from medieval natural science in its aiming at producing concrete results. It was not enough to contemplate nature; the human being was called to perfect it, that is, to participate in its development and bring it to even more effective fruition in ways that could be of direct benefit to humankind. But this was not the will to mastery, for the contemplative side of alchemy was not a means to calculative control; it was an end in itself, indeed the chief end of the *opus*. The participatory perfection of natural processes was not only for the sake of mastering matter or torturing nature to reveal her secrets and then putting this information to the service of mankind; it was above all for the sake of the transformation of the *Magus*: to become himself holy, individuated, a new Christ. Such a work could never be the result of method: The Magus did not transform himself but was transformed by the very divine power that was at work in all natural transformations.

The Renaissance alchemist was above all a hermeneutician of nature. In the Paracelsian doctrine of signatures, the correspondences between things were held to be concretely visible (if encoded). Resemblances showed themselves in visible marks on the surfaces of things, in the outer form of physical things. The work of the alchemist was largely one of deciphering these signatures so as to be able to ascertain the underlying resemblances. "The face of the world is covered with blazons, with characters, with ciphers and obscure words—with 'hieroglyphics'. . . And the space inhabited by immediate resemblances becomes like a vast open book; it bristles with written signs: every page is seen to be filled with strange figures that intertwine and in some places repeat themselves. All that remains is to decipher them."[26] But the hermetic order of the intelligible was not lingual; on the question of universals, hermeticism was a realism, not a nominalism: The reality penetrated by contemplative art was neither an order of extra-mental singular individuals nor concepts existing in the mind alone, but the ontological structure common to both matter and mind.[27] The alchemical imagination penetrated the hieroglyph of nature and immediately experienced some aspect of the infinite structure of the ideal-real; this contemplative knowledge of essences was then applied in concrete alchemical work, an application which in turn revealed new dimensions of structure.[28]

There were compelling politico-economical reasons for the rejection of the hermetic philosophy of nature in the seventeenth century. The epistemology of resemblances, with its penchant for infinity, was unmanageable, limitless, but admitting no certainty, "plethoric yet absolutely poverty stricken."[29] A knowledge that proceeded by infinite accumulation of confirmations could never attain the degree of mastery demanded by techno-science. "By positing resemblance as the link between signs and what they indicate (thus making resemblance both a third force and a sole power, since it resides in both the mark and the content in identical fashion), sixteenth-century knowledge condemned itself to never knowing anything but the same thing, and to knowing that thing only at the unattainable end of an endless journey."[30] In Foucault's account, Faustian man had to outgrow hermeticism if he was to succeed in subordinating all things to his control. The task of modern techno-science is not to contemplate the whole in the parts and the parts in the whole but to determine the properties of things and reduce them to calculable control.

Conclusion

Although Heidegger showed no interest in the Renaissance philosophy of nature, Heidegger's distinction between calculative and contemplative thinking evokes this lost history of modern science.[31] The Heideggerian history of being has been repeated so often that it has become banal. The recycling of slogans such as "onto-theology" or "the metaphysics of presence" in graduate schools all over the planet only obscures the profundity of Heidegger's critique of technology. Heidegger traces the rise of techno-science and global capitalism (for Heidegger, two sides of the one phenomenon) back to three principal sites of *Seinsvergessenheit*: the eclipse of the pre-Socratic experience of *physis* by the Socratic-Platonic demand for grounding reasons; the forgetting of the early Christian breakthrough to factical life in the Scholastic substitution of Greek aesthetic concepts for the dynamic historical categories of early Christian discourse; and last but not least, the Cartesian reduction of being to thought and extension. What is the relation of these three to each other? Is this one movement, as the latter Heidegger argues, or are these three distinct instances of the capitulation of contemplation to control, as such concealing three different faces of the contemplative mind of the West, respectively the shamanic, the apostolic, and the hermetic? This is a question for a further study. According to Heidegger, this progressively deepening obfuscation of being was not anyone's fault, but a destiny: Through these three transformations of thought, *Gestell* (enframing) became the modern default mode of experiencing beings. The future of human flourishing requires, Heidegger says, the retrieval of a lost art of surrendering to the unfathomable granting of being (*Gelassenheit*), a basic reversal of the attitude of critique, calculation, and control, which has granted us apparent dominion over the emergence and withdrawal of beings into presence, but at the expense of a living experience of being itself.

EWN holds Heideggerian *Gelassenheit* to be as ineffective as romantic poetry in the face of the eco-death drive which currently grips us. Morton argues that *Gelassenheit* leaves us bereft of political discernment, placidly accepting "the way things are": Should we also let the BP oil spill be? Should we let the decimation of the Amazon Rainforest be?[32] But *Gelassenheit* was never offered as a *method* of environmental practice; it is, rather, an undermining of techno-scientific-capitalist thought itself, an overturning of its basic assumption, that the human is or ought to be the master of time. To make *Gelassenheit* into a method is to leave the will to mastery at the root of our crisis unchallenged. Morton's mistake is to assume that a contemplative approach to the question concerning nature has been tried and found wanting. On the contrary, it has been found wanting without ever being tried. *Gelassenheit* aims at undoing the will to mastery which Heidegger, among others, identified as the essence of technology.[33] This will to mastery is not only a drive to dominate matter but is also a drive to master the human by reducing all modes of being to one: quantity. Why have we been overtaken by this monistic ontology? Heidegger says it is destiny, but that does not preclude us from thinking further into the matter. What hidden thought connects techno-science, the rise of capital, the ideology of consumerism, and total quantification? Romanticism, Morton's favorite eco-abuser, is symptom, reaction, and rebellion, but hardly a cause of our current problems. Something older is expressing itself in techno-scientific-capitalism, something which, it seems, is also leading EWN by the nose.

New Nature Narratives: Landscape Hermeneutics and Environmental Ethics

Martin Drenthen

Introduction

Philosophical hermeneutics is built on the assumption that people make sense of their lives by placing themselves in a larger normative context. *Environmental* hermeneutics focuses on the fact that environments matter to people, too, because environments embody just such contexts.[1] This is most obvious for cultural landscapes, yet it applies to the specifically natural world as well: Nature can function as a larger normative context with its own narrative dimension. However, there are many different placial and temporal dimensions at play in our relation to the landscape, which can give rise to different normative interpretations of the meaning of a given landscape. Such differences often play a role in environmental conflicts. One such conflict is the clash between those who care for the conservation of cultural heritage landscapes and those who believe that we have an obligation to "rewild" our landscapes, or to "create new nature," as the Dutch like to say.[2] Both ethical positions rely on different *readings* of the landscape, readings that not only reflect a specific ethical relation to the landscape but are also utterly bound to notions of personal identity and sense of place. That is why different landscape readings can easily give rise to deep and seemingly irresolvable conflicts about the landscape, even more so when existing landscape interpretations are challenged by rapid landscape change.

In this chapter, I provide building blocks for a reconciliation of the ethical care for heritage protection and nature restoration ethics. It will do so by introducing a hermeneutic landscape philosophy that takes landscape as a multilayered "text" in need of interpretation, and place identities as built upon certain readings of the landscape. I will argue that, from a hermeneutical perspective, both approaches appear to complement each other. Renaturing presents a valuable correction to the anthropocentrism of many European rural cultures. Yet heritage protectionists rightly point to the value of narratives for Old World identities. I will conclude with a short reflection on how such a hermeneutical environmental ethic can be helpful in dealing with environmental conflicts.

Resurging Wild Nature in Europe's Cultural Landscapes

The European landscape is a contested terrain. European countries are trying to find new, more sustainable attitudes toward nature. The value of wild nature is increasingly being recognized. To compensate for centuries of environmental decline, efforts are made to increase the share of natural areas in Europe.[3] As a result of renaturing projects, designation of new, large-scale habitat areas, and the reintroduction of extinct species, wild nature is literally gaining ground. The establishment of large-scale wilderness areas, the so-called PAN-Parks (Protected Area Network), is meant to create stable refuges for biodiversity, whereas the European ecological network Natura 2000 will connect existing natural areas so that species can migrate more easily and biodiversity loss due to fragmentation is counteracted. These developments are applauded by the general public, but occasionally they meet local resistance, particularly in areas with a long agricultural history, despite the fact that many farmers willingly cooperate when offered financial compensation.

Next to ecological restoration, which is anthropogenic, wild nature also resurfaces spontaneously, notably in abandoned rural areas.[4] The European human population is decreasing, and will continue to do so in the upcoming years. Moreover, Europeans are moving to the urban centers, leaving rural regions abandoned. In some urban zones, too, urban adapters such as fox and stone marten increasingly roam the city centers and suburbs.[5] In general, this means that in many cultural landscapes the human influence on the landscape will become less dominant, and nonhuman species will have the opportunity to occupy new habitats. Lynx and wolf are already repopulating areas where they had gone extinct centuries ago. One of the most spectacular examples is the return of the wolf from Eastern Europe. In the last decade, wolves have already occupied regions in former

East Germany, and they are still moving westward. Ecologists predict that the first wolves will reach the German-Dutch border within ten years. When that happens, wolves will have entered one of the most densely populated areas in the world.[6]

At the same time, other more or less conflicting trends with regard to landscapes are emerging in the European landscape awareness. Increasingly, traditional landscapes are recognized as part of our cultural heritage and worthy of conservation. This revaluation gains ground against the background of a perceived crisis of the European countryside:

> European landscapes are facing a deep crisis. As a consequence of globalization and the economical change associated with it, traditional functions like production agriculture are becoming less important. After the self-evident but inspired landscapes of numerous generations of peasants, monks and landlords, landscape has now largely become a nameless by-product of the global economy.[7]

In the process of globalization, many local landscape characteristics are eroding and are being replaced by interchangeable and "transportable" stereotypes. Traditionally, places served as shared reference points: They both expressed and helped to support regional identities, thus providing human inhabitants with a means of identification and orientation. This traditional connection between people's sense of identity and place has disappeared, leading to experiences of "placelessness"[8] and disorientation.[9] Against this background, many seek to conserve those landscape elements that can still support a feeling of regional cultural identity.[10] Even though the very notion of authenticity has become contested,[11] alleged historically genuine cultural landscapes are broadly valued as places of authenticity amidst an ocean of interchangeable public space.[12]

The tension between these two trends may converge, but more often they are in conflict with one another. Many perceive the emergence of feral nature as a new threat to traditional landscape identity. Some local inhabitants feel that the changes due to ecological restoration undermine their attachment to the landscape,[13] and often imply loss of identity, decrease in character, and a trend toward meaningless stereotypes. When debates about ecological restoration and nature conservation become entangled in issues like these, it can seriously undermine the legitimacy of nature conservation efforts. It is therefore useful to look a bit more closely to an example of such a conflict.

The Case of Land Declamation in the Hedwige Polder

In 2009 a heated public debate took place in the Netherlands about the meaning and purpose of landscape protection in the Dutch province of Zeeland. This conflict is a perfect illustration of how renaturing landscapes can be perceived as a threat to the sense of place, especially to the identity of landscapes and to people's place-identity.

The province of Zeeland lies in the southwest of the Netherlands and is situated around four North Sea estuaries. The history of Zeeland has been marked by the age-old struggle of people against the threat of the sea. Historically, most Zeelanders lived from the sea, but today many of them are farmers who live on land that has been reclaimed from the sea. Zeeland has suffered from many floods. The most recent was the flood disaster of February 1953 that killed nearly two thousand people in one night—it was the immediate cause for the Dutch to initiate the Delta Plan, a nationwide system of sea dikes and water ways designed to protect the Netherlands from major floods in the future.

Figure 5 Map of North Sea Flood in Zeeland, 1953.

The major contributing river to the Zeeland estuaries is the Scheldt. This river once had several estuaries, but the Delta Plan disconnected most of them from the Scheldt. After protests by ecologists and fishermen in the 1970s, it was decided to not close off the Eastern Scheldt estuary entirely (which would have caused the collapse of the saltwater ecosystem), but instead to build a storm surge barrier, with huge sluice-gate-type doors that are normally open but can be closed under adverse weather conditions. The entire Eastern Scheldt estuary became a designated national park in 2002.

The Western Scheldt, on the other hand, was never closed off, for it is an important shipping route to the neighboring Port of Antwerp in Belgium. The shores of the Western Scheldt are heavily embanked to protect the surrounding agricultural land. Close to the Dutch-Belgian border, where the river Scheldt meets the salty waters of the North Sea, is a small nature reserve, *Het Verdronken Land van Saeftinghe* (The Drowned Land of Saeftinghe), named after a town that existed there until 1584. It holds a highly dynamic saltwater ecosystem with high biodiversity, but it is also a treacherous place which cannot be explored without an experienced guide because the tides easily consume large stretches of land in a matter of seconds. Here one can experience the ruthless power of the sea—according to many Zeelanders a reminder why we have to continue to fight nature. The Zeeland flag features a lion struggling to stay above water, the province's motto reads "*luctor et emergo*": I wrestle and emerge.

Figure 6   Flag of Zeeland.

The conflict started with plans to dredge the Western Scheldt shipping channel. According to the Dutch-Belgian Separation Treaty of 1839, the Dutch and Belgian governments are required to carry out "all necessary works" for safeguarding the navigability of the river. In order to ensure future accessibility of the Antwerp harbor for ever-larger ships, the shipping channel in the Netherlands part of the estuary has to be deepened on a regular basis. However, since dredging increases the water flow, this leads to a decrease in food supply for the salt marshes and mud plains along the shore. To counterbalance for biodiversity loss, European nature protection legislation requires compensation measures. After legal pressure from conservation groups, the Dutch Cabinet eventually decided in 2010 that only the flooding of formerly reclaimed land would enable the Netherlands to fulfill its international and European treaty obligations and to ameliorate its relations with Belgium. The land declamation would have to take place in the Hedwige Polder, a small (3 square kilometer) and relatively young (it was reclaimed from the sea between 1904 and 1907) area of land reclamation between the Drowned Land of Saeftinghe and the Belgium border.

The decision led to much controversy. Some heritage protectionists stressed that the Hedwige Polder has a much longer history than is often suggested. Not only does it represent a little-changed early twentieth-century reclamation project but, more important,

> the seemingly young landscape hides a complex layered landscape: under the present surface traces are hidden of an earlier short-lived 17th-century reclamation, of a medieval fenland landscape that was settled and reclaimed from the 10th-century onwards but was lost by 16th-century floodings and, deep below the surface, of a sandy landscape that was used by Mesolithic and Neolithic peoples.[14]

The controversy became more heated when local inhabitants claimed that the Zeeland history of fighting against the sea is of deep importance to the Zeeland sense of identity. A headline from a national newspaper says it all: "Land Declamation Affects the Zeeland Soul."[15] Some of the locals accused nature protection groups for being insensitive to these identity issues. Ecologists would merely answer to an abstract idea of biodiversity without paying attention to the particular history and meaning of this place. The plan of giving back to the sea large portions of this hard-won land was considered by many an insult and ultimately a threat to the regional identity.[16]

In spring of 2011, the newly elected Dutch government, strongly influenced by the strong populist, anti-elitists, and anti-conservationist sentiments in Dutch politics, decided to again recall the decision to flood the Hedwige Polder with salt water.[17] It was decided that anti-renaturing sentiment (along with the economic interests of the five farmers that work in the Hedwige Polder) will be the point of departure for Dutch landscape policy from now on, even if this means that the Netherlands cannot conform to the bilateral agreement with its neighbor Belgium and European legislation concerning nature compensation (e.g., the EU habitat guideline).[18]

Many conservationists consider the Hedwige case as having been crucial for the decreasing support for ecological restoration in Dutch politics. What we can learn from this example is that conflicts between heritage and nature conservation, if not discussed thoroughly enough, can seriously undermine the legitimacy of nature restoration and protection measures.

Culture versus Nature?

Many of the tensions in the debate about landscape can be traced back to the fact that the moral debate on landscape has until quite recently been dominated by two perspectives. On one side, nature conservationists have argued that the growth of industry and agriculture have led to habitat loss and species extinction, and have undermined the ecological integrity of the landscape. They believe that time has now come for humans to take a more modest attitude toward the landscape and to counteract and compensate the devastation of the past by restoring or strengthening existing ecosystems or by the renaturing (agri-)cultural land.[19] Core values of the nature restorationists are biodiversity, scarcity, wildness, ecological resilience, ecological fidelity, and ecological integrity.

On the other side, landscape heritage protectionists start from the idea that landscapes are meaningful reflections of human history. Landscapes can be said to have a biography of their own[20] and can be read as archives that tell a story about the people who dwelled here, how they related to the world and to each other. Cultural landscapes are covered with traces of historic events and remains of past land use. Those concerned with heritage landscapes believe we have to stay in touch with this past "because we owe our existence, our identity, our vision of the world to it."[21] Core values of landscape heritage protection are: landscape legibility, regional identity, sense of place, historical authenticity, and narrative continuity.

This tension between nature restoration and heritage conservation (which more or less coincides with that between the outlook of natural sciences on the one hand, and the humanities approach on the other) has produced a stalemate in the moral debate about the meaning of "new nature" (a term used to denote restoration projects in The Netherlands). Heritage protectionists believe that nature development will inevitably erase valuable and irreplaceable traces of human history, will produce a historically *mute* landscape and thus result in an alienation from the landscape. They regard restoration biologists as nature fanatics who start with a false notion of an authentic landscape and are insensitive toward culture and local human needs. On the other hand, restoration biologists tend to argue that those concerned with cultural heritage are merely conservatives who are unable to acknowledge that certain traditional practices need to be redefined in light of the ecological crisis. Heritage protection would inevitably deify the past, transforming the living landscape into an outdoor museum. The opposition between these mutually exclusive perspectives hinders a productive exchange of ideas about the significance of the landscape.

Today, the need to seek a more productive relation between both perspectives is widely recognized. Many nature conservationists recognize the importance of human perception of nature and the need for public participation in restoration projects; and nature conservation organizations seek to strengthen their social and cultural embedding. Likewise, heritage protectionists increasingly acknowledge the value of biodiversity and the need to make our culture sustainable.

Yet, how exactly both perspectives could be integrated in a comprehensive, more balanced and reflexive view on meaning of place remains unclear. How can we better understand the links between cultural identity and the legible landscape, and contribute to a reconciliation of the perspectives of heritage protection and nature restoration?

Environmental Philosophy on Ecological Restoration and Ethics of Place

The division between cultural landscape protection and wilderness conservation has until recently also marked environmental philosophy. Debates in landscape philosophy have been heavily influenced by the North American bias and its emphasis on wilderness protection.[22] As a consequence, early environmental philosophers had difficulty acknowledging any positive role humans could play in landscape change.[23] Likewise, the debate on ecological restoration focused heavily on issues such as Elliot's critique of the artificiality of "new nature"[24] and Katz's criticism of the anthropocentrism of ecological restoration as such.[25] More recently, environmental philosophers have recognized that ecological restoration can also play a positive role in improving the human-nature-relationship.[26] This is particularly relevant in the "Old World" context of Europe, where culture and nature are indistinguishably intertwined.[27] Therefore, many contemporary environmental philosophers have argued for a better cultural and social embedding of ecological restoration projects.

But despite the growing attention of the cultural impacts of restoration projects, issues of heritage protection as such do not yet play a part in environmental philosophy. Some of the themes that heritage landscape conservationists are concerned with can also be traced back to environmental philosophy. Key terms are "ethics of place" and "sense of place." Some environmental philosophers have pointed out that places are always already filled with meanings[28] and therefore play an important role in structuring people's lives. Others have pointed to the relation between place ethics and land-narratives.[29] Yet, until now, these issues have mostly been dealt with separately. It is about time that environmental philosophers bring together these topics. One way of doing so, I will argue, is to focus on the concept of landscape legibility and its relation to environmental identity.

The Legible Landscape Palimpsest, Place Narrative, and Identity

The legibility of the landscape is a recurring theme in debates about the significance of cultural landscapes in Europe—in popular culture,[30] environmental education,[31] and social environmental sciences.[32] However, this theme is not used as frequently in environmental philosophy and ethics. Originally, the term addresses the relation between landscape perception and human history, yet it could also be used to conceptualize the notion of ecological fidelity that is central in many restoration projects.

The concept of a legible landscape can be used to understand both wilderness protection (conceived of as making explicit the first text of primal nature) and cultural heritage conservation (conceived of making explicit the subsequent historical layers testifying of human interactions in the landscape). The European landscape is like a palimpsest: a multilayered text, consisting of different textual layers written on top of each other. All landscapes contain signs of natural events such as floods and changes in climate, but all European landscapes—even those that may appear pristine—are covered with the signatures of the early human inhabitants. Most landscapes, however, also contain more recent layers that testify of the invention of agriculture, of industrialization, urbanization, and the rise of large-scale agriculture, of the recent rationalization of land use practices, and so on. Today, the designation of large-scale ecological restoration and rewilding projects are inscribing yet another text layer to the palimpsest. Once we recognize the "layeredness" of the landscape text,[33] the legible landscape concept can help connect both perspectives: Cultural heritage conservation is the making explicit of the subsequent historical layers testifying to human interactions with the landscape, whereas landscape rewilding can be conceived of as the unearthing of the primal text of nature.[34]

Moreover, the notion of landscape legibility can also help us to understand the relation between landscapes and human identity. It is by virtue of their legibility that particular places matter to both individuals and communities as embodying their history and cultural identities. People make sense of their lives by placing themselves in a larger normative context. For this reason, environments matter to people, too, because they embody such a larger context.[35] This is most obvious in the case of cultural landscapes: By providing a broader context with which to understand ourselves, they give a sense of orientation and open a perspective on our place in history. This sense of identity is rooted in a narrative understanding of place; these narratives depend on material traces in the landscape combined with the histories that people tell: cultural landscapes are interpreted landscapes. Moreover, because the landscape is always interpreted anew in each era, the cultural landscape's meaning is part of an interpretational history.

Yet these points apply to the specifically natural world as well: The natural world can also function as a larger normative context, with its own narrative dimension.

[N]atural environments have histories that stretch out before humans emerged and they have a future that will continue beyond the disappearance of the human species. Those histories form the larger context for our human lives. However, it is not just this larger historical context that matters in our valuation of the environments in which we live, but also the backdrop of natural processes against which human life is lived.[36]

Both types of reference to textual layers in the landscape hold an implicit moral dimension. The moral dimension of heritage landscape protection does not in the first place refer to some "intrinsic" feature of these landscapes themselves, but rather to the fact that their legible features refer to human history, a history that embodies a meaningful narrative about human relationships with these places and with history. To the degree that the legible landscape serves as a *normative* context that can give some "measure" to the present, one can say that landscape legibility supports a (rather conservative) ethics of place.

Most restorationists, on the other hand, use the concept of deeper, underlying "wild" nature as a moral "baseline."[37] Many will readily admit that it is impossible to reverse history and turn back to an undisturbed past, yet their aim is not to "build" new ecosystems either. Rather, they seek to restore a sense of continuity with a historically deeper past that has been forgotten. Restoration projects should respect the "genius of place" by recognizing (1) the (non-anthropogenic) natural processes and underlying geomorphologic structures that are characteristic of a certain place, *as well as* (2) the (anthropogenic) historical developments of a certain landscape as far as these contributed to the specific character of that area, and (3) the societal functions that have enabled people to interact with these natural processes in ways that are both physically and economically sustainable.[38] Seen from this perspective, ecological restorationists attempt to adjust the anthropocentric place narrative of heritage landscape protectionists and to broaden our sense of human place-history (landscape biography) and our ethics of place. As such, restoration projects could even help revitalize local community's sense of place.[39] By liberating the ancient natural forces that early inhabitants faced and thus consciously reconnecting with the deeper layers of the legible landscape, it becomes possible to reenact some of the forgotten narrative possibilities that these deeper textual layers accommodate.

Both heritage protectionists and ecological restorationists refer to a particular reading of the landscape as legible text that supports

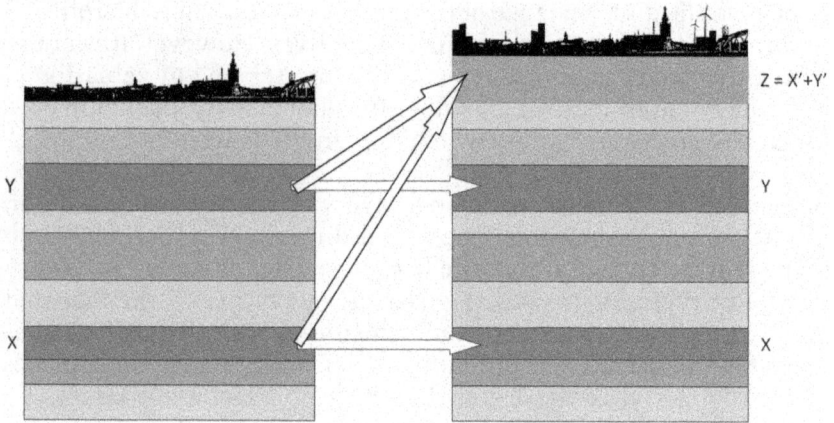

Figure 7 Different layers of the palimpsest support different future place narratives.

particular moral place narratives. Reading the landscape palimpsest in multiple ways can enrich the debate about future challenges and choices, but the readings of past layers in the landscape cannot simply be used as a model for the present because history never repeats itself. Some heritage landscapes may be saved as relicts, but most historic references will at best serve as ideal images or rough guides with regard to our current challenges.

Both perspectives on the meaning of a landscape—heritage and restoration—are one-sided, but together they complement each other. Heritage protection rightly points to the value of history and place narrative. Ecological restorationists must therefore learn to interpret the value of nature protection in narrative terms as well, rather than relying solely on abstract arguments such as biodiversity protection.[40] On the other hand, the restorationists' argument for the value of the natural world should be welcomed as a valuable correction to the anthropocentrism of many traditional cultures of place.

Reflecting on different elements of the landscape biography can provide different narrative possibilities, ways to continue the historic narrative of which we and the landscape are part. Making explicit past human-nature interactions reminds us that we do not start from scratch but find ourselves in a landscape that always already has a natural and cultural history, and thus can enrich our moral imagination. Ecological references broaden and deepen this context for human self-understanding. In the moral debate about the land-scape, both readings can serve a narrative role of guidance for future developments.

Thus the metaphor of landscape as a legible palimpsest can provide a means of thinking through the new developments in the European landscape and their significance for human place identity, and will allow us to develop a pragmatic approach that may help to assess the strengths, weaknesses, and complementary features of the reading practices involved.

Conservation and Restoration Require Reading the Landscape

Restorationists who only refer to abstract values such as biodiversity risk creating the very opposition against nature protection measures that they fear. By arguing solely from an ecological perspective, restorationists risk alienating those who are concerned about the landscape for different reasons.

In the Hedwige case, conservationists at first won the legal and political case, but lost much goodwill among parts of the population along the way. Partly as a result of this event, Dutch conservation groups are currently discussing whether it is wise to rely solely on legal arrangements and European conservation legislation when it comes to finding a new, more balanced relationship with nature. Some have argued that the legal approach, while yielding successful results in the short term, will fail in the long term due to the fact that it distracts attention from the content of the values involved. To get a clearer view of what is morally at stake in this debate, conser-vationists need to articulate a broader and more inclusive vision of what restoration is about. Eric Higgs and William Jordan[41] have both defined ecological restoration as the attempt to not merely heal damaged ecosystems but to heal damaged human-nature relationships as well. As soon as conservationists acknowledge that landscapes also meet deeply human (anthropocentric) needs by providing people with a sense of direction, purpose and identity, they can also recognize the many alternative ways in which one can articulate the meaning of a landscape to show that it is worth protecting.

Not any heritage protectionist story will do, of course. Whatever meaningful and caring relationship with a landscape we wish to foster, in order for it to be sustainable it should meet with the basic ecological facts as well.[42] A historically rooted understanding of the meaning of a landscape can, on second thought, turn out to be misguided if it is based on an incorrect understanding of the role that natural processes play within a landscape system. Ecologists and conservationists could attempt to correct such "mistaken" place identities, for example, by stressing knowledge of the hitherto unknown ways in which the natural system has played a role in the history of a landscape and its inhabitants.

It could be more meaningful, though, to explicitly fall back on an understanding of place-history that appeals to older forms of local traditional ecological knowledge and attempts to provide people with a sense of orientation and deeper understanding of the natural characteristics of the place they care for. Due to modernization processes and the increase in mobility, much knowledge of both natural and cultural history of places has disappeared from the public sphere. Yet, part of this heritage still lives on, in local practices, habits, local songs and narratives, albeit mostly implicitly. The explication and articulating of the meanings inherent in these place narratives—meanings that only unconsciously play a role in a community's sense of place—can help local communities to better acknowledge the deeper significance of the natural systems they rely on.

An example of how this can be done is the Wealthy Waal Project[43] along the Waal River (the main branch of the Lower Rhine River in the Netherlands). Starting from a bioregional framework that combines existing notions of regional identity with basic knowledge of the biotic system, this spatial development project engages water managers, ecologists, local authorities, civil servants, entrepreneurs, and inhabitants to collectively envision what a sustainable future of the region could look like, both economically and culturally. The project greatly benefits from local landscape historians who show the many ways in which natural forces such as rivers have been (and will always be) structuring forces in the formation of the landscape, and—more indirectly—of local cultures dependent on it. Many of the environmental problems that we face today are not much different from those of former inhabitants. We can learn from past experiences by recalling how earlier inhabitants answered to the challenges posed to them by the landscape, and see how this interplay of people and nature has produced the landscape of today. Many of the typical local characteristics of certain places that people identify with can be traced back to specific natural events, such as river floods, and people's responses to them. History can reveal how past human-nature dialogues have had a real effect on the landscape and on the inhabitants who live there today. For example, studying the layout of river banks from the middle ages (legible traces of which can still be seen to the trained eye today[44]) can teach us how river inhabitants who did not yet believe they could subdue the river tried to attune themselves to nature's rhythms. By integrating the heritage story into a deeper and broader landscape history, and revitalizing dormant layers of the local culture that still contain some of the older "ecological" wisdom of local communities, it becomes possible to escape the dualist choice between culture and nature, and between cultural heritage and ecological restoration.

Like all narratives, local histories organize the world and help to understand who and where one is. But, like other narratives, they also create their own audience, as it were. Reframing the restoration issue in narrative terms and complementing the historical self-understanding with deep landscape history can provide inhabitants of a particular landscape with a new story about whom they are and where they came from, and thus create a new sense of community. Place histories can awaken a sense of having a shared burden to take care of the land, its cultural heritage and its ecology alike. A narrative understanding of restoration can thus deepen the sense of place and develop a sense of ecological citizenship and help to find new, more mature relationships to the world we inhabit.

It goes without saying that this overview of how new nature narratives could help us come to terms with resurging nature in cultural landscapes has to be worked out in much more detail. Probably, it can only be done convincingly *in place*, that is, together with the local community that tries to understand its place, together with historians and ecologists and other experts who know a place, and directed at finding a better self understanding of what it means to be living "here" today.[45] In a way, a convincing, meaningful story about a place is told by that place itself.[46]

Coda: Landscape Hermeneutics and Environmental Ethics

We humans are meaning-seeking beings, and the world we inhabit is a reflection thereof. We live in a world that is always already interpreted. The meanings and interpretations of our world are no secondary addition to an otherwise "objective" reality, but rather *form the very fabric* of the kind of world that matters to us. And yet, although the world we live in is an interpreted (and therefore) thoroughly human world, nature presents us with issues that we have to acknowledge in our interpretations of the world.

Historically, one strand of hermeneutics, emerging from Friedrich Schleiermacher's work, advocates that understanding the meaning of a text amounts to knowing the intention of the author. Analogously, a good understanding of nature would amount to an understanding of the meaning that nature itself expresses. This does not mean that we have a direct access the meaning of nature (after all, the intention of the author is also not readily available but must be discovered through interpretation), but it does presuppose that there is such a thing as a true or original interpretation of nature that is more appropriate than others. Hans-Georg Gadamer has criticized this "romantic hermeneutic" view of meaning for failing to appreciate that our understanding of the meaning of the world differs throughout history and within different cultures.[47] Often, the meaning of something which appeals to us is not clear to us. Meaning does not just lie there waiting to be discovered. Meaningful things appeal to us through experiences, but the *meaning* of those experiences only becomes clear once we attempt to articulate them. The experience may precede our understanding of it, but its meaning only exists through our interpretative appropriation, that is, *after* our attempt to "bring home" what it is that beckons to be understood. It makes no sense to talk about the "real" meaning of nature apart from our articulations in a specific cultural form. For environmental philosophy, this means that it makes no sense to refer to nature as having an intrinsic meaning apart from *our* understanding of it. But this does *not* imply, of course, that the meanings we encounter are made by us, the world outside exists, and throws its questions at us.[48]

An adequate hermeneutic of the landscape therefore has to acknowledge that our relation to the landscape is deeply historical, that is, we humans inevitably, and always already, interpret the landscapes we find ourselves in. Past interpretations of the meaning of a particular place in which we find ourselves can play a role in how we act toward and think about certain places. We may be under the sway of certain past interpretations often without being aware of them. But every now and then, the places we find ourselves in beckon to be interpreted anew because they appear to us as somehow meaningful in a new way that we have not yet understood.

We get our meanings from the cultural contexts surrounding us, but that does not mean we are imprisoned in that context, or that we are forced to only conserve the meanings of our cultural tradition. As cultural beings, we are not merely the result of history but make history as well. We may find that we have gotten stuck with stories and interpretations about our world that have been told before, petrified interpretations, or fixed narratives that do not always properly articulate the actual meaning that these places have for us now. In these cases, we will not always be able to adequately articulate what that new meaning actually is. It is at this point that environmental hermeneutics can play a constructive role.

Moreover, an environmental hermeneutics will also have to recognize that the interpretations of the places in which we live in turn provide an ongoing and ever-changing narrative context from which we can understand ourselves.[49] Environmental hermeneutics will therefore have to explicate the interpretational base of our being-in-the-world by articulating those preexisting meanings and interpretations that already play a role in how we act and think, and in doing so force us to have a second look at them. Some of our previous interpretations of the land may prove to be inadequate or outdated once we properly reflect upon them. A hermeneutical environmental *ethics* will ask in what sense these old interpretations can still be considered adequate articulations of how the world we find ourselves in beckons to be understood, or whether we should seek new articulations. Rearticulating these meanings can be laborious but plays a critical part.

The task of a hermeneutical environmental ethics, then, is to articulate and make explicit those interpretations and meanings that are already at work in our everyday practices, to bring them to light and make them explicit, and to confront existing meanings and interpretations with other, less obvious interpretations. Doing so will increase our sensitivity for the many different meanings that can be at stake in our dealings with a particular place, although it will also make the questions of ethics even more complex than they already are. However, by showing how our understanding of ourselves is already emplaced, a hermeneutical environmental ethics can help us to better understand what is at stake in our complex relation with the landscape.

Environments, Place, and the Experience of Time

12

Memory, Imagination, and the Hermeneutics of Place

Forrest Clingerman

Introduction

We Live in the Present

Humans are creatures of the present, and the places that we inhabit oftentimes abet an emphasis on presence. For example, much of our daily interaction occurs in spaces that offer little to discriminate the times of day or season. Artificial lights, heating, air conditioning, walls, and doors maintain a continuous backdrop and regulate the experience of embodiment in space as days and weeks move into the past. Yet we might still find ways to break through mere geometric space, through the anonymity of these situations. In the materiality of these environments, there are fractures and idiosyncrasies that beckon us to brush away the shallowness of space and to read the depth of places.

Suggesting one possible plumb line with which to explore the *temporal* depth of environments is the purpose of this chapter. In other words, I wish to explore one dimension of how we are situated in environments and places, namely, the ways that we experience the hermeneutical dialectic between the time of place and the place of time. The time of place, we might say, is the experienced, lived temporality of a given environment or location. The place of time, in contrast, is the framework of temporality as part of our conceptualization of environment per se. Thus environmental hermeneutics recognizes the complexity of place: Place is not simply a concept but also a lived particularity, which has spatial and temporal materiality. Spatial and temporal materiality is the "stuff" out of which the narrativity of environments is constructed. As Edward Casey has pointed out, throughout the modern period the emphasis has been on time at the expense of space.[1] Thanks to the work of philosophers such as Casey, place has become central to our philosophical understanding of environments. But in the desire to re-place philosophy, environmental philosophers often unwittingly emphasize the spatial (e.g., as seen in discussions of physical embodiment, wilderness, and restoration) with the inadvertent result that nature is static and adrift in time's flow. *Where* is time, we might ask? For many reasons, we must maintain our guard against the overwhelming priority of the spatial, at the expense of a revitalized, vibrant sense of re-placing time. In the contemporary world, our daily built environments erase the temporal dimension, and oftentimes the past and future are effaced in our presence. For a fuller understanding of our world, we must challenge the fact that we frequently live in the present—in a shallow temporality of a radically present "everyday." Even in our everyday lives, there is an occasional inbreaking of temporal depth, which is imperative for our sense of place.

This chapter argues that a hermeneutics of place, among other things, presents an interpretative structure through which to understand the time's depth in place. By way of example: Although if much of our time is spent in temporally sterile, controlled environments, the liturgy of the day thrusts time's mark into our experience. After a day spent in atemporal space, dusk settles and draws us nightward. At night there is silence and darkness; there is a transition marked by personal rituals that serve as the secularized vespers of contemporary life. Sounds and movements echo differently in the evening, insofar as the sounds of the day have dissipated. The bombast of creaking floorboards resounds in ways unheard during the day. Meanwhile, the household is filled with objects that do not seem completely inert: They almost seem to finish their day's work, and only then gradually begin their slumber. And finally, a state of calm is victorious and we encounter the reality of dreaming—and we leave ourselves.

At dawn there is a transition to yet another state. Each morning we become ourselves again, albeit groggy with the change of surroundings from the night before. When we went to sleep, moonlight served as sentry. Now the fluidity of light overcomes the window coverings. Sunshine enters the room, weakly at first as it builds its strength for the day's initial assault. It reaches its tipping point and the daylight pours in, bursting through the panes of glass to etch moving squares of light in the room. Even before this breach of light in the window, the sounds of morning have begun before the last rays of darkness. Without watching the hands of the clock move, we feel and experience time. When situated, we begin to be aware of the placial length of temporality.

If we are even more attentive, this daily cycle is permeated with layers that betray an even greater nuance to the temporality of place. Throughout the summer, the calls of birds, which seem always ready for the morning, replace the insect sounds of evening. The warm breaths of wind against maple and oak leaves, so common in the late afternoon and evening, die down at dawn. Or perhaps a spring fog has settled, shrouding the sharp lines of the landscape and covering blades of grass with dew. Or the icy geometry of frost etches itself on window glass, or we hear the start of the slow drip of icicles. Differences between the vibrant mornings of July and the frozen mornings of February are clear, and they lend themselves to a detection of change and temporality. From the vantage point of a window, we see testaments to the variations of temperature and moisture. In these commonplace details, we feel the traces of autumn, winter, or spring—and finally the suggestion of the place of time as the time of place, as what appears in the everyday environments. In the everyday, we still perceive the palpability of time.

Throughout this illustration of the ordinary, the presentness of place includes three important terms: self, memory, and imagination. To explore these terms, my thesis is this: In finding the trace of the days and seasons—and in the other ways we rediscover the temporal sense of nature—one can reaffirm the depth of the temporality of the self in place. Unlike a radically anonymous present, the meaningful depth of the present (and, perhaps, the presence of a depth of meaning) resounds with the space between the past and future. That is to say, *remembering* how *I* have come to be *here* is an important and complex act that grounds the past to place, despite the envelopment of time. And *imagining* how *we* will be *somewhere* enacts a new possibility of our situation, of the future in place. Thus I argue that these terms are complementary, and that together they illuminate a depth of nature not present in each term individually.

Place, Self, and Narrative

The Hermeneutics of Place

If this chapter seeks a richer understanding of nature, this understanding runs through the meaning of place. Our encounters with the "environment" (broadly understood) occur with the experience of place *as* place. This is the case not only for wildness as paradigmatically "natural" but also for built environments, destroyed or toxic ecosystems, and other examples that we might hesitate to consider "nature." We began with a domesticated example of an environment in evening and morning, which nonetheless clearly exemplifies a sense of place. How so? The introduction of this chapter was not a description of myself exclusively, but rather is a reminder that "I" am surrounded by others, by beings and things that transcend me in many ways. These others—together with the relationship between them—constitute the place in which we dwell, we live. Such a gathering of living creatures and inanimate beings comes to fruition in an interpretation of place.

To reflexively knit together the meaning of environments—to discover the narrative of the "Book of Nature"—is to narrate how the particularities of a place as "place" are multivalent, competing narratives of individuals, relationships among individuals, and the whole itself. But despite this complexity, the way we "live place" is below such an explicit framing of reflection on space and time. The meaning of the place, in fact, is often taken for granted and emerges from the smallest and nearest things: slight tiptoes of little feet on oak floors, the smell of morning coffee, or changing leaves falling from the dogwood outside the window. Moving beyond my own walls, I might situate these things in light of the floodplain of Little Riley Creek, the surrounding farm fields and woodlots, or the village in which I live. The hermeneutical task is not simply to identify experiences, even in a fuller phenomenological sense, but to reflect on how particularities might be understood as meaningful. What we take for granted suggests a broader task for environmental philosophy: We need more than policy discussions or the analysis of concrete ethical dilemmas. Environmental hermeneutics, at its root, offers a reflection on how to engage and find meaning in the world, *so that* there is an appropriate framework for policy and ethics. As Van Tongeren and Snellen write in their contribution to the present collection, "Instead of solving problems, [a hermeneutical ethics] is rather concerned with interpreting and clarifying the conditions and frameworks which determine those problems. . . . By clarification of the conditions and frameworks of our current problems, a hermeneutical ethics may contribute to our self-understanding, and by doing so, to the moral quality of our way of living on the long run."[2]

There are significant benefits to a hermeneutical approach to place in light of this desire for environmental philosophy. Understanding environments (whether "built" or "natural") requires a middle path to overcome certain intellectual obstacles and limitations of perspective. Structurally, a hermeneutics of nature "reads" nature by navigating through the way that our experience is irreducible to explanations of a naïve realist or social constructivist. A naïve realist tends to assume that nature is a simple object that is independent of observer, and that the subject experiencing nature does not affect the meaning of "nature." In contrast, a social constructivist position argues that "nature" is primarily or entirely dependent upon social conventions for its meaning. A hermeneutical view of nature seeks mediation between these two views: Nature has independent existence that can be explained, but it is also influenced and understood in light of the observer's perspective and experience.[3] Furthering this dialectic, the reading of place can be seen to encounter its own version of the hermeneutical circle: Environments emerge in the placial manifestation of the all-encompassing dialogical interactions found between subject and object, sameness and otherness, part and whole. Humane disciplines such as philosophy and theology, in sum, will find a hermeneutical account to be the most adequate way to understand nature; such an account responds to the limitations of other narrative accounts, as discussed by Brian Treanor in the present volume.[4]

At the same time, it should be apparent that reading nature presents unique challenges. Philosophical hermeneutics historically has been devoted to the art and science of the interpretation *of texts*. This is certainly the case in premodern hermeneutics, and even hermeneutical philosophers such as Gadamer and Ricoeur seem to operate in paradigms greatly amenable to the written text. But if the object of interpretation is considered a text, then what is the "text of nature"? The emphasis on place becomes especially important here: The concept of "place" is what allows us to organize our thinking around the meaning of environments and landscapes as texts. A text is not simply a narrative; it includes also a textuality or physical presence. Likewise, "place" includes geometric space and chronological time, but it is more than this. Tim Cresswell writes, "This is the most straightforward and common definition of place—a meaningful location."[5] Similarly, Arto Haapala has noted, "Place is, indeed, the horizon that determines our perceptions and preferences."[6] Following Heidegger, we might further say that place is where humans dwell and encounter the other, human and nonhuman alike.[7]

So, environmental hermeneutics is the hermeneutics of the concept of "place" and actual places. To create a model of place as a text, it is possible to use a reflexive mediation of three interconnected ways of thinking (something I have worked out in greater detail in other works). First, our interpretation of place starts with the thinking of the concept "place," which serves to demarcate and categorize landscapes, urban environments, homes, and other spatio-temporal locales. Much like "art" identifies a cultural creation as an artistic work, "place" serves to guide what constitutes an acceptable or truthful interpretation of an environment, for the concept of "place" identifies the complex unity and differentiation of space and time in lived experience. But this leads to a second quality of understanding place—namely, thinking of specific manifestations of place; places are particular, experienced sites and locales. A particular place, together with its objects, inhabitants, and visitors, can serve as an element of environmental philosophy. Finally, interpreting place includes an element of reflexive thinking: Being situated in place mediates the correlation between specific manifestation and general concept. Environments are interpreted through an understanding of how we are emplaced—emplacement being a spatiotemporal echo of Ricoeur's "emplotment" of a narrative.[8] In other words, we are challenged to think about thinking when we are emplaced in place. Humans (as natural and cultural beings, as well as inhabitants and observers) find themselves in particular places; we understand human embodiment in light of our experience and conceptualization of place.[9]

Thinking of place therefore is a variation of the hermeneutical circle: Place is a structure through which to understand the interaction between individuals who dwell or otherwise are in particular places, and at the same time these places gather the individuals together. Echoing Paul Ricoeur and his discussion of relationship between "emplotment" and narrative, a hermeneutics of place establishes its understanding through our reflexive "emplacement" in the textuality of built and natural environments. Discovering the textuality of place is not simply an objective reflection upon nature "out there" but equally reflexive thinking of how we are as subjects understanding place through a placial pre-understanding.

If place and the emplaced are in a hermeneutical relationship, it means that environmental hermeneutics must adapt and deepen some of the concepts of hermeneutics in an effort to understand material, spatial, and temporal existence. For example, some of the more significant terms that fit into the discussion of environmental hermeneutics are metaphor, narrative, the idea of the "legible landscape," and environmental identity. Metaphors (and related terms such as *models*) illuminate the linguistic pre-understanding of nature and place. Jozef Keulartz has helpfully suggested that metaphors work on three levels in environmental philosophy: cognitively, discursively, and normatively. These three things together show how metaphors "determine our attitude towards entities in the world." [10] He suggests, for example, that seeing nature as divine text leads to a passive approach (something I think he overstates), while seeing it as a machine leads to seeking mastery. In turn, metaphors suggest that our understanding of place is narrative. Anton Markoš,[11] Brian Treanor,[12] Martin Drenthen,[13] and others have developed significant resources in showing the narrativity of place and nature. For example, Treanor argues that narrative offers us a significant way of knowing nature, such that we are not stymied by a limited scientific objectivism or a postmodern constructivism or subjectivism. A related concept to narrative is the "legible landscape," taken from Dutch landscape activists and discussed by Martin Drenthen. Drenthen shows how hermeneutics can approach the legible landscape in more dialectical ways, rather than through a subjective bias as is found in more semiotic approaches.

Environmental hermeneutics, in sum, engages the conceptualization of "place," the manifestation of individual places, and the self-identity of inhabitants and visitors to places through the structures of metaphor and narrative. The Book of Nature is one fruitful metaphor for manifesting the narratives of nature and environment; environmental philosophers seek to understand this Book, taking up the heritage of medieval and early modern science. As I noted in an earlier work,

> The need to return to a more inclusive tradition of the Book of Nature arises from the fact that understanding nature necessitates turning toward a hermeneutically complex and yet potentially unifying metaphor. Seeing nature as a text allows us to reconcile and weave together the various narratives through which we interpret nature. As a single entity that gathers the separate, heterogeneous parts of what we casually call 'nature,' the Book of Nature is a collection of metaphors and models.[14]

We construct the sense of narrative in place, insofar as the Book of Nature serves as an object to experience nature with a hermeneutical lens. At the same time, this reflexive book includes the reader, opening the possibility of an embodied narrative identity—an environmental identity situated in place.

The Self and Narrative

Our concern with place implicitly acknowledges that the hermeneutical structure of place is intimately tied to the meaning of self. To explain: The self cannot be fully known or manifested without being distinguished from the other. This knowledge requires a robust reflexivity. For if a quest for the self begins with the quest for the other, this quest includes not simply other selves but also a detour through the otherness of self as self. Self-identity arises from out of the question of place, insofar as place is an other-than-self that envelops the self. At the beginning of this chapter, I offered examples of the self in dialogue with place: in the quiet of the evening, or the way that an autumn day begins. At first glance, what I described was simply a vignette of everyday experience. But just as it offers a context to discuss place, it was readily apparent that the hermeneutics of place raises the question: *Who* starts this day? Bracketing the sound of overzealous sparrows outside the window, or other inhabitants of this place, who instantiates the meaning of "I am awake"? This is a particularly difficult question because of the uniqueness of the human in place: We are creatures with a presence inside and outside nature. Or, stated otherwise, we are present when we absent ourselves from an absolute immersion in the natural world. As humans, we transcend nature and are situated participants within it. We are agents who act upon nature, and objects within the natural world. But how can we present a unified description of the self in light of these paradoxical qualities?

Here the concept of a narrative identity becomes a persuasive way of answering the question "who?" at the boundary of nature and culture. As advocated by Paul Ricoeur and others, narrative identity is attentive to the need to bring concordance to the discordant elements of the self. Insofar as place is a collection of discordant elements, narrative identity is a useful framework of investigating how to situate self in place. Further, a narrative view of self allows us to interrogate the temporality of place as involved in the temporality of self.

Of course, Ricoeur does not see the answer of the question of the self in the hermeneutical understanding of place. Rather, his answer to the question of human identity—the question, "Who?"—comes through the recognition that humans name and narrate. This ability to narrate responds to a specific discordance found in our self-identity: How is it that we change, and how do we stay the same? The answer for Ricoeur is found in the fact that there are two main uses of "identity." First, there is "sameness" (*idem*). Identity as sameness can be numerical identity, close resemblance, continuity, or permanence over time. This last sense of sameness causes problems when the second type of identity, selfhood, is also present. Selfhood (*ipse*) can be understood as a response to the question of who, rather than what, and relates to the ascription of an agent to an action.[15] Dieter Teichert succinctly explains, "To be a person and to gain one's identity—in the sense of identity as selfhood—means to be a being which does not possess a stable, closed and fixed identity. Identity as selfhood is not simply there like an objective fact. To possess an identity as selfhood means to be the subject of dynamic experience, instability, and fragility. . . ."[16] The result, Ricoeur argues, is that there is an ontological break between *ipse* and *idem*. Selfhood changes over time, whereas sameness denies such change. Which is the appropriate model for selfhood? Thus insofar as we change and stay the same, we are constant and novel across time, and we are both the same and different over the course of a life, neither model of selfhood can be dispensed with in the end.

Ricoeur suggests that these two forms of identity can be gathered together through narrative (a concept already shown to be important for place). "According to my thesis, narrative constructs the durable properties of a character, what one could call his narrative identity, by constructing the kind of dynamic identity found in the plot which creates the character's identity."[17] A plot mediates between permanence and change, making narrative identity correspond with the "discordant concordance" of the story—in other words, mediating between sameness and selfhood. Thus the need for a narrative identity, defined as "the kind of identity that human beings acquire through the mediation of the narrative function."[18] Ricoeur further explains, "It is thus plausible to endorse the following chain of assertions: self-knowledge is an interpretation; self interpretation, in its turn, finds in narrative, among other signs and symbols, a privileged mediation; this mediation draws on history as much as it does on fiction, turning the story of a life into a fictional story of a historical fiction, comparable to those biographies of great men in which history and fiction are intertwined."[19] This has implication on the relation of the "everyday self." Narrative not only configures but refigures the self—"In the course of the application of literature to life, what we carry over and transpose into the exegesis of ourselves is this dialectic of the self and the same."[20]

There is a significant lacuna in Ricoeur's discussion of narrative identity, at least from the perspective of environmental thought. Narrative identity does not adequately account for places and environments as *more than* backdrops for agents. Places are situated otherness, the identity of otherness that the self encounters as other than self-identity. While Ricoeur's view of narrative identity is predicated on ethical concerns, nature itself has no agency or role relative to place. But this limitation is not endemic to a theory of narrative identity. In *Oneself as Another*, Ricoeur notes in passing that the existential embodiment of the narrative marks the difference between literary fictions and lived narratives, and makes related remarks in *Memory, History, Forgetting*. Furthermore, S. H. Clark notes that "[a] novel, with comparatively rare exceptions, unfolds a textual world of its own, whereas in a life-history, each story is necessarily bound up with those of others."[21] Interestingly, Clark here intimates that narrativity opens the possibility of a multiplicity of worlds—I would argue that this same dynamic would apply to places—bound up together. As our discussion of place intimated, the gathering of places and life histories *is* the basis for our reading of nature. By *re-placing the self* and remembering place, nature becomes more than a backdrop; *it is a participant in the narrative, an other that embodies memory, an other that locates imagination, and thereby an other that provides a constitutive element of selfhood.*

Naming the import of the foregoing discussion, environmental hermeneutics supplements narrative identity with a sense of "environmental identity." David Utsler has defined environmental identity as a "self-understanding in relation to the environment. Neither the eco-centric self nor the anthropocentric self is privileged over the other; rather, each is a constitutive element of one's identity."[22] The self is present in our experience of the wild, just as much as the built environment. The self holds together the start of our day, and it carries through time. It walks along wooded pathways, on city streets, and beside superfund sites. With our starting point discussed previously, we have already presupposed certain things about the self that wakes up, lives in place, and interacts with others. Moving further than Utsler, I suggest environmental identity is grounded on the self encountering place as an other with its own sense of identity. The self describes and redescribes itself in response to the confrontation of place; place coheres in its own work of identity and encounter with otherness. In other words, in a reflexive manner place similarly is manifested and reconfigured in response to the selves of its inhabitants.

Memory and Imagination

Embedding Place in Time

Place situates the self, and the self brings meaning to the place. Place and self are indelible in the formation of their intertwined, reflexive identity. A hermeneutical reading of nature is based on the narratives that dwell within individual places as place. "Reading nature," in other words, means investigating how metaphor, narrative, and environmental identity are emplaced in place. This depiction of nature sees place as a uniquely textual text, in that the reader reflexively reads the text as and from within the world of the text. The world of the text is the embodied and material text of the world; it is a material narrative that lends itself to hermeneutical reflection. In the present section, I argue that any interpretation of place as text—that is, any knowledge of the metaphors and narratives of nature—temporally rests on memory and imagination. To give this further nuance: The *place of time* is at the intersection of self and environment, and thereby the *time of place* is discovered through the work of memory and imagination.

The time of place is ongoing: We can see minor variations, slight divergences, and subtle distinctions that mark this day as unlike any other. The relationship of self and place is embedded in both space and across time. The self is someone who *remembers*, who *imagines*, and who is *here*. And so other terms enter into the discussion: Memory and imagination become important because *we are always already in the lived temporality of place, and will continue to be so*. This chapter began with a description of the end and beginning of a day. However, it was not this *particular* day—it was not *today*. This initial illustration showed the inevitable thrown-ness in time; there is an inevitable placial "alreadyness" in our self-identity. Such "alreadyness" is encountered through a weaving of the past in a relational structure, which is built upon the task of memory. Likewise, we continue to be rooted in place the site of our imagination, the locale of who and what we seek to be. The work of the imagination reconfigures place on the foundation of memory.

Place is a text of *instantiated* temporality. That is to say, place is a site wherein time is mediated as something both within us and beyond us; place negotiates the identity found in the difference between "clock time" and "lived time," and between the naïve realist and the socially constructed. Temporality, then, contains this reflexivity of abstract and concrete when viewed hermeneutically. Furthermore, there is an implicit embodiment of temporality in our experience of nature—places are the *matter* of time. Environments, places, and landscapes float in the currents of time, even apart from our inhabitation. Places are defined by the ways that they grow, change, and perhaps even die while interlocked with time's arrow. In the case of environmental hermeneutics, we can note that the meaning of place—wherever and whenever that place *is*—frames itself through the traces of time that physically constitute that which is "nature." Furthermore, metaphors of landscapes and environments work only in and through time. And finally, the environmental self (particularly in light of the dialectic of selfhood and sameness that rests in the presence of environmental identity) is bound in time as well. We might argue that many—most, even— environmental problems are the consequences of forgetfulness of nature's time. Sustainability is a prominent example of how environmental concerns re-place the self in both space and time, living more deeply in the broad narratives of place.

To read the text of nature, we attend to the interlacing of geometric space and chronological time into the place of meaningful material presence. Because of this, the question of memory and imagination might be stated in this way: If "nature" is understood through place, then how should the temporality of nature be understood in response to lived experiences of environments? Certainly any narrative has a sense of time. But place *instantiates*, place *embodies*, place *manifests* time through its traces and physical presencing. Insofar as place is a gathering of self and other-than-self, such physical presences work as the locations of not only human time but a temporality that moves apart from us. So if place is truly the point of mediation of self and nature, how might our sense of environmental identity encounter the *time* of place as a *place* of time, and vice versa?

Our answer receives insight from Saint Augustine. In *The Confessions*, Augustine meditates on the complexity of being embedded in time, especially given the "vast storehouse" of memory. The past is no longer, but yet it still remains within us (and, I would argue, within others: human and nonhuman, living and nonliving alike). Similarly, the future has not yet come to be, but we can envision and predict it. Thus Augustine writes,

> What is by now evident and clear is that neither future nor past exists, and it is inexact language to speak of three times— past, present, and future. Perhaps it would be exact to say: there are three times, a present of things past, a present of things present, and a present of things to come. In the soul there are these three aspects of time, and I do not see them anywhere else. The present considering the past is the memory, the present considering the present is immediate awareness, the present considering the future is expectation.[23]

Augustine's general reflection holds true for the particularities of place: Memory holds the past of place in our presence, and expectation (in the form of productive imagination) holds the future of place in our presence as well. Thus we are given the inevitable temporality of placial narrative. Instead of remaining locked in mere presentness, memory and imagination are pivotal elements in stretching out time, of allowing us to live in time's duration. In other words, memory and imagination are avenues for deepening the hermeneutics of place emplaced. *Memory and imagination are instantiations of the personal and collective mediation of emplaced time—a mediation of temporality and embodied experience.*

To recapitulate, memory and imagination deepen our reflection on nature by providing us with a key to the mediation of time's duration in place. Experience of place includes more than mere presence; through physical and intellectual traces, we find the past of the place, just as we can discover a sense of the future through imaginative variations of what the "where" is. Places make present the past and future of individuals as individuals residing in place. Simultaneously, places manifest the past and future of the place as *reflexive narratives of environmental identity*. Therefore in material, reflexive, and multivalent ways, *places* are present past and present future. To dwell in this past and future, it is necessary to acknowledge the memory of place; to put it plainly, places contain the trace of memory. Likewise, environmental imagination is essential to the narrative of place; environments hold a depth of time that opens a vision of the "otherwise." Memory and imagination ground the possibility of moving from abstract, calculative time toward this reflexive sense of placial, meaningful temporality.

Memory

Narratives of place are meaningful only when grounded on living presentness of the past, which individual and collective memory offers.[24] *The place of memory in environmental hermeneutics is to connect the past of the place with its presence or presentness.* In discussions of environmental philosophy, we might perhaps forget what is most obvious: that experiences of environments are sometimes—often—infused with memory.[25] The variation of landscape, sound, light, and temperature are all manifestations of the temporality of self and place. This temporality remains; traces of the past are included in the present, even if only as decaying leaves, the temporary presence of one species or another, the absence of erstwhile dwellers or recognition of the changes we are unwitting participants of. Our memories allow us to know the places in which we dwell, and to acknowledge who is dwelling—and who migrates. When we feel autumn as distinct and different than summer and winter, we *remember*.

Presently, we are interested in how memory informs our interpretation of place. To offer a short description: In the case of the memory of place, memory is a form of knowing that recalls past dwellers and configurations of place. This means more than simply retrieving objective information about past events and objects. The memory of place is more complex: It demands that someone or something remembers; through memory, this someone or something arguably comes to be. Further, forgetting is an important part of memory. Thus electronic storage (e.g., computer hard drives) is not a good analogy for memory: Successful memory of place requires us to selectively save, interpret, organize, and perhaps even alter information about the past. Third, memory is an embodied form of knowing. It is not a subjective feeling divorced from perception, but neither is it simply a replaying of perceptions. Rather, memory is an awareness that interacts with all of our senses in a way that structures them in conversation with our meaningful reality. Finally, memory can be—and in the case of place, *is*—something that occurs both individually and collectively. Personal memory is informed by our collective memory, while collective memory is more than a mere aggregation of individual memories. Thus Janet Donohoe argues that places serve to make collective memory possible, and the intersubjectivity of experience shows how closely intertwined collective memory, individual memory, and place are.[26] Insofar as our ethical responsibilities toward environments account for the temporal dimensions of place, environmental ethics is an ethics of memory on both individual and social levels.

Memory, as part of the time of place, is contained in both the reader and the text; we see cognitive and physical presences of our placial memory. A key concept for explaining the way that memory manifests itself is the "trace." Edward Casey gives this definition:

> [T]he ordinary notion of the trace is that of a mere mark left by an entity or an event of which it is but the finite and fragile reflection. Its nature seems to consist in a self-surpassing operation whereby its meaning or value lies elsewhere—namely, in that of which it is the trace, that which the trace signifies by a self-suspension of its own being or happening.[27]

On one hand, there is the trace of the past *in our cognition;* this is what is commonly debated as a "trace"—a signifier that stands in for the original. On the other hand, nature offers us a form of "trace" *in its own physical embodiment.* The rings of a tree, decomposing leaves, watermarks, and erosion patterns all offer a trace in this manner. These two types of traces form a more complete memory of nature, for together they explain the ways in which memory has a presence within our cognition and within nature itself.[28]

The foregoing description of memory has repeatedly suggested its normative element. Certainly there is a desire to understand memory as descriptive of past events. However, our memories are not mere recordings of the past, but are renewed and changed through the present. Insofar as we are interested in moving beyond an anonymous presentness and toward a renewed sense of temporal depth, therefore, we seek a *right* memory of place. Through a hermeneutical reflection on how places are aggregations of both dwellers and the traces of time, right memory is *the work of holding a place in time, and of holding time in place.* Our encounters with nature dive into a temporal flow that is captured in recollections and remembrances. We seek to understand our present experience in light of the traces of the past, whether these traces are mental or physical. At the same time, the trace highlights something else: We are led to forget certain elements of place, just as we remember others, as inevitable parts of our experience. As Drenthen has shown using the metaphor of the legibility of landscape, we forget some strata of landscape and highlight others. This influences not only our interpretation of place, but also our ethical response to it. Yet the past is a voice that calls us out of the radical present. Memory might be rooted to a sense of tradition and the past, but it is not limited to these. In fact, without the possibility of hope and expectation, we are unable to extricate ourselves from the oppressiveness of the past to encounter the future. Right memory, therefore, leads to imagination.

Imagination

If held captive to the past, memory cannot finally emerge as a right memory of place—that is to say, right memory includes possibility and temporal openness to what place *might be*. This is where imagination becomes important for environmental discourse. In particular, imagination provides us with the capacity to envision alternatives to present situations and worldviews; for this reason I am equating Augustine's sense of expectation with the productive imagination. In fact, such a connection is found throughout environmental thought; we can describe the place of imagination as *the experience of grasping the possible futures of place in time's presence.* Thus Dylan Trigg, in his reflections on place-memory, the uncanny, the temporality, writes, ". . . what is immediately apparent is the future-directed orientation of the imagination. Whereas memory gathers place, imagination appears to disperse the origin of place by throwing it outward and ahead of time."[29]

What is the imagination in the context of environmental philosophy? Sara Ebenreck writes, "The work of imagination is quietly present in much writing about environmental ethics. Discussions of future, more sustainable societies call upon our capacity to imagine cultures different from the one in which we now live."[30] Ebenreck does not provide a complete definition, but she does offer us with a useful starting point for reflecting on the imagination:

The imagination is that power which allows us to (a) creatively envision a reality different from the one in which we are immersed, and thus to formulate purposes, goals, or ideals; (b) participate in another's perspective by constructing a sense of what that perspective is; (c) creatively envision an action that embodies the compassion or respect called for by ethical principle; (d) both construct examples for ideas and articulate paradigm cases that we allow to illuminate our thought; (e) grasp or articulate in an image relationship which embody paradoxical qualities that are difficult to express in linear logic; and (f) approach the description of reality through the creative 'naming' of a metaphor or story. The work of imagination may be embodied in metaphor, in image and symbol, in story, in envisioning an action or situation, or may be at work in the mode of awareness that gives rise to those embodied results.[31]

Imagination, according to Ebenreck, is a reflective enterprise that allows us to metaphorically and perhaps even literally "see otherwise."

Ebenreck's discussion of the imagination can be augmented by Paul Ricoeur's sense of the imagination. Ricoeur's sense of imagination is much narrower than Ebenreck's. For Ricoeur, discussion of the imagination as image must be discarded in favor of seeing imagination as akin to "semantic innovation" and metaphor. Ricoeur writes,

> The ultimate role of the image is not only to diffuse meaning in the various sensorial fields but to suspend signification in the neutralized atmosphere, in the element of fiction. . . . But it is already apparent that imagination is indeed just what we all mean by the word: the free play of possibilities in the state of noninvolvement with respect to the world of perception or of action.[32]

In other words, it neutralizes our conventional "theory of the world." When we tie imagination to language rather than vision, we are offered a way to try out new things within a fictional world. Such a view of imagination sees its task as participating in the redescription of reality. It is more than descriptive, however. Rather, it is upbuilding and centered on a project; a "productive imagination" (rather than the more typical "reproductive" imagination)[33] allows us to imagine variations and new ways of acting based on our symbolic, fictional, and narrative sense of the world. "It is imagination that provides the milieu, the luminous clearing, in which we can compare and evaluate motives as diverse as desires and ethical obligations, themselves as disparate as professional rules, social customs, or intensely personal values."[34] Like Ebenreck, for Ricoeur the imagination opens up an alternative that can be enacted in the future.

Because of this projective sense of imagination, there is a clear place for both individual and social imagination in environmental thinking. We can imagine the places and environments of our world in new ways that are in keeping with notions of sustainability, restoration, simplicity, and the like—that is, in terms of environmental "virtues" that all rest on a sense of *kairos*. In terms of the personal, we imagine the environments and places of our lived experience as altered by climate change or empowered by a desire for increased biodiversity. Resting on the ground of memory, our individual imagination looks to the future as otherwise. Similarly communal imagination allows us to "try on" new views and alternative worlds. The social imagination presents us with ways of understanding ourselves and our world; Ricoeur argues that our social imagination allows us to mediate tradition and innovation, thereby letting us have ties with each other in the midst of change. He writes, "The truth of our condition is that the analogical tie that makes every man my brother is accessible to us only through a certain number of *imaginative practices*, among them, *ideology* and *utopia*."[35] Treanor has explained how the social imaginary and the dialectic between ideology and utopia are at play in environmental discourse; he concludes that "[e]nvironmentalists should hold fast to a utopian vision of a sustainable future, articulated in compelling narratives, while remaining on alert for unconventional allies and unexpected opportunities for forwarding our agenda."[36] This claim for narratives of sustainability, I would argue, depends upon the complex work of the imagination, insofar as such narratives seek to understand places as reflexively embodied places of space and time.

For both individual and social imagination, we see the possibility of the imagination opening up a renewed sense of the world. That is to say, imagination allows us to form a bridge between tradition, the present, and possibility. Through the innovation of imagination, we discover new ways of understanding, envision new ways of acting, and uncover better possible futures. Tracey Stark, in a discussion of Richard Kearney's work, recognizes that imagination forms precisely such a bridge:

> Kearney argues that one of the most pressing tasks facing the future is to ensure a creative relationship between tradition and the historical future. The trouble is that the indispensable interplay between past and future is becoming increasingly threatened in our time. . . . This is our 'ethical duty'—namely, to designate concrete steps toward an emancipatory future while liberating untapped potentialities of inherited meaning.[37]

To echo this, we might say that the task of imagination is the *work of opening the possibilities of time to place, and place to time.*

Conclusion

We have reached the conclusion of this chapter. Perhaps it is dusk, or midmorning. Times and places are inflected in the reading of this text, just as they have been lightly woven into its writing: It might be the settled chill of late fall, or the lazy humidity of summer. Through the particular qualities of time and place, memory and imagination work to deepen the presentness of place—or, alternately stated, they allow us to ground the present in the past and possible future as we experience the presence of place. I began by suggesting that one of the pressing problems that must be confronted by environmental philosophy is the fact that contemporary Western society lives in the present, shorn of temporal depths. What I have argued is that environmental hermeneutics confronts this radical presentness through a reflection on how our lived experience of the temporality of place moves beyond the present through memory and imagination. We fully understand self and other in place by questioning *when, who,* and *where.*

Such a temporal deepening is the aim of a hermeneutics of place. Perhaps a fitting conclusion, then, is to once again remember and imagine how hermeneutics is not merely descriptive but normative as well. For example, competing individual's imaginations of place (e.g., developers vs. conservationists) are limited by the recognition of the complex individual and social memory of place. Likewise, the harm of merely static memories of place is overcome through the process of environmental imagination. A depth of memory challenges and complicates the consumerist, throw-away culture by calling us to remember other forms of living in place, as well as seeking an imaginative response of living otherwise. This deepening results in a fuller sense of our emplacement in place: The meaning of place emerges in response to change, attachment, reminiscence, and the values present in envisioning new possibilities. It offers insights into aesthetics, ontology, theology, and other practical reasoning. Thus the present discussion is, truly, in a present that includes the presence of the past and future—it is neither end nor beginning, but an arrow toward a broader strand of who, where, and when we are.

CHAPTER

13

The Betweenness of Monuments

Janet Donohoe

Often when we think about the environment, we think about natural places and the negative impact of the human being upon those places. We think of global warming, melting ice caps, mountain topping, extinction of animals, and other threats to nature. With the increasing public and social emphasis on environmentalism, we are encouraged to think of our impact on the natural environment by recycling more, using water less, and reducing our environmental footprint. The environment we are supposed to be concerned with and thinking about is "out there" beyond the perimeters of our cities where we only go for a deliberate experience of nature. So, we frequently address concerns about the natural environment in isolation from questioning or thinking about the built environment. I would like in the following to propose a broader understanding of what we mean by environment. The human, built environment within which we experience the everyday, where we live, work, bring up our children, and lay to rest our dead is just as much our environment—if not even more so—than the so-called natural world. By focusing my discussion on one aspect of that built environment, I hope to show, in part, how to approach our built environment hermeneutically and to challenge us to do so with just as much attention and just as much rigor as we might consider the issues facing the natural environment. For, if we do not know what we are doing in the built environment, we will have a much harder time conceiving how that built environment necessarily overlaps and conflicts with the natural environment.

Hannah Arendt argues in *The Human Condition* that the built environment is one of the fundamental contributions of the human being to the condition in which we find ourselves.[1] The production of buildings and monuments is positive insofar as they contribute to the public sphere in their everlastingness. Because buildings and

264

monuments that we erect are made of lasting materials, they indicate our desire to produce something that outlives us, something that contributes to a cultural world that will continue. She lauds this contribution as a passing along of tradition and heritage to future generations. Monuments are part of that which creates a lasting world.[2]

From this description of monuments, one can conceive of an encounter with one as an encounter with temporality—erected in the past by a prior generation, a monument projects itself in its permanence into the future. In the experience of a monument, we are reminded of our own transitoriness in the face of the permanence of the past through the monument and its preservation of that past into the future. But perhaps, too, the monument produces in us sweet dreams of such immortality, as Arendt suggests. Through the monument, we can be pitted against nature—nature in its cyclicalness is violated by the permanence of the stone of the monument—we have triumphed over the death nature brings.

What Arendt overlooks in her claims, however, is that while monuments are at once attempts at staving off temporality, at transcending our time, they are always caught back up within history, never escaping the temporality altogether since they are always within the subjective phenomenological experience. In spite of pretensions to immortality, or historical objectivity, monuments can only ever be experienced within a transitory present by a transitory individual, from within a certain historical perspective. It is this tension between the eternal and the temporal, as well as between the natural and the built, that I would like to explore. Drawing upon the work of Martin Heidegger and Paul Ricoeur, we can come to conceive of monuments as elements of our built environment that are not to be revered and unassailably preserved, but that are living elements of our temporality, historicity, and broader environment. Precisely this tension between the eternality and historicality of monuments leads to the need to approach them hermeneutically. A hermeneutic approach appropriately repositions them, reducing their ideological power without denying their role in transmitting to us a tradition, opening for us a world. But it also allows us to see the place of monuments as existing in the tension between the presence of the past and the presence of the future, between memory and history, between testimony and archive, between one and another, between nature and built environment. In the following, I would like to trace out this position of the betweenness of monuments.

In *Memory, History, Forgetting*, Ricoeur describes the tension between history and memory: "On the one hand, history would like to reduce memory to the status of one object among others in its field of investigation; on the other hand, collective memory opposes its resources of commemoration to the enterprise of neutralizing lived significations under the distant gaze of the historian."[3] I will elaborate on Ricoeur's position and suggest that monuments are important in occupying this space between memory and history, between testimony and archive, and in maintaining the tensions between them.

Monuments and the Presence of the Past: Memory and Testimony

On a general and fairly surface level, we can speak of monuments as narratives written in stone—both literally through the inscriptions on many monuments and figuratively through their bas-relief elements, the sculptures associated with them, the symbols of the state, or cultural references that grace many of them. But we recognize, too, that this narrative must be encountered by one who comes to it with a forestructure out of a cultural horizon that shares elements with it, while at the same time coming to the narrative from out of his or her own era of history. The monument in its transference of tradition across generations appeals to a shared tradition and produces a collective cultural memory. Unlike other elements of the built environment, monuments set themselves off from the everyday, encouraging us to reflect upon a collective history. I have elaborated elsewhere on the way in which the sedimentation of layers is embedded in the monument.[4] But here, I would like to raise a further issue with respect to the relationship between collective memory, tradition, and monuments that arises when we consider monuments in their productive role. As narrative, they produce memory and history.

In production of collective memory, monuments function as testimony in recording the experience, in giving image or shape to something that frequently cannot be thought or remembered otherwise, and in making that memory available to the community as a whole. They serve as living reminders that appeal to others to remember. The monument, as W. James Booth suggests, is a "signifier, pointing to a past event, and sometimes also a call to remember, latent and awaiting a witness's voice to cast light on it."[5] It illuminates and shapes the meaning of the past. A monument bears witness to the past by preserving it within the present. Booth also suggests that bearing witness is a gesture of defiance and resistance. Monuments, in being built of stone, marble, or other durable material, put up a resistance to the passage of time and have pretensions to immortality. But they are also acts of resistance against the fragility of memory and the silence of the past by ushering that past and those memories into the present.[6]

Take, for example, the small memorial placed on the spot where Amy Biehl was killed by a mob in Cape Town, South Africa, on August 25, 1993. Amy Biehl was a Stanford University graduate and Anti-Apartheid activist in South Africa on a Fulbright fellowship. She was giving three South African friends a ride home to the Gugulethu Township when she was pulled from her car by an angry

Figure 8 Monument to Amy Biehl in Gugulethu Township, Cape Town, South Africa, 2001. (Photo courtesy of Dr. Steven Gish.)

mob and stabbed to death. The simple stone cross memorial serves
as a reminder of Amy Biehl's tragic and violent death, but it also
serves as a historical marker of a period of 1990s South African history
marked by many such killings and violent unrest.[7]

The memorial stands between the historical past and the call to
remember. It is historical and archival in that it sits on the spot
where Amy was killed. But it also testifies to her death and life. It sits
barely off the road near a busy gas station in the middle of the township. It does not have a lofty position, was not erected by the State, but
serves as a constant presence of the past in the lives of Cape Town
citizens in a way that many more exalted monuments do not. The
monument serves as a salve to free the members of the township
from a past that threatens it with shame or conflict. But it also
serves as a call to remember a particular individual.

Preserving memory against those who would otherwise be
inclined to erase it altogether out of guilt or indifference is to bring
that which is immemorial, other, and which surpasses us into a
manageable sphere where it can be accommodated, experienced,
grasped. This is after all what we must do—we must place into
narrative that which cannot be spoken. We must accommodate that
which is radically other.[8] The monument serves as a command:
"Remember this!" It speaks of shared responsibility, of a debt to
the past. Yet the process of understanding that past is always at the
same time memory of the present, in the present, through the present.
It is only we who can interpret the experience of the monument for
ourselves and our time, through the lens of our own traditions and
prejudices. Thus, our interpretation of the monument always runs
the risk of reduction, sameness, hubris, elevation of our time and
our meaning. As Booth notes, commemoration "can yield silence by
replacing one memory with another."[9] It perhaps replaces the testimony of the witness with our memory now. We question, then,
whether it is utterly self-serving, now-serving, anti-frailty to place
into stone that which we remember.

Ultimately, the meaning of the memorial or the monument can
only be determined by those who encounter it. This does not mean,
however, that we are at liberty to use monuments simply to reflect
ourselves back to ourselves. The monument serves as a point of
contact between present and past and as such it shows forth the
ongoing life of a community. It serves to call us to responsibility for
preserving the memory and inscribing us within the history of the
community. Booth argues that it calls us to bear witness and

"bearing witness can be understood as a 'reciprocal obligation' between generations of a community and an affirmation of their deep identity."[10]

Booth's analysis of the function of the preservation of concentration camps and the memorials that have been erected there can also be useful to us here. He argues that such places have become central in the recollection of European Jewish life in and of the twentieth century insofar as they "rescue the hour of their destruction from the silence of the rustic settings that might have shrouded Auschwitz and other killing centers."[11] The preservation disrupts the silence that might otherwise ensue in the growing over of a site of an atrocity. However, preserving this site of victimhood also leaves untold and forgotten the civilization and the culture that had been established prior to the destruction of millions. Extrapolating from Booth's position, we could say the same of Amy Biehl and perhaps of monuments generally. In the testimony to Amy Biehl's violent and untimely death, she has become a symbol of resistance to Apartheid and of the power of reconciliation due to the circumstances surrounding her death and her parents' subsequent reconciliation with the four men convicted of her murder. But what about Amy Biehl, the Stanford student? What about Amy Biehl, the California child? These are shadowed by the monument itself, by the testimony.

Thus if we only view monuments in terms of our duty to remember, we run the risk of reducing the historical to the testimonial. Instead, the monument holds both testimony and history in tension in order not to run the risk of either idolatry of the dead or ideology of the state. In spite of the shadow on Amy Biehl's early life, the memorial to her death does seem to balance the tension between testimony and history by its placement in the midst of the everyday lives of the people of the township, and its human-scale presence that serves as both a reminder and as a historical marker.

In their role as markers of events of the past and representations of the past through the interruption of the everyday, then, monuments can serve as testimony. Annette Wieviorka argues that testimony is more than simply the recounting of facts or even a narrative that provides objective knowledge of the past. Instead, testimony attempts to keep the meaning of the past alive through a presentation of the truth of the past not as objectively and scientifically determined, but as remembered through narrative and through symbolic representation. And as Booth elaborates, testimony "seeks to guard the truth against effacement or oblivion. But it is a relationship to the truth about the past that is part of a mesh of identity, justice, and debt. Its governing imperative is a mix of debt to the now voiceless past, to preserve the voice of justice against forgetting or falsehood, and to the needs of continuity in identity."[12] Monuments cannot, however, be reduced to testimony, for they also serve as archives and as projections of the now into the future. They draw together temporality and become the locus for the meeting of the past and future, the "I" and the other, the testimony and the archive.

Monuments and the Presence of the Future: History and Archive

If we consider Arendt's position again here, she suggests that the production of monuments is important for the transference of a story to future generations to allow for the survival of history. She writes, "acting and speaking men need the help of *homo faber* in his highest capacity, that is, the help of the artist, of poets and historiographers, of monument builders or writers, because without them the only product of their activity, the story they enact and tell, would not survive at all."[13] Arendt places stipulations on how this needs to happen by suggesting that the monuments need to be of a sort that they be "fit for action and speech." She writes, "in order to be what the world is always meant to be, a home for men during their life on earth, the human artifice must be a place fit for action and speech, for activities not only entirely useless for the necessities of life but of an entirely different nature from the manifold activities of fabrication by which the world itself and all things in it are produced."[14] She acknowledges that the projection into the future by the monument is not necessarily projection into a certainty since it risks the reinterpretation by the future, the re-valuation of the stone marker by a generation that cannot be controlled or foreseen. What becomes clear from these passages from Arendt, and is underscored by Ricoeur's reading, is that the world is made for humans during their life on earth. Monuments are produced for the living, not for the dead. Perhaps we are called by the dead to produce the monument as testimony, but it is erected in the world of human living. The living read and understand the story of the monument. Its message is for the living. It may direct us to that which is not of the living, to that which is radically other, but it must direct *us* and those to come, not those who are no longer. Its focus, then, is on the future. It is archival preservation for future generations.

Ricoeur views the archive as written to be consulted as the deposited testimony which is itself the narrative of a declarative memory.[15] The physical place of the archive is a shelter for the trace of memory. This shelter, this preservation of the trace, however, is not entirely innocent. Because it takes up its position within the social and the public, it is open to whomever approaches it. Without a designated addressee, the monument as archive takes on a certain authority over those who view it. It is most frequently seen to be giving us the truth of the event or the person it glorifies or represents. Like the process of producing an archive, decisions are made as to what is to be represented in monuments and memorials. Such decisions are made with the expectation of future interpretation. How we in the present want an event to be remembered in the future, or even determining which events shall be remembered, are the motivating desires for the production of the archive that is the monument.

And if we think of the power of the monument with respect to its surrounding world, we can see that it organizes the surrounding world, with itself as a focal point such that the other elements fade into the background. It conquers the landscape in a very physical sense in which its substance (usually cement or marble) withstands nature, does not allow for the growth of anything in its place, and puts up resistance to the winds and the weather.[16] In its physicality, it is future oriented, erected for the future, and meant to carry a tradition and meaning to the future.

In the encounter with the kinds of monuments that Arendt's passage evokes, we might be tempted to view monuments as primarily productive (and I do not mean this in a positive sense) in creating an environment of national power, collective ideals that attempt to persuade the individual of the priority of the nation or the culture over the individual. We kill in the name of the United States, for instance, and can rest assured that our names will ultimately be glorified in a war monument. War memorials easily come to mind as creating the national archive of participation in violent conflicts.

If we think of the typical glorifying monument to soldiers in war, we experience a moment that contributes to the ongoing narrative of a nation. What transpires in this kind of monument is an abstraction. We are not to pay attention to a particular soldier. The soldier represented in the monument is frequently anonymous. It is not a soldier, it is representative of soldierliness. It is an abstraction that allows us to view war as glorious and is about heroes as opposed to being about the horrors of the destruction of flesh, the gruesomeness of mutilation, and the death of particular individuals. In this respect, the abstraction goes hand in hand with the ideology of the monument. This is not a historical narrative, telling the story of what happens in battle, this is the production of a narrative of a nation as war-winners, as gloriously on the right side of the battle. But it is precisely in the abstraction that the distortion takes place and the particular deaths, the particular victims, are put aside, forgotten, dismissed.

Even those monuments that bring into question the glorious war have been opposed and limited, their narratives restricted. In France, for instance, after World War I, there was a movement to erect pacifist war memorials not celebrating or honoring the fallen soldier but pointing to the horrors of war through the presentation of widows and orphaned children, the grieving and the lost. One of the most famous of these is at Gentioux-Pigerolles with its inscription "Maudite soit la guerre" (cursed by war). This memorial, made of stone with a sculpture of an orphaned boy standing cap in hand fist raised toward the inscription of names of the dead, was only officially inaugurated in 1990 after many, many years of obstructionist behavior on the part of the French military. In the effort to preserve the power to shape and produce the archive as desired by the state, the testimony to the effects of war was to be covered over. The monument in its testimonial power was to be denied in favor of a constructed history.

It seems, then, that what monuments and memorials can do, in holding the past and the future in tension, in speaking as both archive and testimony, is to provide narrative of the past, to bring it into the present, and to project for the future the memory of those who are no more. In order to achieve this, monuments cannot be restricted to archive and the archival element cannot be strictly anonymous history, but must include the testimony of the other within its narrative.

How do we hold the testimonial element in tension with this almost overpowering element of securing our vision of our role, or the role of our generation for future generations?[17] How do we allow for the testimony in light of the ideological decisions necessary for archives? As Ricoeur points out, ideology is "linked to the necessity for a social group to give itself an image of itself, to represent and to realize itself, in the theatrical sense of the word."[18] Among the characteristics that Ricoeur attributes to ideology is one that is particularly apt for monuments: Ideology is "simplifying and schematic." It gives an overall view of a group and that group's history in relation to the world as a means of justifying not only the group itself but many of the group's actions. "For it is through an idealized image that a group represents its own existence, and it is this image that, in turn, reinforces the interpretative code. The phenomena of ritualization and stereotype thus appear with the first celebrations of the founding events."[19] This ritualization itself is a necessary element in the cohesion of a group and is not of itself negative. When ideology becomes something of which to beware is when we recognize that ideology functions at the level of formulating thought rather than inspiring thought. It is uncritical and nonreflective in its opacity.[20] Monuments in their inherent temporality are in some respects automatically tied to a certain level of ideology. "All interpretation takes place in a limited field, but ideology effects a narrowing of the field in relation to the possibilities of interpretation which characterized the original momentum of the event. In this sense we may speak of ideological closure, indeed of ideological blindness."[21] Ultimately for Ricoeur, then, "[i]deology is the error that makes us take the image for the real, the reflection for the original."[22] It is important to note that the ideological role here is not necessarily negative. It becomes negative when it is narrowed and fails to incorporate an element of critique. With regard to monuments, then, if the monument is instrumental in securing a social bond, it is ideological, but it can secure that social bond in ways that are open to the reinterpretation of the past as well as to creative projection into the future. It is that position in between that must be preserved for monuments and that we intend to explore in the following.

Ricoeur on the Between

We adopt Ricoeur's understanding of the twofold use of the word "history" here as "the set of events (facts) past, present, and to come, and as the set of discourses on these events (these facts) in testimony, narrative, explanation, and finally the historians' representation of the past."[23] Ricoeur suggests that our making of history is something that comes about because we ourselves are historical. Monument making is just one way among many that we make history. It is in this way, too, that we justify our continued monument making in spite of the difficulty that monuments pose for the transmission of culture across generations. This does not eliminate the possibility that monument making could happen in a very different fashion in the future. We should consider how human temporality and historicality contributes to our role as monument makers.

Ricoeur raises concerns about Heidegger's use of the fundamental structure of care and its temporality as a way into a revised understanding of the temporality of humans. For Ricoeur, Heidegger is mistaken when he conflates the totality and the mortality of the human being. In spite of Heidegger's claim to the openness of Dasein to its possibilities, the insistence upon Dasein's own most possibility as being that of death seems to have the effect of closing off Dasein. It is precisely this closing off that Ricoeur finds problematic in its focus on the imminent threat of dying that masks "the joy of the spark of life."[24] Ricoeur instead wants, along with Hannah Arendt, to stress the theme of natality which he describes as underlying the Arendtian categories of labor, work, and action, which compose the *vita activa*.[25]

Moreover, Ricoeur asks, "Does not the jubilation produced by the vow—which I take as my own—to remain alive until . . . and not for death, put into relief by contrast the existentiell, partial, and unavoidably one-sided aspect of Heideggerian resoluteness in the face of dying?"[26] As a response to this question, Ricoeur finds value in the dialogue between the historian and the philosopher regarding death. It is also at this juncture that I find the role of the monument. Initially, Ricoeur wants to explore the relationship between the embodied desire for life and its association with death. He suggests that in fact, this knowledge of death, as promoted by biology, is still heterogeneous to the desire to live. "It is only at the end of a long work on oneself that the entirely factual necessity of dying can be converted, not to be sure into the potentiality-of-dying but into the acceptance of having to die."[27] For Ricoeur, even when such an acceptance of death has taken place, it does not replace the anguish or the fear of death precisely because of the "radical heterogeneity" of death with respect to the desire for life. The second tack for Ricoeur is in the plurality of death. He challenges Heidegger's lack of attention to the death of the other as having the same impact as the angst of one's own being-toward-death. For Ricoeur, "[w]hat it is important to plumb instead are the resources of veracity concealed in the experience of losing a loved one, placed back into the perspective of the difficult work of reappropriation of the knowledge about death."[28] Through the death of the other, we learn loss and mourning. With respect to the loss of someone close, Ricoeur likens it to a loss of an integral part of one's self-identity. The loss of the other with whom we communicated constitutes a step toward the loss of self. In mourning the loss of the other, we anticipate the mourning that others will go through at our own future death. This redoubles the anticipation of one's own death with which we struggle to reconcile ourselves.

Ricoeur recognizes, however, that this is only on the order of the loss of someone close. How, then, do we accommodate the loss of the strangers whose names we frequently read on monuments? Do monuments play a role in this process? The answer to this question is a decisive "yes."[29] Ricoeur draws upon Levinas, too, when he thinks through the way in which the death of another, not a close other, but an other, reflects my own fear of death back to me. For Ricoeur, the most extreme example of this is the violent death of murder. What we have in the form of a war memorial or monument in particular is the violent death of countless others in war. Death by the state? Death by the national enemy? But violent death all the same. Ricoeur refers to this as the problem of death in history.

What does this mean for monuments? It provides us with a way of understanding the role of monuments vis-à-vis memory and history that can be helpful in shattering the ideological pretensions of monuments and in providing us with a way of approaching the monuments with an eye toward our own lives. I have argued before that some monuments are successful in bringing us to mind of our own being-toward-death while other monuments represent an attempt to allow us to escape that grasp. I do not back away from this claim, but I do want to qualify it by recognizing that part of what makes those successful monuments successful is due in large part to their role in making us feel the death of the other in its particularity. This brings us face to face with our own death but also face to face with our own life or, as Ricoeur says (drawing upon Levinas), our "being against death."[30]

The monument to Amy Biehl calls us to remember her violent death at the hands of fellow human beings. In such a memory is also the shadow of life. This simple confrontation on a dusty street corner in a township brings one to mind of one's own mortality, life cut short, but also brings one to mind of the very aliveness of one's living with the dust in one's eyes and the heat on one's neck. It is testimony to Amy Biehl and her violent death, but as testimony, it is also confrontation with the death of the other that calls to mind one's own death. At the same time, it testifies to one's own being alive since the monument was placed as a reminder to the living. In its archival element as a reminder in the township of an era of violence, but also of reconciliation, it serves, too, as a call to continue that heritage of reconciliation. There is a power in the monument that many more ideological monuments lack.

The attention to life as well as death marks the value of the monuments at the individual level of memory, but what of the other side of the equation, the value of the monument at the historical level. If the monument is bridging a gap between individual temporality and history, on what grounds does it do that? Ricoeur attempts to reconcile these questions through, on the one hand, the concept of attestation as discovered in Heidegger, and on the other hand, the concept of "heritage-debt." On Ricoeur's reading, attestation in Heidegger is the locus of the unity of our historical condition through its three temporal phases. It is the testimony of the past, the anticipation of our historical condition, and the "I can" of the present that is felt in the capacity for speech, action, and narrative. Heritage-debt likewise "constitutes the existential possibility of standing for" which remains dependent on the retrospective orientation of historical knowledge.[31] These are the two prongs between which monuments can be understood to mediate. Both of these prongs are an effort to exist in the tension between the absence of the past and the representation of that past in the presence of history. What Michel de Certeau attributes to the writing of history, I likewise attribute to the monument (not unlike what he calls the sepulcher). He identifies two different movements in this writing. It "exorcizes death by inserting it into discourse," while in the same moment it "performs a 'symbolic function' which 'allows a society to situate itself by giving itself a past through language.'"[32] For Certeau, Ricoeur asserts, the writing plays the role of the past. Ricoeur also refers to Ranciere who indicates that death in history is not usually the death of anonymous people but is the death of the king or other important personages. The writing of these deaths attempts to make the deaths redeemable by history. We need only look at a brief history of monuments themselves to see this.

Early war memorials, in particular, honored great military leaders and their victories. The Romans imported obelisks from Egypt to celebrate their victories. Common soldiers, however, were not among those honored and celebrated. It was not until after World War I that monuments and memorials became more democratic honoring the common soldier. These monuments were often erected to unknown soldiers—the ones whose remains were never recovered, or whose remains were recovered but were unidentifiable. Only in the most recent past have the names of the known, common soldiers been included in the commemorative process. The victims of atrocities, too, have only recently been commemorated or marked by a monument. What might this tell us about the process of commemoration? Why has it become a necessary and important thing for us to attend to? And must we commemorate every national conflict?

The duty to remember does not ensure that such a memory is put to effective use. Tzvetan Todorov suggests that it is not easy to determine the good uses of the past from the bad ones, but that this does not mean that we should give up only in favor of the particularities of the past.[33] Todorov makes a distinction between the literal use of memory and exemplary use of memory. Literal use of memory is where an event is preserved in memory as fact, not necessarily true, and is never moved beyond. In such a case, the initial event is associatively connected to subsequent events but leads "nowhere beyond itself."[34] Exemplary use of memory, on the other hand, opens the event in question—trauma, for instance—to generalization thereby distancing it from oneself, using it as a principle for present action, and making it an example from which to learn. Todorov upholds exemplary memory due to its ability to generalize and bring justice which is impersonal law, impartial judgment. In arguing for exemplarity, he charges that one must move beyond victim to fight against injustice. Monuments to the victims of atrocities can be seen as one attempt to do just that—to provide a narrative that acknowledges the injustice, that serves as testimony and witness for the victims, and attempts to bring justice by using their memory against obscurity through time and as a reminder of a lesson to be learned. This is a fine line, however, for in the making of an event and example, one must guard against the shadow that can cover the particularity of the experience of the victims. Thus the conservation of the memory for history can only truly be effective, I would argue, if in the experience of the monument is one confronted with the particularity of the event and of the particular deaths that allows one to see beyond the particularity at the same time to the example it represents. This is the position in between that the monument takes up.

This fine line is accomplished through mimetic narrative of a successful monument. While monuments themselves may not be so clearly seen as fully developed narratives, they share the qualities of mimesis as outlined by Ricoeur and which Brian Treanor has explained more fully in his essay within this volume than we can do here. In brief, they contribute to the narrative identities of the groups who erect them as well as the groups for whom they are erected. They are narratives in their symbolic depictions of victims or heroes and in the use of cultural symbols that emphasize the valor or the pain they commemorate. In this way, they are instrumental in promoting a prefigurative understanding of one's own nation, for instance, as a certain kind of player on the world stage. They also contribute to the emplotment of a national story by enveloping a particular event into the general story and help to refigure the event within our own lives perhaps allowing us to walk away from a monument or memorial with a different perspective on that which it commemorates.

This commemorative narrative, however, cannot be a reduction of history to memory, nor can it be a reduction of memory to history. Instead, narrative, as transfigured in monuments, must be the duty to hold memory and history in tension with one another. If we view the monument as merely archival, we could see monuments as simply an injunction to remember a particular event of the past, leading to the ready question of how one could be required to remember something that was not even of one's own generation. Remember the holocaust when it happened before I was born? Surely, I cannot be expected to remember such a thing, so how is it that memory could be involved here? This is precisely where memory and history are held in tension through the physical manifestation of the memorial and its role as both archive and testimonial. And what precisely is one being called to remember? Remember those who were killed? Remember the perpetrators? Exactly who were the perpetrators? Can we know that exactly? And who were those who were killed? Can we even know that? We know names, perhaps. We have numbers, perhaps. But what does this mean? How do we remember the untold millions who died on The Middle Passage, whose names were unrecognized and unrecorded? Ricoeur describes the duty as a duty of justice where justice is understood to be related to a component of otherness. It is, in other words, a call to do justice, through memory as both archive and testimony to those who are other than the self. As he writes,

> [W]e are indebted to those who have gone before us for part of what we are. The duty of memory is not restricted to preserving the material trace, whether scriptural or other, of past events, but maintains the feeling of being obligated with respect to

these others of whom we shall later say, not that they are no more, but that they were.[35]

Ricoeur rightly argues that this imperative to remember has been too easily separated from history in the sense that it has been taken up as a frenzy to commemorate. If we simply divorce the duty to remember from a place in history, we end up with a rush to memorialize that lands us in an ideological position that is not about the duty to the other but is about securing our own place as victim.[36] The duty of memory, then, might more properly be understood to involve the duty to remain in the tension between memory and history. This is where commemoration must reside, but it is a point that monuments and memorials so frequently fail to find.

Conclusion

Monuments are responsible for making the dead speak, for bringing the past into the present, and for projecting that past and our present into the future. As the monument is always experienced in the now from the perspective of a current generation, its interpretation is the transmission of a dynamic meaning across time. The monument takes up a position between past and future, between "I" and other, between temporality and history, between testimony and archive.

Of course, any encounter with a monument must recognize the need to be open to the tradition, the history, and the memory that the monument evokes. And in that openness is precisely the responsibility of the experiencer to recognize the relationship between history and temporality that makes the experience possible in the first place. As Ricoeur reminds us, "the temporal constitution of the being that we are proves more fundamental than the simple reference of memory and of history to the past as such. In other words, temporality constitutes the existential precondition for the reference of memory and of history to the past."[37] Successful monuments take up the position in between allowing for us to experience both the possibility of testimony, memory of those who are no more, while at the same time allowing for the possibility of archive and projecting that history and tradition for the future. To reduce the effect of a monument to testimony is to misunderstand its position within history, but to reduce the monument to archive is to misunderstand its role in bringing us to be mindful of the other and of ourselves with respect to the absent other. Its position in between makes it complex in its meaning and allows us to take on a position of critique of its possible ideological and historical assertions. In calling both of these aspects to mind in the experience of a monument, we are ever more so aware of our own historicality and our own position in the in-between of the present. This kind of hermeneutic approach is not to be reserved for monuments but is an approach that can be taken up in general with our built environment and can help us to engage in the critique of environment and the human-environment relationship more generally.

My Place in the Sun

David Wood

In this chapter, I pursue the thought that it is via temporality, especially history, that place is distinct from space. I show that this claim survives our moving away from a naïve naturalistic understanding of the past to one constructed and constituted so as to include narrative, intentions, and projections even when these form the basis for serious contestation of what we take to be the past.

Space and Place

The distinction between space and place is not difficult to grasp. On the one hand, geometry, on the other location, permeated one way or the other with meaning. We identify space with measurement, with the neutral, objective way in which extended things are separated from each other.[1] Space, in this way, is understood as "external," in classical contrast with mind, with "thinking stuff" (*res cogitans*) which is internal, a distinction which itself sounds spatial but is not. This sense of the objectivity and neutrality of space is complicated somewhat both by theories of perspective and by the theory of relativity, in which space is subjected to perspectival considerations—perhaps ineliminably—suggesting that the position of the subject, orientation, is an essential ingredient, even to "external," "objective" space. But the outcomes are still calculable even when we take into account the requirements of perspectival representation. Adding a subject just brings in a new "angle."

Place, however, seems very different. When American poet Gary Snyder talks of the need (finally) to put down roots, he is making a move even within a literary and cultural tradition that celebrates mobility and quest.[2] Moreover, the distinction between space and place can quickly become charged, even politically, tied up with territorial conquest, mapping, and property. In both the United States and Australia, spaces and places were nullified, neutralized, by surveying practices that located even natural features on geometrical grids, with some residual respect given to rivers and peaks.

"Place" operates at many levels and registers. "A place for everything, and everything in its place" refers to a well-designed kitchen or workshop in which a range of useful things is easily accessed. One place (a kitchen) will provide subordinate places for smaller things. "Come back to my place" and "Your place or mine?" are invitations to a home, a dwelling, a locus of personal life and meaning. To win first place in a competition is the mark of accomplishment in which one is differentially ranked in relation to others. And there are many situations in which we evoke the value of place without saying, "This is (a) place," as when people talk about their favorite cities, restaurants, or vacation destinations. We do, it is true, speak of place in more subjective ways ("I was in a bad place for months after she left."), but this seems like a metaphorical extension of the more usual sense of a location whose meaning is shared with others. Indeed if one had to choose one word to distinguish place from space it would be meaning.

At this point, one could launch into a neo-Aristotelian description of the conceptual geography of place, aided and abetted by, for example, Dilthey's hermeneutics. Or (with Ed Casey) we could creatively explore the phenomenology of place in its extraordinary variety.[3] However, these are both valuable exercises for another occasion.

Place as Site of Contestation

I propose here to consider those situations in which "place" is the site of contestation and where history plays a central role in that contestation. I want to highlight and explore the connection between the existential sense of right attached to one's bodily existence, the values associated with dwelling and home, and the political and ecological consequences of the ways in which the scope of this sense of right is interpreted. Place and its history and promise are central to these issues.

Here is the argument: We each take for granted as the ground of every other right and privilege our individual right to exist, our "place in the sun." This is rarely contested, and it is undoubtedly defensible. We are speaking of my (mere) right to exist, not as a bloated plutocrat, one of the 1%, but as a mere man. The image of "my place in the sun" makes it clear that we are not talking about a privileged spot on a private beach but a place on earth that does not deprive others of theirs. Not a zero-sum game.[4] After the holocaust, it must have seemed to many that merely being alive was a special privilege for which one might feel guilty. Levinas' meditation on "my place in the sun" would be a good example.[5] But if (to be very quick) the conditions under which this was or seemed like a zero-sum game (my survival = your death) were the work of the devil, then one is surely buying into evil in endorsing this position, even in feeling this guilt. The moral of this thought is that even if one's right to exist is something like an unconditional background, it is one that can, at least apparently, be suspended or modified. The idea of an "unconditional" right may not ultimately fly. It may be that there are conditions, such as lifeboat situations, under which unconditional rights would be suspended. If there were unconditional rights, the right to the minimal means of existence, given abundance all-round, seems like a strong candidate, and on a very different level to the right to free speech, education, health care, and so forth. It seems to be a condition for these other rights that one continues in existence. And yet when one says, "Given abundance all-round," it sounds like a condition. What if that were not satisfied?

And what consequences does this have when one takes seriously the idea that continued bodily existence can only artificially be separated from those interactions with the world that provide sustenance—what we often call labor? Is there then a right to work, and where does that take us?

When John Locke famously connected the right to property to whatever one had "mixed one's labor with," he included an important caveat—that the exercise of such property rights would not significantly affect the supply of material for others to do the same.[6] And we might add, not at the time, nor subsequently. If we relegate to the cognitive margins the existence of native inhabitants and nonhuman creatures, this looks like a splendid way of justifying and motivating the colonizing of the New World, and indeed they did, and it was.

The question I am posing is this: Is there a slippery slope from distinguishing place from space in the existentially uplifting way evoked by Gary Snyder to the attribution of a narrative value to place that would legitimate possession, exclusive use, privilege, and so forth? And if there is in fact a slippery slope, can we mark the point at which such slippage becomes dangerous?

If one has something like a natural right to one's own corporeal existence (hence the right to self-defense), the obvious next step is whether that right implies or can properly be extended to the locus of one's life—one's home, for example. Some sort of extension seems inevitable once we admit, as we have suggested, that bodily existence is impossible without its relation to some sort of sustaining ground. We need air, food, water, shelter, and space to exercise just to preserve our biological existence. Only thereafter are information, companionship, and so forth considered.[7] Bodies are not just "things" but indices of our essential relationality, both in a physical and social sense. Ears, eyes, mouth, hands, and genitalia all shout this in different ways. But it is not clear what implications this has for any particular situation one might find oneself in. The burglar who finds the beer (or Perrier) in your fridge cannot justify his drinking it on the grounds that we all need to drink. Juridically weighed statements of our collective responsibility for providing minimal conditions of survival even to social outcasts come up with the jail cell—sometimes overcrowded or (possibly worse) in solitary confinement. But no one supposes that a prisoner has a right to some particular cell (#353 at San Quentin), even if he has gotten used to it. Similar considerations apply to the provision of public housing for the poor. Once it is accepted that we often and appropriately live in groups, such as families, local sociocultural norms will generate minimal housing units—apartments, small houses. In each of these situations, one's right to continuing existence implies *some* place, not some particular address.

These cases of social provision make visible only the most basic layers of place—physical sustenance, and a locus for shared space— what we might call (a) "home."[8] Minimal it may be, but we should not underestimate its significance. The plight of the homeless, the refugee, and those who lose their homes in a hurricane cannot be overestimated. In the home is centered both a whole set of utilities (from which one can venture out into the world, and to which one can return), but also a whole slew of meanings.

Mixing One's Meaning

The hermeneutics of place begins, one suspects, with the body and its physical and social needs. But with considerations of meaning it flowers, and beyond the point at which minimal needs are being satisfied. Whether we endorse Locke's understanding of property, there is something compelling about this account ("mixing one's labor") for understanding this flowering of meaning. In the case of involuntary incarceration, meaning is at a premium. A cell mouse may become a prisoner's closest companion. In the case of the social provision of housing, maintenance neglect is a common concern. In each case, there is a deficit of opportunity for mixing not just labor but meaning with one's surroundings, for the investment of time, effort, energy, planning, and so forth. Where such opportunities abound, meaning flourishes. Herein lies the attraction of the pioneer spirit—where one would find, occupy, and develop a piece of land, build a house, and so on. And more standardly, it explains the legal protection offered (under normal conditions) by property rights. They enable long-term planning and investment, the laying down of associations and memories with a place, and so on. These are the sorts of considerations that lie at the heart of objections to people like Proudhon ("property is theft"). But the seeds of his position's appeal are not hard to find.

The condition Locke laid down for legitimating connecting property to labor was that one should not deprive others of the same opportunity—much as we make parallel remarks today about freedom and sustainable development. The clearest cases are those in which there is an infinite (or practically boundless) supply of the material in question. Here we applaud those who freely bestow deep rich complex layers of meaning on the spaces they occupy.

Locke's premise was that property is not a zero-sum game. True, if I own something, you cannot, at the same time, also be the exclusive owner (though we could share). But if creating property were like writing poetry, it would be largely benign. Admittedly, if I write a poem, you cannot write the very same poem and unproblematically claim it as yours.[9] We do not think of this as a restriction because the combinatorial possibilities of language seem infinite.

It turned out that the premise on which Locke's definition of property rested was more problematic than anticipated. When applied to the American colonies, not only were there original Native American occupiers of the land but what was meant by "mixing one's labor" was not at all clear[10] (running cattle?). It soon became clear that there were choice locations (springs, soil, minerals like gold, ocean views) and a limited supply of them. On the ground, mixed in with Locke's agrocentric compatibilism, was violence and land grabbing. Moreover, if we look more broadly at the way place and property comes into being, it is hard not to conclude that forcible expropriation of existing rights is the norm.

If we take this thought seriously, we would be talking not about mixing our labor with the land but mixing our blood. Or the blood of others: "enemies," sons, fellow creatures.

At this point, I want to step back by opening up a new question: Is the meaning structure of a place *sui generis*, essentially sustaining itself, or does it rest on something quite different? I have already, in effect, addressed this question in connection with those meanings associated with property, connecting them back to bodies and work. Does such a grounding demonstrate a wider truth?

Place and Territory

Consider the idea of territory. This term ranges across the spectrum from politics, through economics, to the ecological. Britain has (or used to have) Crown Territories, or "possessions" it administered without their being part of the United Kingdom. Marketing and sales people have distinct territories where they can focus their efforts without competition from their colleagues. And nonhuman creatures have territories, too, that they mark, patrol, and defend. These are clearly places, spaces that sustain and distribute meaning. But these meanings are tied back to imperatives—interests, forces, histories. Colonies were conquered and subdued for their exploitable resources (including human bodies) and their strategic location. Commercial territories are allocations that encourage the investment of energy, time, and attention by guaranteeing exclusive access and control. Animals maintain territories to control mating or to protect their food supply.

If territory is a kind of "place," its skein of meanings (significant subordinate places or sites, its history of important events, ways its boundaries are marked—local dialect, color-coded maps, urine) derive from, and rest on, what we could call vital interests—profit, life/death, reproduction, and so on. Predicated on these factors, lines and boundaries arise.

Place and Contestation

If, for a moment, we took territory to be exemplary for thinking place, it would have dramatic consequences for our basic model. Mixing one's labor (or one's meaning) seems like a relation that a subject has with its world, independently of what others are doing— hence the importance of the "infinite supply of material" caveat.

We may suppose that battles over special sites would be exceptions or marginal cases, but what if they were in fact the norm? Or that we came to see my (or our) relation to place as always in principle contestatory, even if I only want to name it. Can I cultivate a garden or participate in your garden cultivation without excluding or helping to exclude others from making those same investments. Turning Locke on his head, what happens to mixing one's labor under conditions of serious scarcity?

In extreme situations, it is often said that "it's every man for himself!" It is not entirely clear what this means. It could mean that in the absence of the opportunity for a coherent group response, each must seek his own survival in the best way possible, even if that strategy would not, under better conditions, optimize the greatest number of survivors or indeed maximize one's individual chance of survival. Or it could mean that the moral niceties of collective life have ceased to apply; walk over your grandmother if that helps. The latter, in turn, could be interpreted as the breakdown of all order, or a return to the primitive "law" of survival. Such a law would justify, for example, killing the other man in a lifeboat stocked with only enough food for one to survive.

If there is some sort of natural right to ensure one's survival even if, *in extremis*, under conditions of genuine scarcity, it involves taking the life of another, and if some minimal sense of property can be justified as an extension of the right to ensure one's survival, this would seem to offer a justification for the expropriation of another's property, even if they themselves needed it to survive. If I am starving I can justify stealing from your garden even if that would mean you would starve, or even stealing your garden. I would not need to dispute the legality of your title. I would need only to be genuinely desperate to acquire, in whatever way necessary, the means to survive.

I do not say that there is or could be a legal system that would endorse such a course of action.[11] There could be a system that would treat desperate need as a mitigating circumstance in the case of theft, even violence, for example. And that mitigation would evaporate if the situation was not in fact desperate, or if I could just as easily have asked for some food.

The point I am making is this: Even if we cannot legally codify this, we cannot rule out being able to justify violence at the origin of possession and dispossession of property. But just as Locke's account of property through mixing one's labor rests on not depriving others of the same opportunity, so, too, even this limited justification of violence is conditional, with conditions not unlike those for a just war. You have to be genuinely desperate, have exhausted other means, and the violence must not go further than necessary. We must also take into consideration the question of time and history. If you are away from home in the winter and I survive the cold by helping myself to your garden vegetables, this is not a justification for permanently taking your whole garden. By the spring, things will be looking up. Suppose, however, you do not come back and over the years I make your garden mine, and perhaps your house, too. I fix the leaking roof, add an extension, and raise a family. Then you come back.

There is an old story of a poacher caught red-handed at dawn with salmon stuffed down his trousers. The squire tells him he is trespassing, that *he* owns this land. The poacher asks him how it is that he owns it. "My forefathers fought for this land!" "Well," said the poacher, "I will fight you for it now."

I suggested previously that the desperate man who kills for place, for a plot of land for his family to live on, could "justify" doing so, even though it might not be legal. One way of understanding this would be to say that he could tell a story, a narrative, in which his actions would be understandable, perhaps even reasonable. "What choice did he have?"

But when my neighbor returns after many years to reclaim his property he may have an equally good story. He was falsely imprisoned by a brutal dictatorship and unable to make contact.

We have talked about place as a site in which labor and meaning are mixed. We have supplemented this two-term relation with a third term—the other—with whom I may have to compete for possession of this place, or even to stamp my meaning onto it, perhaps to the point of violence. I suggested at the outset that what makes place distinct from space is its temporality. Most likely temporality, especially history, is how meaning and perhaps the sacred gets added to the mix.

The Historical Constitution of Place

Let us now talk about the historical constitution of place. I live on a large farm in Tennessee. Some four hundred million years ago, what is now limestone was the bed of the sea. Numerous coral fossils pepper the surface of wide slabs of rock, often many feet wide. In the creek not far away, Native American arrowheads and scrapers still surface, anything up to ten thousand years old. At the top of the hill, there is a Civil War installation of undetermined function from the 1860s. Google maps shows outlines of fields long ploughed under and farming patterns dating back to 1900. Then there are derelict sheds and barns, roads overgrown and newly cut, and skulls of dead creatures, wild and domesticated. And numerous living beings with their own lives and times—from fleeting dragonflies to slower snakes. Some of this time is current—limestone dissolving in the rain, frogs splashing in the pond, sunlight shimmering in the spider webs. These concurrent rhythms interweave to make something like the fabric of the present. But the sedimented layers of the past are equally real and constitute a fairly different dimension. We may imagine that the reality of the past of a place lies only in what remains today—fossils, blurred outlines, relics, skulls, the contours of the landscape. On this model, events that occurred "here" once upon a time are, as they say, history—done, over with, finished. It is not clear how to change the paradigm here, but what I am proposing is that its past be treated as a real part of what we understand a place to be, so that all the events, processes, and changes that a place had "witnessed" would be part of that place. A place has, as it were, temporal roots. This seems especially plausible in the case of a battle-field, a graveyard, ground zero, a historic town—where well-known events of human significance are recorded and commemorated. The more general claim is that unknown events, events in geological time, are also "part" of what constitutes a particular place. They do not need to have been recorded somewhere, but they did happen here, and so they are part of what "here" is. This is a realist view of place. Interestingly, it may supply a ground for a rather different ontology.

Near an English Civil War battlefield at Edge Hill in Warwickshire, there is a mound called Blood Copse where, it is said, many of the casualties from that battle were buried. To this day, there are reports of ghosts of dead soldiers walking by and dogs that get too close howling at night. I do not know what to say about the dogs, but ghosts are, arguably, a visualized acknowledgement of the reality of the past on the part of those who still bear its memory, within an ontology that wants things to be present if they are to count as real. Derrida's evocation of the specter "of" Marx (Marx as specter, and Marx's evocation of the "specter of communism") makes an analo-gous point. To speak of a ghost or a specter is to speak of the way the dead "live on" through memory, but not just as memories.

Memory gives us access, however imperfect, to their past lives. In the case of soldiers who fell tragically in a short indecisive battle, or a great thinker who many have tried to "bury," we are dealing with what we might call unsettled memory, memory of events or people that challenge or disrupt normalizing narrative. A long discussion of trauma could begin here!

The background here is a certain realism. Places literally contain their past.[12] But our grasp of that past can generate strange phenomena. Of course when these are repeated (e.g., ghost sightings), the growing record of such "sightings" itself becomes part of the past.

The Place of the Future

It might be thought obvious that the past would have a privilege over the future in determining place. The past consists of real sedimented layers of events, while the future is as of yet undetermined. And yet it would be a mistake to cling to too simplistic an account of this privilege. Not only is the developing meaning of (a) place tied up with often contested possibilities of fulfillment of contemporary promise, both rhetorically and causally, but the same is true of the past. The layers of past events that constitute place themselves project possible futures, which may or may not have subsequently been realized. This is true both of explicit projections (plans, programs, etc.) but also of more gestural indicators (things must change, enough is enough, we cannot go on like this). Many past events themselves projected temporal horizons of significance, inheriting and witnessing the past, anticipating and adumbrating the future. A certain naturalism is tempting at this point, even as it includes intentional phenomena. The *soixante-huitardes* believed that the future could be shaped by imagination: "L'imagination au pouvoir!" The cobblestoned streets of Paris bear this history, with all its unrealized aspirations, as facts, even if those hopes have been largely forgotten. And as much as past projections mattered causally in shaping action, whether successfully or otherwise, the same is true today. But there remains an asymmetry. The past, we suppose, with all its complexity, actually happened. Our anticipations of the future, our desires, and our plans are all equally real and in one way or another capable of being efficacious. But if we take as our standpoint a dynamic emergent sense of place, or place as a continuous sedimentation of meaning, the future *as such* is not part of the meaning of (a) place. Hegel was right—*Wesen ist, was gewesen ist.*

Being is what has been. The future in that sense is a different order of being, essentially unrealized.

Once we break with the illusion of presentism, this generous naturalism seems plausible. It is not that the past *is present*, but that, as past, it is a component of the real. It does not "live on," except in and through its effects. It does not need to. It really happened. It has not gone anywhere, it is rather *somewhen*. We may find a spatial analogy helpful, as in the celebrated analogy of the iceberg, much of which floats below the surface.[13] But this is in no way to reduce time to space. The past is the past.

I have tried to elucidate here a certain intuition about the temporally constitutive thickness of place, and how that contributes to its meaning. This relies on a simplistic accommodation of the intentional within the natural. In addition to what happened there is also what we wanted to happen, what we feared might happen, and so on. These are not posited as possible worlds but as subjective or intersubjective attitudes that can sit casually enough alongside the first-order facts.

Ontological Musings

This objectivism is, however, problematic. It is not just that we have no access to the past except through our narrative constructions. In important ways, nothing ever happens at all without being articulated one way or another. This is perhaps the point of Derrida's *Il n'y a pas de hors text*. It seems to get close to a certain idealism, one of a transcendental flavor. It affirms the centrality of a certain complexity for things or events to have any distinctive identity. If something happens, if an event occurs, we are not just saying that the cosmos perdures, or continues. *Something* happens. This rather than that. It rained this morning. The rain started and it stopped. It was rain not snow. It rained this side of town not that. A mass of raindrops fell, each with a common cause (clouds). It started as a heavy downpour, and later eased up.

We need an ontology that acknowledges distinctions, differences, on-the-ground, "out-there," as well as our constructive and selective activity in creating meaningful shapes on the basis of intrinsic material differentiations. And furthermore, we need an ontology that acknowledges the difficulty of identifying those intrinsic differentiations without already putting our constitutive powers into gear. Moreover, we need to recognize that our meaning-construction activity is underdetermined by intrinsic differentiation. Think of the different rhythms one can construct as a passenger from the sounds of a train running on a track. We also need to recognize that while there are clearly opportunities for individual, personal meaning construction, *place* particularly invites shared collective synthesis through ritual, repetition, narration, and so on.

If we accept that human meaning-construction is a process of differentiation and synthesis laid over a natural world in which difference and connection are already happening (in which we are already involved and from which we cannot separate ourselves), it must be further acknowledged that the extra we bring is not (just) truth in some ungrounded acontextual sense but an articulation of interest, and interests that we might call dwelling. Implicit in this formulation, however, is first the possibility of a fracturing of the "we," and the further possibility that this fracturing might not be a mere possibility but at the heart of the matter. "We" can fracture into specific groupings of humans, set against each other. But it can also imply "we humans," set against the other creatures with whom we share the planet.

The central implication of these observations is that the naturalistic/objectivist account of the thickness of meaning intrinsic to place will not do. Does the intuitive appeal, such as it is, of the past being embedded in place depend on it? If we start to include memories, attitudes, narratives, don't we lose the solidity that seems to be provided by the naturalistic account? Analogues to geological strata are one thing, but the swirling complexities of people's transient beliefs about what's happening are quite another.

We are now in a better position to consider place as a site of contestation. What this implies is that the articulation of meaning in and through history that gives place its depth can be expected not merely to testify to some group's (a "people") affection, depth of memory, identity investments and so on, but also to its attempt to exclude or legitimate domination over other contenders for the same space. The meanings of place are rarely free of territoriality. But that implies that the rhetoric of place is inextricable from the kind of claim to property rights we discussed previously, with reference to Locke. Developing such a relation to place (mythological, sacred, narrative) can be treated as mixing one's symbolic labor with a piece of the earth. Yet if analogous caveats and conditions obtain here, the legitimacy of such entitlement would rest either on an unlimited supply of similar sites, or on dire necessity. I suggested that *some sort* of justification could be made even for killing one's neighbor if the alternative really was one's own death or that of one's family. One could "understand" someone doing that, even if one could not approve of it. There are also special new conditions, hard though they might be to specify, such as that the events being knitted together can plausibly be said to have actually happened, if they are not openly acknowledged to be myth. And that the actors in these events have some plausible connection to "us." If the inhabitants of a land weave their history together with Viking invaders, it is important both that there were Viking invaders and that "we" have some special connection with them. If other contenders equally identify themselves with the same invaders, we have a problem. The plausibility of the narrative construction will be a mix of objective factors (whether what is being claimed as a continuity meets a minimal standard of plausibility, for example, logical consistency), whether "we" want to believe it, and perhaps whether it is in our interest to believe it.

Consider two extreme cases: First, the status of Jerusalem, and second that of "Man's Place in Nature."[14] Each of these "places" is heavily contested though in different ways.

A Tale of Three Cities

Jerusalem both personifies and symbolizes the "sanctity of place" for all religions deriving from or responding to biblical scripture.[15]

Jerusalem is claimed as a sacred site by three religions. What follows are a series of freely available non-scholarly accounts of how/why Jerusalem is important to each of them.[16]

A city divided between east-central Israel and the Israeli-occupied West Bank. Jerusalem was founded as far back as the fourth millennium b.c. and was ruled by the Canaanites, Hebrews, Greeks, Romans, Persians, Arabs, Crusaders, Turks, and British before being divided in 1949 into eastern and western sectors under Israeli and Jordanian control. In 1967, Israeli forces captured the eastern sector from Jordan, later declaring the city as a whole to be the capital of Israel. The legal status of Jerusalem, considered a holy city to Jews, Muslims, and Christians, remains fiercely disputed. Population: 729,000.[17]

Around the year 1010 B.C.E., King David defeated the Jebusites in Jerusalem and decided to make the city his administrative capital. When he brought the Ark of the Covenant to the city, he stripped the Twelve Tribes of the spiritual source of their power and concentrated it in his own hands.[18]

Jerusalem is also very important to Christianity, as Jesus Christ lived and died here. The Christian quarter alone houses some 40 religious buildings. . . . One of the most prominent and important sites . . . is the Via Dolorosa . . . Jesus' final path, which according to Christian tradition led from the courthouse to Golgotha Hill, where he was crucified and buried. [. . .] The Church of the Holy Sepulcher is a pilgrimage site for millions of Christians from all over the world.[19]

Jerusalem is considered a sacred site in Sunni Islamic tradition, along with Mecca and Medina. Islamic tradition holds that previous prophets were associated with the city, and that the Islamic prophet Muhammad visited the city on a nocturnal journey. Due to such significance it was the first Qibla (direction of prayer) for Muslims and the prophet Muhammad designated the Al-Aqsa for pilgrimage. [. . .] Muhammad is believed to have been taken by the miraculous steed Buraq to visit Jerusalem, where he prayed, and then to visit heaven, in a single night in the year 620.[20]

These accounts already anticipate ways of privileging one version over another. Length of (semi-) continuous occupation would favor the Jews. Spiritual significance seems to be shared, as well as evidence of a canny awareness of what needs to happen (to be built) to strengthen the case for distinctive legitimacy. But how can we best thematize the role of history here in constituting the meaning(s) of this place? If all meaning is differential, whether one is aware of participating in that process, Jerusalem seems to take this to another level. It seems, at least, that it is not merely the case that the status of Jerusalem is fought over now but that this very question of legit-imacy-dispute itself has a long history, a second-order truth that the various contenders for exclusivity might arguably agree on, even as they would disagree about the verdict. We could go further: Jerusalem appears here because it is *exemplary*, combining an extreme his-torical determination of meaning, *some* shared agreement about the facts, and disagreement on how to interpret them. Many of these disagreements are performative in character—reaffirming the party line. Some will be more or less disinterested. It may be, however, that Jerusalem is too good an example, to the point of not being at all typical. It may be a really good example of a continuing battle-ground in which the contestation of history and the history of contestation are inextricably interwoven. It would be too good an example because it goes beyond the claim that the history of a place is an intrinsic part of it. If every history reflects a certain interest, the history of Jerusalem, weaving together and/or counterpoising so many different interests, both confirms and belies that generalization. It is not clear, for example, how to count the Jerusalems. One? Three? But even if Jerusalem is "too good" an example, it nonetheless bears witness, albeit problematically and in excess, to the historical constitution of place.

Man's Place in Nature

Man's place in nature challenges the "we" in a different way. At the heart of the phrase is some sort of distinction between man and nature, even as we seek to bring the two back into alignment. The idea of evolution, promulgated by Darwin [1859] only four years before Huxley's book of this name, inserted man back into nature through our biological *history*, but also reasserted our species privilege as the latest achievement of that process. The word "place" does not function as the name of a literal space, but the way it does function confirms the normative sense we have been arguing for, made explicit in such a phrase as "first place." To have a place is to belong, somewhere at least. (Cf. "A place for everything and every-thing in its place.") To speak of man's place in nature is to reaffirm that we *have* a place in the face of the death of God and the broader displacements of secularization. Contestation is implicit here. What is at stake is whether we (still) have a place, and, if so, where is that place and how is it to be conceived. In the light of our earlier remarks about the privilege of the past over the future, it is interesting to note that Teilhard De Chardin extends his own brand of evolutionary thinking into the future. Spirituality is man's distinctive contribution to evolution, and it opens up a future in which that spirituality will be developed and exercised. This is quite compatible with our reser-vations about the future defining place in the ordinary sense of place. Projections (current and previous) of possible futures properly play a role in our understanding of who and where we are. Such projections embody and fulfill the very normative understandings of man that can anchor our sense of having a place (in nature). It is not implausible, for example, that developing a certain selfless wisdom will make it possible for us to rescue the planet, and by extension ourselves, from the consequences of a myopic and inadequately evolved model of agency. Continuing to externalize the toxic and dystopic byproducts of our seemingly productive activities is unsustainable. Recognizing this, and projecting a different path, is necessary for the changes we need to happen in a managed way. We do not have to subscribe to Teilhard's spiritual language to acknowledge a certain aporetic dimension to our relation to nature, as when we may be said to be in the world but not entirely of the world. Perhaps Heidegger is echoing this when he recommends at least a *soupçon* of the uncanny as a condition of being truly "at home" in the world. Our proper place would be to be slightly displaced.

Conclusion: Our Place in the Sun

The sun bathes the earth in light and warmth. There would seem to be more than enough to go around. Having such a place does not seem like a zero-sum game. Yet we battle for the well-lit room in the house, the best spot on the beach, the building site on the ocean, not to mention the shady spot in a summer parking lot. To be placed in a race is very much a zero-sum game. We have argued throughout that while the intuitive plausibility of history being essential to place rests on a somewhat naïve naturalistic grasp of that history as including everything that ever happened here, the thesis survives the inclusion of intentional attitudes, and narrative constructions, even contradictory and conflicting ones, albeit in a more problematic (and interesting) form.

15

How Hermeneutics Might Save the Life of (Environmental) Ethics

Paul van Tongeren and Paulien Snellen

Can a hermeneutical approach be helpful to environmental moral philosophy? Can it help to deal with the main issues of this applied ethic,[1] that is, the improvement of the disturbed relation between humans and their natural environment, the way this relation ought to be (conceived of), and the moral status of the nonhuman world? And if so, what—if any—would be the limits of this environmental hermeneutics?

In order to answer these questions, we will first ask why environmental moral philosophy would be in need of this hermeneutical move at all: Why not stay within the scope of mainstream ethical theories? To do this, we will first introduce and summarize Bernard Williams' criticism of "ethical theory" that, according to us, also applies to environmental ethics. Second, we will point out why the limits of moral philosophy bring us to hermeneutics, in particular to hermeneutical ethics. In conclusion, we will ask what the limits of this hermeneutics are, especially in the framework of a hermeneutical ethics of the environment.

The Limits of Philosophy with Regard to (Environmental) Ethics

In this first section, we will give a summary of the argument of Bernard Williams' well-known book *Ethics and the Limits of Philosophy*,[2] and relate it to the issues with which an environmental ethics deals. Williams himself does not speak of environmental ethics (or any other form of applied ethics whatsoever) but is interested in limiting or restricting the presumptions of philosophy—"the powers of philosophy"[3]—with regard to what he calls "ethics." For the moment, but we will come back to this, Williams' conception of ethics can be best—although vaguely—described as the collection of all the elements that are relevant for answering Socrates' question, that is, the question of how one should live.

A problem arises immediately when we want to apply this conception to environmental ethics—first, because the term "ethics" does not refer to the same thing in the expression "environmental ethics" as in William's notion of ethics. The first points to a particular kind of philosophical theory, whereas the second points to what we will call "ethical experience." Moreover, Socrates' question does not admit the division into the separate fields of applied ethics; it is about one's life as a whole. Environmental ethics is either directed at life in a much broader sense than only human life, or, if limited to human life, at solely a particular aspect of it: the way ethics deals with the natural environment. This brings to light that possible answers to the main questions of environmental ethics are not necessarily applicable to human life as a whole. Hence, it is only with much precaution that we—for the sake of the argument— provisionally rephrase Socrates' question as "how we should deal with our natural environment" and conceive of environmental ethics as the part of philosophy that tries to answer that question.

The reasoning Williams presents is aimed at more specific (but very common) approaches to Socrates' question. "Ethics," when defined as mentioned previously, should according to Williams be distinguished from two other uses of the term: first, the philosophical discipline which he calls "moral philosophy" or "ethical theory"[4] and second, a particular subset of the considerations that are of importance to Socrates' question, namely considerations that are typically expressed in terms of a particular kind of obligation, an obligation that makes one feel guilty if unable to fulfill it. Williams calls this subset "morality" and in the last chapter of his book he explains what morality implies and "why we would be better off without it."[5] We may categorize most theories of environmental ethics with one of these two uses of the term, at least to the extent to which they claim to be a philosophical discipline and attempt to outline what our obligations are with regard to our natural environment.

The question that Williams poses in his book concerns what philosophy is capable of with regard to ethics. Phrased in terms of Socrates' question, Williams asks whether philosophy can answer the question of how one should live. Does one have to philosophize in order to answer this question? It seems that Socrates' answer to the question was affirmative—"the unexamined life is not worth living for men"[6]—and without doubt many philosophers have claimed that their philosophy provides the answer.

Williams' answer is skeptical, although his skepticism is "more about philosophy than it is about ethics."[7] He does not hold that an answer to the ethical question is impossible, but rather that philosophy cannot contribute much to it. It should be noted that the term "philosophy" has a specific meaning, which becomes clear when he writes: "[W]e can think in ethics . . . but . . . philosophy can do little to determine how we should do so."[8] His skepticism is aimed at philosophy considered as a theoretical foundation, justification, or test of the answers given to Socrates' question. Here his criticism becomes relevant for environmental ethics, for we think this is precisely the role philosophy takes on in traditional environmental ethics: to found, justify, and test the answers to (not Socrates', but) environmental ethics' question: how we should deal with our natural environment—for example, to find a criterion on the basis of which we can ascribe intrinsic value to (part of) the nonhuman world.

According to Williams, philosophy cannot do much in this regard. However, the fact that philosophy can only play a very modest role in finding a foundation for answers to ethical questions is no problem at all according to him, since both the importance and the effect of such a philosophical justification should not be overestimated.[9] For traditional environmental ethical theories in which the ultimate aim is to help to find a solution to concrete environmental problems, this observation does pose a problem. For how then could environmental ethics help to improve the relation between humans and the nonhuman world? Hence the limitations of ethics Williams points out will have serious consequences, not just for ethical theory as such but also for environmental ethics in particular.

Williams develops his skepticism by exploring two questions. The first question is whether philosophy can provide a rational foundation for any answer to Socrates' question, that is, whether it can demonstrate what answer should be given by means of a rationality that has not already committed itself to a particular answer. He distinguishes two traditional attempts to do so, teleology and deontology, and relates them to Aristotle and Kant respectively. In both cases, Williams shows that their justificatory power relies on the prior acceptance of the presuppositions of the answer.

That it is—as Aristotle puts it—in one's own interest to live a virtuous life, or to relate to the natural environment in a virtuous way, will only be persuasive to those whose upbringing has already interested them in a virtuous life. That is, it is persuasive to those for whom the pursuit of a virtuous life already fits in with their life as a whole. One cannot prove that a certain way of living is in one's best interest without using arguments that already presuppose this perspective. Thus there is no external access providing an insight into the ethical goodness of the virtuous life. There is no "outside view" on the basis of which the claim for the virtuous life can be proven to be true.[10] Aristotle himself acknowledges this. He recognizes the importance of proper moral motivation, that is, the motivation to act in a virtuous way, to ensure virtuous behavior; but he knows that he cannot persuade people with the help of rational argumentation to form the proper moral motivation if they do not already in a way possess it.

Kantian deontological theory focuses on the rights and duties that are implied in our rationality, and can therefore not easily be extended to what we owe to nonrational nature. But apart from that, the Kantian "fact of reason" (*Faktum der Vernunft*), that is, that it is rational for everybody to prevent others from interfering with their freedom—and thus that everybody has good reasons derived from their own interest in freedom—does not automatically imply that in reality everybody has good reasons to accept what others demand. Deontological theory speaks about what any given person "would reasonably do if he were a rational agent *and no more*,"[11] but the question that ethics poses is related to real rational agents:

> Unless you are already disposed to take an impartial moral point of view, you will see as highly unreasonable the proposal that the way to decide what to do is to ask what rules you would make if you had none of your actual advantages, or did not know what they were.[12]
>
> [P]ractical deliberation is first personal, radically so, and involves an *I* that must be more intimately the *I* of my desires than [the Kantian] account allows.[13]

Williams' conclusion is that "there is no route to the impartial standpoint from rational deliberation alone."[14] In other words, both with regard to Aristotle and to Kant, the answer to the first question is identical and in the negative: There is no external access to ethics.

Hence someone will only be persuaded by (specific kinds of) arguments by environmental ethics, if he is already engaged (in that specific way) with the natural environment.

This point of criticism is crucial for environmental ethics. Environmental ethical theories attempt to provide a rational foundation for an answer to the question of how we should deal with the natural environment. But this foundation cannot be based on rational argument alone (for there is no external access to "ethics"). Instead, the search for a rational foundation is itself already motivated by the very same motivation that ethical theories try to produce. The problem is, however, that if people do not share the presuppositions of the rational foundation, the argumentation is unlikely to appeal to them, even when they hold the arguments in themselves to be correct. The mere recognition that an argument is logically legitimate does not lead people to change their behavior. This becomes apparent in our daily practice. Most of us are familiar with arguments that show that and how we should deal with the natural environment in a more sustainable way. But although we know, myriad of us do not do it.

Williams' observations thus point out that the core problem for ethics is a problem of moral motivation. This might be even more obvious with respect to the way we should deal with the natural environment than in other cases. And precisely this problem of motivation is not solved by the philosophical theories in which ethics takes its refuge. The same point comes to the fore again in Williams' treatment of his second question.

The second question Williams poses takes his claim for granted that an external access to ethics is impossible. He is prepared to assume that we are "in some general sense committed to thinking in ethical terms." Nevertheless, he asks whether an ethical theory— even if its point of departure are actually given moral convictions— can show "how [we] should think,"[15] that is, whether it can test our factual "ethical beliefs."[16] Williams is again skeptical about the possibility "that philosophy can determine, either positively or negatively, how we should think in ethics."[17] His skepticism is aimed against two "ethical theories" that are nowadays considered to be the most important forms of teleological (consequentialist) and deontological ethics and that for that reason also provide the most prevalent forms of environmental ethical theories: utilitarianism and (Rawls') contractualism.

Williams has a reputation for his shrewd criticism of utilitarianism. Here, we will only present one aspect. Utilitarianism prescribes its solutions from the standpoint of an impartial observer, who is—in the words of Roderick Firth—"omniscient, disinterested, dispassionate, but otherwise normal."[18] Williams notes that this impartial observer would not act at all without at least some motivation, and that these motivations will always be "mine," that is, that the actor will be attached to them in a strong way, something utilitarianism denies or requires that we forget:

> The belief that you can look critically at all your dispositions from the outside, from the point of view of the universe, assumes that you could understand your own and other people's dispositions from that point of view without tacitly taking for granted a picture of the world more locally familiar than any that would be available from there.[19]

Rawls' attempt to test his deontological ethics in a contractualist manner is criticized in a similar way: The primary goods that Rawls proposes cannot be produced "by the machinery of rational choice deployed under selective conditions of ignorance" but are only understandable "in terms of a fundamental ethical conception of the person."[20] Rawls' model of a reflective equilibrium between theory and intuition is in fact liberal, rationalistic, and prejudiced in several other respects. In itself, this should not be an objection, but it becomes one when the theory declares itself applicable to fields that do not share these prejudices.

Williams' criticism can be characterized by two interconnected lines of thought. First, he criticizes in principal the attempt to give an ethically persuasive answer to the ethical question from an external standpoint; second, he unmasks the particular intuitions behind the claims for an impartial standpoint and he refutes the claim that ethical theory would be capable of giving "some compelling reason to accept one intuition rather than another."[21] Ethical theory cannot provide a foundation for any answer to the ethical question, and it can never supply us with an ultimate test of our ethical intuitions—neither from a point of view external to these intuitions, since from this position the criteria could not be internalized, nor from the inside, due to a lack of criteria external enough to judge the intuitions. To the extent to which this criticism holds for all prevalent ethical theories, the criticism also affects environmental ethics. Does this mean that ethics has been shown to be impossible, environmental ethics no less than ethics as such?

Can one follow Bernard Williams in his claim that philosophy is of no use whatsoever for our dealing with Socrates' question?

The image of philosophy that arises from Williams' criticism of philosophical claims with regard to ethics bears heavily on scientific rationality. Science is concerned with a notion of the world regardless of our own place in it. Williams would have no problem acknowledging that science can tell us something about our natural environment, for example, about global warming or the role of biodiversity in certain ecosystems. However, he denies that ethics can use rational and "impartial" facts and arguments as a foundation for the answer to the question of how we should deal with our natural environment (and how we should live our lives in general). This is a challenging observation for environmental ethical theories, which often refer to scientific results (e.g., about the extinction of species or the alteration of ecosystems) to show how pressing the issues are with which they deal and how much we are in need of a change in the way we behave with respect to our natural environment.

Right from the start, Williams makes it clear that his skepticism only concerns the claims of this theoretical type of philosophy with regard to ethics, and not of ethics as such, not even of another way of critical reflection on ethics. This other kind of reflection is, according to Williams, not "ethical theory" but "ethical thought," the aim of which "is to help us to construct a world that will be our world, one in which we have a social, cultural and personal life."[22] Therefore, Williams' book can be read as a plea for the vindication or rehabilitation of practical philosophy. This might bring him— unexpected as this may be— closer to hermeneutics, which after all, according to Gadamer, is essentially practical philosophy.[23]

From a Philosophical Ethics to a Hermeneutics of (Environmental) Ethical Experience

We will show that what Williams calls "ethical thought" or "reflection" can be characterized as hermeneutics, although he does not use this term himself, let alone apply it to environmental philosophy. In his book on the limits of philosophy (and as far as we know in his whole oeuvre) the term "hermeneutics" does not appear. Nevertheless, what Williams proposes as an alternative, and what he refuses to call an "ethical theory," comes down to what we would call "hermeneutical ethics" or "ethics as hermeneutics of ethical experience."

Thus Williams' challenging observations might be met by a turn to (environmental) hermeneutics. Williams puts it this way:

> There could be a way of doing moral philosophy that started from the ways in which we experience our ethical life. Such a philosophy would reflect on what we believe, feel, take for granted; the ways in which we confront obligations and recognize responsibility; the sentiments of guilt and shame. It would involve a phenomenology of the ethical life.[24]

However, he continues, "This could be a good philosophy, but it would be unlikely to yield an ethical theory."

Let us briefly specify some characteristics of a hermeneutical ethics as we conceive of it,[25] apply these to an ethics of the environment, and point out the similarities with Williams' "phenomenology of the ethical life." A hermeneutical ethics is a philosophy that not only conceives of its own task as interpretation but that also regards moral reality, that is, the world as it appears in ethical experience, as interpretation. In the expression, "ethics as the hermeneutics of ethical experience," this "ethical experience" is not only the interpretandum—some sort of "reality" to be interpreted—it is itself already an interpretation. There is no nature that is not already interpreted. Whether we speak of nature as environmentalists, or as scientists, or from another perspective, we are speaking about an interpretation. Nature as we experience it, whether in our technological manipulation of nature or in our wonder about nature's beauty, is always already an interpretation, and this interpretation is always already part of a temporal structure of interpretations, that is, a tradition which is not an object to which I relate as a subject but of which I am a part, with my ideals, expectations, and experiences. For example, the rise of the average temperature or the disappearance of species would not mean anything if we would disconnect it from the web of interpretations of which it is a part.

In his criticism of utilitarianism, Williams discusses the idea of a "World Agent" in whom all preferences are brought together and which is an operationalization of the thesis that what is morally right is that what maximizes the satisfaction of as many preferences as possible. On this, he writes,

> This idea is often criticized in terms of its ethical results, but the fundamental objection is that it makes no sense as an

interpretation of the world. It is because of this that it makes
no sense of ethics.[26]

Concerns, interests, and preferences are not neutral things in the
world but are always personal and singularized. They are eyes and
ears, and so forth, through which we are in the world: they are interpretations. Utilitarianism tends to confuse the theoretical interpretation of certain dispositions with the interpretation of the world
based in these dispositions.[27] Thus, Williams reproaches utilitarianism
for its naïve denial of the hermeneutical character of the reality of
preferences that are its point of departure. By overlooking the fact
that these preferences are themselves interpretations, utilitarianism
interprets them incorrectly.

A first requirement of practical philosophy is that it should give
an adequate explanation of the interpretative nature of ethical experience and ethical reality. As such, when Williams distinguishes his
"ethical reflection" from the "ethical theory" and "moral philosophy"
that he criticizes, he does indeed emphasize the interpretive character
of this reflection. It does not, so he says, lead to "ethical knowledge"
but produces something else: "understanding."[28] A hermeneutical
environmental ethics thus should primarily be concerned with
acknowledging the interpretative nature of our experience of the
natural environment and especially of the moral aspects of this
experience. Martin Drenthen's work on "legible landscapes" forms
a scarce but fine example of how this could be given shape. He
discusses how we can interpret a landscape as a "multi-layered text"
and uses this interpretation to reconcile different positions in the
traditional protection-restoration debate by referring to the different
ethical experiences that underlie both positions.[29]

Second, this reflection takes into account the argument that there
is no starting point outside the interpretation of our ethical experiences. Ethical experience cannot be put between brackets and should
not be concealed by a universalistic and rationalistic method. Quite
the reverse, this experience should be the point of departure for our
interpretation—it is through our experience that we understand the
world. In the expression "ethics as the hermeneutics of ethical
experience," ethical experience is not only interpretation or interpretandum but is also the interpreting activity, it is the interpreter
itself. A hermeneutical environmental ethics would not attempt to
take a neutral stance in an effort to convince those who have a different view of nature or the environment. It would instead start from

our moral experience and try to explain it. Making abstraction from our moral experience of the world would abandon the only starting point we have for moral reflection.

Ethical theory, as criticized by Williams, attempts to provide, with as little presupposition as possible, an explanation of and a theoretical foundation for what it contends. His alternative is a "reflection that asks for understanding of our motives, psychological or social insight into our ethical practices." Indeed, "*our* motives . . . *our* ethical practices"—for these motives and practices are always deeply "ours" or "mine." The same holds true for my understanding of the so-called "thick ethical notions" "such as *treachery* and *promise* and *brutality* and *courage*,"[30] of which Williams speaks, and to which we, from our environmental perspective, could add such notions like "responsibility" and "care," or "wonder." The personal or internal nature of these notions is, just like Gadamer's "prejudices," not an impediment but rather a precondition of the kind of practical philosophy that Williams proposes. Instead of taking a point of view that resembles as much as possible no point of view at all, we should rather take a human point of view. To use Williams' words, "[T]o see the world from a human point of view is not an absurd thing for human beings to do."[31]

Earlier we referred to the vagueness of Williams' notion of ethics. This vagueness has to do with the fact that his reflection has its point of departure *in* ethical experience. This is not a trace of carelessness—the last thing of which one can accuse an analytical thinker such as Williams—but a deliberate attempt to evade a rigid definition. Earlier we described ethics as "the collection of all the elements that are relevant for answering Socrates' question, that is, the question of how one should live." This now has to be stated more accurately, since issues other than ethical considerations can be relevant as well to answer Socrates' question. Any further distinction, however, comes at the cost of circularity—ethics is concerned with all those things that are *ethically* relevant to answering the question of how one should live; an environmental ethics is concerned with all those things that are *ethically* relevant to answering the question of how one should deal with the natural environment, whether it is nature's beauty or nature's brutality. In other words, Williams' ethical reflection recognizes that besides ethical considerations, other aspects might be relevant, but acknowledges at the same time that the relative importance of these other aspects cannot be determined from the outside. It is precisely ethical life that "can see that things other than itself are important."[32] Hence exactly when we take the moral aspects of our experience of the natural environment as a starting point, can we come to understand that the natural environment is important independently of the human world.

Third, if ethics can only commence with ethical experience, ethical theories that attempt to begin elsewhere, or "outside," can be and should be reread and reinterpreted as being internal interpretations of ethical experience themselves. Such an approach enables or even compels us to reread the history of moral philosophy and ethical theory as a collection of interpretations of aspects of ethical experience, that is, as a range of hermeneutical drafts. If looked upon in this light, the positions criticized by Williams—Aristotle, Kant, utilitarianism, and Rawlsian contractualism—can be vindicated as "partial" interpretations of ethical experience.

The same holds true for environmental ethical theories (that often draw on the theories Williams criticizes). Sometimes this can be quite clearly the case. For example, in an article on intrinsic worth and pragmatism, Ben Minteer starts out with a description of his personal experience. One day when he was ten years old, he was killing garter snakes in the back garden. When his mother saw this, she confronted him and pointed out that what he did was cruel (for, so she argued, he was killing living things). Minteer says, "It *was* wrong, and I knew it. I knew it not because of my mother's undeniable authority (. . .), but because clearly the snakes *were* alive." He then goes on to tell us his reason for sharing his experience: "I think that it is suggestive of a common intuition many of us have toward the value of elements of the natural world."[33] But instead of sticking with the experience and trying to understand it, he attempts to construct a rational argument and foundation for the intrinsic value of all living things. Even though it is not always as apparent as in Minteer's article, environmental ethical theories can be vindicated as "partial" interpretations of ethical experience as well.

It makes less sense to criticize theories because of their pernicious effects on our relation to the natural environment than to try to understand what aspect of our moral experience they present. To achieve this, the basic hermeneutical rule should be followed, which prescribes that one should "reconstruct the question to which the transmitted text is the answer."[34] When this rule is applied to the most well-known ethical theories, we realize that it is illusory to think that they each only give different answers to one and the same question. Kant and Aristotle, as well as Mill and Rawls, pose different questions, since they aim to explain different aspects of ethical experience. A hermeneutical environmental ethics should in this way develop a differentiation of questions regarding our relation to the natural environment, questions about our duties, our responsibilities, our desires and ideals, our reliability, and so on.

Bernard Williams does not provide his own examples of such a vindication of ethical theories. His criticism of utilitarianism has become famous, and he states "that we would be better off without"[35] the so-called "morality system," of which utilitarianism is "a marginal member"[36] and of which Kant's deontological ethics of duty and Rawls' contractualism are classic examples. Nevertheless, we think that Williams primarily opposes the idea that this "morality system" comes "to dominate life altogether"[37] rather than wanting to deny its meaning entirely.

This brings us to a fourth point. Characteristic of a hermeneutical ethics is that it does not reduce the plurality of ethical experience but rather aims at enriching it. On the contrary, the morality system typically reduces the ethical to what Williams considers to be only one aspect of it: "It misunderstands obligations, not seeing how they form just one type of ethical consideration."[38] Ethical theory as it is *criticized* by Williams attempts—as we said before—to provide an explanation and a theoretical foundation of its contentions with as few presuppositions as possible. The purpose of his alternative ethical reflection is indeed the reverse; instead of ridding ourselves of ethical ideas and considerations, or at least reducing their number, we should collect and articulate as many as possible. The relevant considerations are multifarious and diverse—unable to be reduced to one teleological or deontological denominator. This complexity of ethical considerations and motives reflects the complexity of the tradition to which we are the heirs. The interpretation of this tradition is itself an ethical venture in the sense that it is a means of finding our way around, and of making us feel at home (ètheios, èthikos) in this tradition.

However, this does not mean that hermeneutical ethics cannot be normative.[39] Let us explain. It will be clear that hermeneutical ethics cannot be normative in the same sense as the "morality system." It might be due to the point that Williams refuses to speak of another "moral philosophy," but instead modestly offers his "ethical reflection" as opposed to the "ethical theory" that he criticizes. A hermeneutical environmental ethics would also have to moderate its normative claims. But in what way *can* it be normative?

Hermeneutics is not only the technique of interpretation but equally and also more importantly the theory that tries to show what it means when we say that human life is determined by interpretation. If we say that moral experience is in need of interpretation, we claim that there is something in our moral experience that makes an appeal to us. There is something in experience—for example, in wonder, indignation, concern, or whatever our natural environment may evoke in us—that presents the world as meaningful and that claims to be meaningful itself. The world as we experience it has something to say to us, and our understanding of what it has to say makes a call on us—at least it demands that we respond to it appropriately. Moreover, while making this appeal, our experience does not simply communicate what it has to say, but rather withdraws from our understanding and demands to be appropriated and interpreted. Therefore, it cannot be something that coincides completely with our immediate opinions, for then we would know in advance everything experience has to offer. Interpretation would not be necessary. Since, however, we *are* in need of interpretation, that is, since we don't know yet the meaning of that which presents itself as meaningful but are trying to grasp that meaning, we inevitably presuppose a criterion (even if we do not dispose of it) to which we can compare our opinions. Every interpretation is a proposal, and one that asks for confirmation or—if necessary—refutation. Hence, as Williams also admits, there is no reason why the form of reflection he suggests could not be critical: "Much explanatory reflection is itself critical, simply in revealing that certain practices or sentiments are not what they are taken to be."[40] Thus environmental hermeneutics can uphold the critical stance an environmental ethics traditionally has, albeit in a more modest way.

In conclusion, we can say that Williams' skepticism with regard to ethical theory is not directed at ethics in general but only at a particular kind of philosophy and philosophical claims. The limits of that particular type of philosophy—"ethical theory"—can be transcended in a hermeneutical ethics. Contrary to the philosophy that he criticizes for its attempts to provide a foundation after the fact, hermeneutics focuses on the interpretation of "pre-understandings" that necessarily precede their explication. Modifying Kant's formula, we could say that with regard to ethics, Williams—albeit unintentionally—limits philosophy in order to make room for hermeneutics.

The Limits of (Environmental) Hermeneutics?

The beginning of hermeneutics is that we are addressed by something. Things have something to say to us, even if they do so in silence. In the case of environmental hermeneutics, nature places us before the question of what it has to say and what we should do about it. In this beginning lies the universality as well as the practical nature of hermeneutics. Not only because, as Gadamer writes with regard to the art work, "it does not leave us the freedom of pushing it away" but also because it shows that we are always already involved in this process of understanding, which includes misunderstanding and lack of understanding; and since we are always already involved in it, the elaboration of our understanding will also be a development of our self-understanding. In Gadamer's words, "*Verstehen . . . ist immer auch Gewinn eines erweiterten und vertieften Selbstverständnisses. Das heißt aber: Hermeneutik ist Philosophie, und als Philosophie praktische Philosophie.*"[41] Williams also admits that the type of reflection he points at can be practical, at least in the sense that it informs our ethical practice: "[E]thical reflection becomes part of the practice it considers, and inherently modifies it."[42]

This does not mean, however, that a hermeneutical (environmental) ethics has unlimited possibilities. Although it is plausible to assume that human experience and interpretation indeed form the only entrance if we want to deal with the ethical question, a hermeneutical approach to environmental ethics cannot easily meet its aim to solve existing environmental problems. Although environmental hermeneutics can point out to people that they are situated in an environment and can help them to understand what that means, a radical reform of their habits, let alone a real improvement of the fate of the earth, are not easily achieved. The seriousness of environmental problems and their global scale aggravate this limitation.

However, the limitation of a hermeneutical ethics might include an advantage as well. It is true that a hermeneutical ethics does not help much, at least not quickly, to solve practical problems. Instead of solving problems, this type of ethics is rather concerned with interpreting and clarifying the conditions and frameworks that determine those problems. This does not mean that a hermeneutical ethics could not have practical relevance. However, its importance for practice will rather be related to preventing future problems than to solving the present ones. By clarification of the conditions and frameworks of our current problems, a hermeneutical ethics may contribute to our self-understanding, and, by doing so, to the moral quality of our way of living in the long run.

We would like to point out one interesting example of this emphasis on prevention, which is, after all, always a long-term matter. If we regret this "long-term-character," we display the same activistic impatience, which probably is one of the causes for the problems we are facing. Activism is the attitude that considers reality dependent on our making and such an attitude is inevitably impatient: There is, after all, no reason why we should wait to have what is in our power to acquire. Activism and impatience might not only be a characteristic for our technological culture but also one for the kind of moral philosophy that attempts to solve the problems produced by that culture. Such problem-solving moral philosophy, although it may prescribe norms that put limits on our activity, runs the risk of repeating the very same impatient activism that it wants to restrict. A preventive moral philosophy, on the contrary, can only patiently prepare what it can foster, but not enforce.

Therefore, it could be important to elaborate an environmental ethics of patience. Patience is, after all, not resignation but a combination of resolute resistance with the power to endure what one fights against. In an old (Hellenistic) tradition, the virtue of patience is compared with natural processes. One of the terms we find for the concept of patience is *hypomonè*. This indicates that something is being maintained (*menein*), even ripens, *in* what is going under (*hypo*). Like the seed ripens by going under in the soil, so patience may be considered as a power that is becoming stronger in what may look like destruction. Environmental ethics might have to become more patient than it sometimes seems to be at present.

There is another limitation that is rather specific for a hermeneutical ethics of the environment. To put it in the form of a question: isn't "Nature" as such a limit to interpretation? Hermeneutics takes up a human point of view. This raises the question whether it can be extended to topics of environmental ethics that stretch to the nonhuman world. Williams claims that there is no outside view, no external access to the answer of what is morally right. An inside view cannot provide objective criteria either, but it can bring understanding. However, as humans we cannot understand the nonhuman world independent of ourselves.

In that sense, anthropocentrism, which traditionally many environmental ethicists tried to avoid, is inevitable. Since—as we stated before—there is no nature for us which is not always already an interpretation, it is impossible for human beings to "step outside" the interpretative sphere and fully grasp the "real" worth and needs of the natural environment in itself. But that a strict non-anthropocentrism is not achievable does not mean that we are sentenced to an egocentric stance. Rather, an environmental hermeneutical approach can try, among other things, to understand the ethical experience of persons who, like Minteer, hold that the natural environment has intrinsic value and that something other than oneself is important. Moreover, whereas "environmental ethical theory" has to make an effort to justify the talk of intrinsic value of the natural environment in the first place, environmental hermeneutics can simply take it as an ethical experience that is in want of understanding.

Conclusion

With the help of Williams' criticism of ethical theory, it has become apparent that environmental ethics, just as ethical theory in general, cannot give an objective foundation for its theories and encounters the problem of moral motivation: If people are not motivated in a certain way, arguments will not persuade them to change their ideas and behavior.

An environmental hermeneutics can meet these challenges. It does not try to step outside human experience and thus has an eye for what already motivates people. However, this means that environmental philosophies have to alter their ambitions. Environmental hermeneutics requires modesty and patience in regard to normative claims. Therefore, it has trouble dealing with the urgency and global scale of existing environmental problems.

However, Williams' criticism of ethical theory shows that traditional environmental ethics has these problems as well. But unlike traditional "environmental ethical theory," environmental hermeneutics *can* fruitfully deal with the issue of moral motivation and its insights are less likely to clash with a person's life as a whole. In addition, a patient approach to environmental issues can clarify the conditions and frameworks of our current problems and can thus contribute to our self-understanding and the moral quality of our way of living in the long run. Lastly, an environmental hermeneutics does not have to argue that its topics are of importance, but can use given ethical experiences as a starting point. Hence, even though it knows its limitations, a turn to environmental hermeneutics seems indeed a helpful one to take.

Notes

Introduction: Environmental Hermeneutics
David Utsler, Forrest Clingerman, Martin Drenthen, and Brian Treanor

1. See F. Nietzsche, *Kritische Studien Ausgabe* (KSA), Herausgegeben von G. Colli und M. Montinari (Berlin: DTV/De Gruyter, KSA 12.315, 1980).

2. For a more detailed description of this history, see Jean Grondin, *Introduction to Philosophical Hermeneutics* (New Haven, Conn.: Yale University Press, 1994); Richard E. Palmer, *Hermeneutics: Interpretation Theory in Schleiermacher, Dilthey, Heidegger, and Gadamer* (Evanston, Ill.: Northwestern University Press, 1969).

3. Robert Mugerauer, *Interpreting Environments: Tradition, Deconstruction, Hermeneutics* (Austin: University of Texas Press, 1995), xxvii.

4. It should be noted that there is a kinship between these descriptions and the field of ecosemiotics, which is also an emerging field but which has a slightly different focus. Ecosemiotics is interested in the role of signs and signifiers within the realm nature itself whereas environmental hermeneutics asks about the meaning the environment has for "us" as readers.

5. Kenneth Liberman, "An Inquiry into the Intercorporeal Relations between Humans and the Earth," in *Merleau-Ponty and Environmental Philosophy: Dwelling on the Landscapes of Thought*, ed. Suzanne L. Cataldi and William S. Hamrick (Albany: State University of New York Press, 2007), 38.

6. Hans-Georg Gadamer, *Truth and Method*, 2nd rev. ed., trans. Joel Weinsheimer and Donald G. Marshall (New York: Continuum Press, 2004), xxvii.

7. Gadamer, *Truth and Method*, xxvii; emphasis added.

8. David E. Linge, "Editor's Introduction," in Hans-Georg Gadamer, *Philosophical Hermeneutics*, trans. and ed. David E. Linge (Berkeley: University of California Press, 1976), xii.

9. Gadamer, *Philosophical Hermeneutics*, 7.

10. Gadamer, *Truth and Method*, 295.

11. Ibid., 470.

12. Cf. J. Baird Callicott, "Environmental Philosophy *Is* Environmental Activism," in *Beyond the Land Ethic: More Essays in Environmental Philosophy*, ed. J. Baird Callicott (Albany: State University of New York Press, 1999), 27–43. In much of this chapter, Callicott is critiquing the supposed divide between environmental philosophy and activism.

13. Gadamer, *Philosophical Hermeneutics*, 31.

14. Ibid., 30–31.

15. Gadamer, *Truth and Method*, 307.

16. See Charles S. Brown and Ted Toadvine, "Eco-Phenomenology: An Introduction," in *Eco-Phenomenology: Back to the Earth Itself*, ed. Charles S. Brown and Ted Toadvine (Albany: State University of New York Press, 2003), ix–xxi. See also Adrian Parr, *Hijacking Sustainability* (Cambridge, Mass.: The MIT Press, 2009) for an excellent analysis of "green washing" in terms of the language of sustainability.

1. Environmental Hermeneutics Deep in the Forest
John van Buren

An earlier version of this essay, written for an interdisciplinary conference on woodlands, "Land Life Lumber Leisure," organized by the geographer Anne Buttimer at the University of Ottawa, was published in *Environmental Ethics* 17:3 (1995) and appears in revised form in the present volume with the acknowledgment of the journal.

1. Paul Ricoeur, *The Conflict of Interpretations*, trans. Willis Domingo et al. (Evanston: Northwestern University Press, 1974).

2. Martin Heidegger, *Being and Time*, trans. John Macquarrie and Edward Robinson (New York: Harper & Row, 1962); Gadamer, *Truth and Method*; Paul Ricoeur, *Hermeneutics and the Human Sciences*, trans. John B. Thompson (Cambridge: Cambridge University Press, 1987); Jürgen Habermas, *Knowledge and Human Interests*, trans. Jeremy J. Shapiro (Boston: Beacon Press, 1971); Richard Rorty, *Philosophy and the Mirror of Nature* (Princeton, N.J.: Princeton University Press, 1979).

3. See Robert Frodeman, *Geo-Logic: Breaking Ground between Philosophy and the Earth Sciences* (Albany: State University of New York Press, 2003); Paul Rabinow and William M. Sullivan, eds., *Interpretive Social Science: A Reader* (London: University of California Press, 1979); David Seamon and Robert Mugerauer, eds., *Dwelling, Place & Environment: Towards a Phenomenology of Person and World* (New York: Columbia University Press, 1985).

4. Regarding the term "environmental hermeneutics," cf. Robert Mugerauer, "Language and the Emergence of Environment," in Seamon and Mugerauer, *Dwelling, Place & Environment: Towards a Phenomenology of Person and World*, 51–70; and Robert Mugerauer, *Interpreting Environments:*

Tradition, Deconstruction, Hermeneutics (Austin: University of Texas Press, 1995).

5. *Poetics* 1450b. Cf. Jim Cheney, "Postmodern Environmental Ethics: Ethics as Bioregional Narrative," *Environmental Ethics* 11 (1989): 117–134. For the role of narrative in hermeneutical theory in general, see Paul Ricoeur, *Time and Narrative*, 3 vols., trans. Katherine McLaughlin and David Pellauer (Chicago: University of Chicago Press, 1984).

6. Holmes Rolston, III, "Storied Residence on Earth," in Rolston, III, *Environmental Ethics: Duties to and Values in Natural World* (Philadelphia: Temple University Press, 1987), 341–354. For the general point that our self-understanding is interpretive and narrative, see Donald E. Polkinghorne, *Narrative Knowing and the Human Sciences* (Albany: State University of New York Press, 1988), 146–155.

7. See Ricoeur, *Hermeneutics and the Human Sciences.*

8. See the study of Ricoeur's hermeneutic arc in Dan R. Stiver, *Theology after Ricoeur* (Louisville: Westminster John Knox Press, 2001), chapter 2.

9. See Bill Devall and George Sessions, *Deep Ecology: Living as if Nature Mattered* (Salt Lake City: Gibbs Smith, 1985).

10. These and other features, as they relate to the interpretation of diverse environments, have been explored in, for example, Seamon and Mugerauer, eds., *Dwelling, Place & Environment: Towards a Phenomenology of Person and World* and Don Ihde, *Technology and the Lifeworld* (Bloomington: Indiana University Press, 1990).

11. "Knowledge" or "episteme" is to be taken here in Michel Foucault's sense of an underlying interpretive "worldview," where "knowledge" can therefore be used in the plural—"knowledges." See Michel Foucault, *The Archaeology of Knowledge*, trans. A. M. Sheridan Smith (New York: Pantheon Books, 1972), 191.

12. For this classificatory scheme and that of "people," "officials," and "academics," see Anne Buttimer, John van Buren, and Nancy-Hudson-Rodd, eds., *Land Life Lumber Leisure: Local and Global Concern in the Human Use of Woodland: An Interim Report* (Ottawa: Department of Geography, University of Ottawa, 1991). For complimentary classifications, see Ray Raphael, *Tree Talk: The People and Politics of Lumber* (Washington: Island Press, 1981), as well as the special issues on forests in *The Trumpeter: Journal of Ecosophy* 6, nos. 2–3 (1989) and 7, no. 2 (1990).

13. Cf. Linda H. Graber, *Wilderness as Sacred Space* (Washington: Association of American Geographers, 1976); Jay H. C. Vest, "Will-of-the-Land: Wilderness among Indo-Europeans," *The Trumpeter: Journal of Ecosophy* 3 (1986): 4–7; John C. Miles, "Wilderness as Healing Place," *The Trumpeter: Journal of Ecosophy* 3 (1986): 11–18.

14. Alasdair MacIntyre makes this point throughout his *After Virtue: A Study in Moral Theory* (Notre Dame, Ind.: University of Notre Dame Press, 1984).

15. Cf. the classifications of forest-related values given in Holmes Rolston, III, "Values Deep in the Woods," *The Trumpeter: Journal of Ecosophy* 6 (1989): 39–41; Donald VanDeVeer and Christine Pierce, eds., *The Environmental Ethics and Policy Book* (Belmont, Calif.: Wadsworth, 2003), 519–545.

16. World Commission on Environment and Development, *Our Common Future* (New York: Oxford University Press, 1987), 43.

17. Cf. J. Ronald Engel and Joan Gibb Engel, eds., *Ethics of Environment and Development* (Tucson: University of Arizona Press, 1990); the special issue on sustainable development in *The Trumpeter: Journal of Ecosophy* 5 (1988); Ingrid Leman Stefanovic, *Safeguarding Our Common Future: Rethinking Sustainable Development* (Albany: State University of New York Press, 2000).

18. For the general notion of truth as interpretive fittingness, see Paul Ricoeur, "On Interpretation," in *Philosophy in France Today*, ed. Alan Montefiore (Cambridge: Cambridge University Press, 1983), 175–189. Robert Mugerauer gives in effect the example of applying the biophysical criterion of fittingness to conflicting views of the Grand Canyon by explorers, tourists, native peoples, and government officials; see his "Language and the Emergence of Environment," 57ff.

19. For this historical criterion in hermeneutical theory, see Gadamer, *Truth and Method*, 274ff.; G. B. Madison, "Method in Interpretation," in his *The Hermeneutics of Postmodernity: Figures and Themes* (Bloomington: Indiana University Press, 1988), 25–39.

20. See Habermas, *Knowledge and Human Interests*, 301–317.

21. See Jürgen Habermas, *The Theory of Communicative Action, vol. 1: Reason and the Rationalization of Society* (Boston: Beacon Press, 1984).

22. Cf. John S. Dryzek, *Rational Ecology: Environment and Political Economy* (New York: Basil Blackwell, 1987) and his "Green Reason: Communicative Ethics for the Biosphere," *Environmental Ethics* 12 (1990): 195–210.

23. Cf. Cheney, "Postmodern Environmental Ethics: Ethics as Bioregional Narrative," 117–134.

24. Cf. Dryzek, "Green Reason: Communicative Ethics for the Biosphere," 205–207.

25. For this notion, cf. Richard J. Bernstein's *Beyond Objectivism and Relativism* (Philadelphia: University of Pennsylvania Press, 1983).

26. See the essays in Seamon and Mugerauer, eds., *Dwelling, Place & Environment*.

27. Cf. Cheney, "Postmodern Environmental Ethics: Ethics as Bioregional Narrative," 117–134; John Llewelyn, *The Middle Voice of Ecological Conscience* (London: Macmillan, 1991).

28. Martin Heidegger, *Martin Heidegger: Basic Writings* (New York: Harper & Row, 1977), 193.

2. Morrow's Ants: E. O. Wilson and Gadamer's Critique of (Natural) Historicism
Mick Smith

1. See Mick Smith, "Edward Hyams: Ecology and Politics 'Under the Vine.'" *Environmental Values* 20 (2011): 95–119.

2. Edward Hyams, *Morrow's Ants* (London: Allen Lane, 1975).

3. This particular understanding of "Natural History" will be capitalized to distinguish it from both biological natural history and specific uses like the social-theoretical understandings of Adorno and Benjamin. See Beatrice Hanssen, *Walter Benjamin's Other History: Of Stones, Animals, Human Beings and Angels* (Berkeley, Calif.: University of California Press, 1998).

4. Rémy Chauvin, *The World of Ants: A Science-Fiction Universe* (New York: Hill and Wang, 1969). Edward Hyams and George Ordish, *The Last of the Incas* (London: Longmans, 1963).

5. Edward O. Wilson, *Sociobiology: The New Synthesis* (Cambridge, Mass.: Harvard University Press, 1975), 379.

6. See also Bert Hölldobler and Edward. O. Wilson, *The Superorganism: The Beauty, Elegance and Strangeness of Insect Societies* (New York: Norton, 2008).

7. Wilson shared a Pulitzer Prize for Bert Hölldobler and Edward O. Wilson, *The Ants* (Cambridge, Mass.: Belknap Press, 1990). He also has another Pulitzer for Edward O. Wilson, *On Human Nature* (Cambridge, Mass.: Harvard University Press, 1978).

8. Marshall Sahlins, *The Use and Abuse of Biology: An Anthropological Critique of Sociobiology* (Ann Arbor: University of Michigan Press, 1977).

9. Issues treated at length in Stephen Rose, Richard C. Lewontin, and Leon J. Kamin, *Not in Our Genes: Biology, Ideology and Human Nature* (Harmondsworth: Penguin, 1985); and Elliot Sober, *Philosophy of Biology* (Boulder, Colo.: Westview Press, 1993).

10. Wilson, *On Human Nature*, 208.

11. Edward O. Wilson, *Biophilia* (Cambridge, Mass.: Harvard University Press, 1984).

12. Edward O. Wilson, *Consilience: The Unity of Knowledge* (London: Little, Brown & Co., 1998), 279.

13. Ibid., 296.

14. Wilson quoting Hilbert, ibid., 47.

15. Wilson, *On Human Nature*, 208.

16. Wilson, *Consilience*.

17. Wilson, *On Human Nature*, 204.

18. Ibid., 207.

19. Ibid.

20. Ibid., 208.

21. Wilson, *On Human Nature*, 208; emphasis added.

22. Ibid.

23. This also potentially undermines Wilson's influential argument for preserving nature on the basis of our inherent biophilia, for the "people" of tomorrow might just engineer their way around this penchant for natural diversity, too.

24. Wilson, *On Human Nature*, 207.

25. Interestingly, Wilson was bought up as a Southern Baptist and at "the age of fifteen was 'born again' at a revivalist meeting." Michael Ruse, *Monad to Man: The Concept of Progress in Evolutionary Biology* (Cambridge, Mass.: Harvard University Press, 1996), 516.

26. Wilhelm Dilthey, *Introduction to the Human Sciences* (Detroit: Wayne State University Press, 1988), 15.

27. Jean Grondin, *The Philosophy of Gadamer* (Montreal: McGill-Queen's University Press, 2003), 68.

28. I will simply adopt the term "human sciences" despite the odd ring it has in English, since this is how *Geisteswissenschaften* is usually translated. Dilthey actually adopted this term to translate John Stuart Mill's "moral sciences"—but this might sound even stranger.

29. For example, Steven Shapin, *A Social History of Truth: Civility and Science in Seventeenth-Century England* (Chicago: The University of Chicago Press, 1994); Bruno Latour, *Pandora's Hope: Essays on the Reality of Science Studies* (Cambridge, Mass.: Harvard University Press, 1999).

30. Caryl P. Haskins, *Of Ants and Men* (New York: Prentice-Hall, 1939).

31. Ruse refers to Wheeler as Wilson's "intellectual grandfather," *Monad to Man*, 513. A. J. Lustig, "Ants and the Nature of Nature in Auguste Forel, Erich Wasmann, and William Morton Wheeler," in *The Moral Authority of Nature*, ed. Lorraine Daston and Fernando Vidal (Chicago: The University of Chicago Press, 2004).

32. Matt Ridley, *The Origins of Virtue* (London, Viking, 1996), 12.

33. Ibid., 17.

34. Ibid., 33; emphasis added.

35. Wilson and Hölldobler's definition of a superorganism is "a society, such as a eusocial insect colony, that possesses features of organization *analogous* to the physiological properties of a single organism" (*Superorganism*, 513; emphasis added). Just what level of social integration and how closely the analogy has to hold is, they admit "subjective" (514), hence presumably not scientific by Wilson's own definition.

36. "When we study the works published by Gadamer throughout the 1950s, we realize that they are dominated by this Diltheyan problem of historical consciousness and truth in history. . . . We can also say that the whole of Gadamer's work begins with Dilthey, to which his hermeneutics is the reply" (*Philosophy of Gadamer*, 67). Gadamer, *Truth and Method*.

37. Gadamer, *Truth and Method*, xxxv.

38. See Forrest Clingerman's chapter in this collection.

39. Alrlo Vanderjagt and Klass Van Berkel, *The Book of Nature in Antiquity and the Middle Ages* (Leuven: Peeters, 2005). Klass Van Berkel and Arlo

Vanderjagt, *The Book of Nature in Early Modern History* (Leuven: Peeters, 2006). Forrest Clingerman, "Reading the Book of Nature: A Hermeneutical Account of Nature for Philosophical Theology," *Worldviews* 13 (2009): 72–91.

40. Although there is, of course, a debate about whether evolution might, despite occasional mass extinctions and cataclysmic events, be understood in general as bringing about increasing levels of biological and ecological *complexity* over time, this is very different from the kind of "progress" Wilson's consilience envisages. That said, it is probably not coincidental that, as Ruse notes, "[T]he most overt, living Darwinian enthusiast for absolute progress is E. O. Wilson. He is convinced that there has been such progress that we [humanity] have won the race." Michael Ruse, "Evolution and Progress," in *The Philosophy of Biology*, ed. David. L. Hull and Michael Rose (Oxford: Oxford University Press, 1998), 620.

41. Francis Fukuyama, *The End of History and the Last Man* (New York: Free Press, 1992).

42. Gadamer, *Truth and Method*, xxiii.

43. Grondin, *Philosophy of Gadamer*, 68.

44. Gadamer, *Truth and Method*, 239.

45. Wilson, *Consilience*, 45.

46. Ibid., 43.

47. Gadamer, *Truth and Method*, xxi–xxii.

48. In which case, a real question is whether and how Wilson's way of recognizing individual finitude makes a difference to self-understandings (and hence understandings of ethics, politics, and society) that have always recognized that finitude in other ways. This, of course, might be exactly the kind of "existential" and socio-theoretical question that someone like Foucault might chose to raise, and that Wilson thinks we should simply walk away from.

49. This is the title of part 3, chapter 3 of *Truth and Method*.

50. Martin Heidegger, "The Way to Language," in *Basic Writings* (London: Routledge, 1993), 424.

51. Ibid., 423.

52. Gadamer, *Truth and Method*, xxxi.

53. Words do not re-present things more or less accurately (whatever that might mean) in some immaterial realm of thought; they (incompletely) expose the history of a things appearance in the world.

54. Dermot Moran, *Introduction to Phenomenology* (London: Routledge, 2000), 282.

55. Mick Smith, "Lost for Words? Gadamer and Benjamin on the Nature of Language and the 'Language' of Nature," *Environmental Values* 10, no. 1 (2001): 59–75.

56. Gadamer compounds this by sometimes speaking of the "fusion of horizons" as leading to a greater *universality*, which might be misread as suggesting some overall or absolute epistemic progress.

57. Hannah Arendt, *The Human Condition* (Chicago: The University of Chicago Press, 1958), 178. See also Mick Smith, "Ecological Citizenship

and Ethical Responsibility: Arendt, Benjamin and Political Activism,"
Environments 33, no. 3 (2005): 51–64.

58. Wilson, *Sociobiology*, 383.

59. Edward O. Wilson, *Anthill: A Novel* (New York: W. W. Norton, 2010).

3. Layering: Body, Building, Biography
Robert Mugerauer

1. Robert Mugerauer, "Deleuze and Guattari's Return to Science as a
Basis for Environmental Philosophy," in *Rethinking Nature: Essays in
Environmental Philosophy*, ed. Bruce V. Foltz and Robert Frodeman, 180–204
(Bloomington: Indiana University Press, 2004).

2. Ted Relph, *Place and Placelessness* (London: Pion, 1976); Karsten
Harries, "Thoughts on a Non-Arbitrary Architecture," *Perspecta* 20 (1983):
9–20; Robert Mugerauer, "Toward a Phenomenology of Midwestern Yards,"
Places 2, no. 2 (1984): 31–38.

3. Christian Norberg-Schulz, *Genius Loci* (New York: Rizzoli, 1979);
Kim Dovey, "Dwelling, Archetype, and Ideology," in *Dwelling*, ed. Robert
Mugerauer, 10–19 (Austin: University of Texas Press, 1993).

4. Robert Mugerauer, "The City: A Legacy of Organism-Environment
Interactions at Every Scale," in *The Natural City: Re-envisioning the Built
Environment*, ed. Ingrid Stefanovic and S. Scharper, 173–194 (Toronto:
University of Toronto Press, 2011).

5. Robert Mugerauer, "Language and the Emergence of Environment,"
in *Dwelling, Place and Environment*, ed. David Seamon and Robert
Mugerauer, 51–70 (Dordrecht: Nijhoff, 1985), see esp. 68.

6. See, in this volume, Brian Treanor, "Narrative and Nature"; Nathan
Bell, "Environmental Hermeneutics With and For Others"; and Martin
Drenthen, "New Nature Narratives."

7. Mugerauer, "The City."

8. Robert Mugerauer, "Anatomy of Life and Well-Being: A Framework for
the Contributions of Phenomenology and Complexity Theory," *International
Journal of Qualitative Studies of Health & Well-Being*, July 2010, <5:5097-
DOI: 10.3402/qhw.v512.5097>.

9. See, in this volume, Forrest Clingerman, "Memory, Imagination, and
the Hermeneutics of Place."

10. Robert Mugerauer, "Toward a Phenomenological Hermeneutics of
Rivers: A Way to Integrate Design, Ecology, and Politics," paper presented at
the 15th annual meeting of the International Association for Environmental
Philosophy, Philadelphia, October, 2011.

11. Merleau-Ponty, *Nature: Course Notes from the Collège de France*,
trans. Robert Vallier (Evanston, Ill.: Northwestern University Press, 2003), 227.

12. Matthew Biro, *Anselm Kiefer and the Philosophy of Martin Heidegger*
(New York: Cambridge University Press, 1998), 194–199.

13. Merleau-Ponty, *Nature,* 305.

14. Ibid., xvi, 204, 212.

15. Ibid., 208.

16. Ibid.

17. Ibid., 212, 219. Though he repeats the motif of layering and leaves in doing so (perhaps another chiasm, criss-crossing?), in *The Visible and the Invisible* (1959–1960) Merleau-Ponty corrects the idea of language as "sedimented positivities" and the body as made of two leaves:

> body . . . flesh of the visible . . . carnal being, as a being of depths, of several leaves or several faces, a being in latency. . . . We say therefore that our body is a body of two leaves, from one side a thing among things and otherwise what sees them and touches them; . . . [but] one should not even say, as we did a moment ago, that the body is made up of two leaves, of which the one, that of the 'sensible', is bound up with the rest of the world. There are not in it two leaves or two layers; fundamentally it is neither thing seen only nor seer only. . . . And as such it is not in the world. . . . To speak of leaves or of layers is still to flatten and to juxtapose, under the reflective gaze, what coexists in the living and upright body. . . . If one wants metaphors, it would be better to say that the body sensed and the body sentient are as the obverse and reverse, or again as two segments of one sole singular course, which goes above from left to right and from below from right to left, but which is but one sole movement in its two phases. (*The Visible and the Invisible,* trans. Alphonso Lingis [Evanston, Ill.: Northwestern University Press, 1968], 135, 137–38.) Compare with Lefebvre, see note xxxviii.

18. See Mugerauer, "The City"; "Anatomy of Life and Well-Being."

19. Biro, *Anselm Kiefer and the Philosophy of Martin Heidegger,* 172–174; see also William Anderson, *Green Man: The Archetype of our Oneness with the Earth* (San Francisco: HarperCollins, 1990).

20. Monty, 1979, cited in Harold J. Morowitz, *Beginnings of Cellular Life: Metabolism Recapitulates Biogenesis* (New Haven, Conn.: Yale University Press, 1992), 33.

21. Morowitz, *Beginnings of Cellular Life,* 32–33. In hearing that such mats "have to be cemented or they will rot" (33), we find another case of the language of self-organization, here seemingly parallel to Manuel DeLanda's interesting explication of Deleuze and Guattari's double-action of assemblage by means of "sorting and cementing" (Manuel DeLanda, *A Thousand Years of Nonlinear History* [New York: MIT Press, 1997], 57–62). Similarly, we are told that "[s]cores of fragile alliances between amino acids act in concert to stabilize the folded protein, resulting in a three-dimensional configuration . . ." (Debra Niehoff, *The Language of Life: How Cells Communicate in Health and Disease* [Washington: Joseph Henry Press, 2005], 13).

22. Morowitz, *Beginnings of Cellular Life,* 55.

23. Niehoff, *The Language of Life*, 13; Dee Silverthorn, *Human Physiology: An Integrated Approach*, Second Edition (Upper Saddle River, N.J.: Prentice Hall, 2001), 25.

24. Silverthorn, *Human Physiology*, fig. 2-11.

25. Ibid., 23; fig. 2-8.

26. Ibid.

27. Ibid., 24.

28. Ibid., 26.

29. Ibid., 29, 48.

30. Of course, the self-organization of three-dimensionality is a complex topic. We would need to add more on the structures of protein molecules: the secondary structures of the ά-helix and β-pleated sheet; the genuinely three-dimensional tertiary structures of fibrous proteins and globular proteins (and how the latter two are, respectively, insoluble and soluble in water) (Silverthorn, *Human Physiology*, 32–33). Similarly, there is the role of the weak hydrogen bonds occurring among neighboring molecules in the folding back of proteins on themselves (25). Then, too, there are the historically important advances in molecular biology such as the work by Monod on the primary and globular structures of proteins in "molecular ontogenesis" (Mugerauer, "Deleuze and Guattari's Return to Science").

31. Silverthorn, *Human Physiology*, fig. 3-5.

32. Niehoff, *The Language of Life*, 14–15, 65–67.

33. Ibid., 13.

34. Ibid., 46–47.

35. Silverthorn, *Human Physiology*, fig. 12-3.

36. www.history.com/search.do?searchText=underground+cities (accessed in September 2007).

37. Grzegorz Micula and Magdalena Micula, *Cyprus: Eyewitness Travel Guides* (New York: DK Publishing, 2006).

38. Henri Lefebvre, *The Production of Space* (New York: Blackwell, 1991), 86–87. It needs to be pointed out, however, that Lefebvre warns us that "[f]igurative terms such as 'sheet' and 'stratum' have serious drawbacks: being metaphorical rather than conceptual, they assimilate space to things and thus relegate its concept to the realm of abstraction. Visual boundaries, such as walls or enclosures in general, give rise for their parts to an appearance of separation between spaces where in fact what exists is an ambiguous continuity." He prefers the analogy of hydrodynamics (*The Production of Space*, 87); cf. Merleau-Ponty, *The Visible and the Invisible*, see note xvi.

39. Anelia Jaffé, "Symbolism in the Visual Arts," in *Man and His Symbols*, ed. Carl Jung (New York: Dell, 1970), 169.

40. Ibid., 170.

41. Gaston Bachelard, *The Poetics of Space* (New York: Orion, 1964), 18.

42. Alfonso Ortiz, *The Tewa World: Space, Time, Being, and Becoming in a Pueblo Society* (Chicago: The University of Chicago Press, 1969); David

Saile, "Many Dwellings: Views of a Pueblo World," in *Dwelling, Place and Environment*, ed. David Seamon and Robert Mugerauer (Dordrecht: Nijhoff, 1985), 159–182, see esp. 165–166.

43. Sebald, *The Emigrants* (London: Harvill Press, 1996), 160–162.

44. Another artist-architect who works by layering is Christos Hajichristos, where the attitude-strategy especially pervades his paintings and media work.

45. Christine Lampert, *Frank Auerbach: Exhibition, Hayward Gallery, London, 4 May-2 July, 1978* (London: The Council, 1978), 13.

46. Isabel Carlisle, "People," "Landscapes," and "Drawings," in *Frank Auerbach: Painting and Drawing 1954-2001*, ed. Christine Lampert, Norman Rosenthal, and Isabel Carlisle (London: Royal Academy of the Arts, 2001), 62.

47. Christine Lampert, "Auerbach and His Sitters," in *Frank Auerbach: Painting and Drawing 1954-2001*, ed. Christine Lampert, Norman Rosenthal, and Isabel Carlisle (London: Royal Academy of the Arts, 2001), 18–33, see esp. 19.

48. Lampert, "Auerbach and His Sitters," 25.

49. Carlisle, "Drawings," 124.

50. Christine Lampert and Kate Austin, "Chronology," in *Frank Auerbach: Painting and Drawing 1954-2001*, ed. Christine Lampert, Norman Rosenthal, and Isabel Carlisle (London: Royal Academy of the Arts, 2001), 146–151, see esp. 147.

51. Mark McCulloh, *Understanding W. G. Sebald* (Columbia: University of South Carolina Press, 2003), xv–xvi.

52. Hanna Arendt, *The Human Condition* (Chicago: University of Chicago Press, 1958).

53. Carlisle, "Landscapes," 100; Martin Swales, "Theoretical Reflections on the Work of Sebald," in *W. G. Sebald—A Critical Companion*, ed. J. J. Long and Anne Whitehead (Seattle: University of Washington Press, 2004), 23–28, see esp. 26.

54. See Robert Mugerauer, "The Double-Gift: Place and Identity," in *Back to the Things Themselves: Architectural Experience, Memory, and Thought*, ed. Iris Aravot (Haifa: Technion University Press, 2011).

55. Clingerman, "Memory, Imagination, and the Hermeneutics of Place."

4. Might Nature Be Interpreted as a "Saturated Phenomenon"?
Christina M. Gschwandtner

1. When I speak of "natural" phenomena here (especially as distinguished from the human), I do not mean to imply that there is no continuity between the human and the nonhuman or that humans (and human artifacts) are somehow not "natural." Social ecology in particular seeks to overcome these distinctions between "natural" and "social" and to establish a greater

continuity between these realms. Indeed, several essays in the present volume address this issue explicitly or implicitly. My intent here rather is to overcome precisely the implied contention in Marion's work that "saturated phenomena" are primarily or even exclusively human or cultural (associated with history, art, or religion) and to "extend" this depiction in some sense to include the nonhuman and our experience of the nonhuman other. See also the comments about anthropocentrism at the end of this chapter.

2. Jean-Luc Marion, *Being Given: Toward a Phenomenology of Givenness*, trans. Jeffrey L. Kosky (Stanford, Calif.: Stanford University Press, 2002), 126. To be exact, there are actually two brief mentions of trees here. In the first, Marion speaks of the "intentional objects" tree or triangle, which are objects in the flux of lived experiences like the objects of mathematics and indicates that a distinction between natural, temporal, or purely theoretical objects "is not pertinent here" (126). Shortly after that, he briefly suggests the notion of "habitual phenomena" (an idea to which he never returns) and mentions a tree in the desert (which most interestingly "opens a world," left entirely unexplored) together with a taxi as two phenomena that "impose" themselves on me as I am in need of them. It is very difficult to ascertain exactly what status these examples (and many others, primarily of "technical objects") have here, as they are used to develop the "anamorphosis" of the given phenomenon which later becomes applied to saturated phenomena. In the conclusion to the section, Marion indicates that instead of a firm "either/or" distinction between poor and saturated phenomena, phenomenality might be envisioned to increase in "degrees of givenness" and he suggests that these objects mentioned earlier might hence be a very lowly first degree of givenness. He never returns to explore this notion further, something Anthony Steinbock criticizes more generally in "The Poor Phenomenon: Marion and the Problem of Givenness," in *Words of Life: New Theological Turns in French Phenomenology*, ed. Bruce Ellis Benson and Norman Wirzba (New York: Fordham University Press, 2010).

3. A recent dissertation-turned-book examines the possibility that creation might be conceived as a "gift," drawing on Derrida's and Marion's phenomenological accounts of the gift. The author tries to develop what he calls an "oscillational eco-ethos" by suggesting that a "double movement of acceptance and return" can get beyond the aporia of the gift or at least hold it in continual tension. In that sense, all of "what-is" can be conceived as a gift that should be loved. Mark Manolopoulos, *If Creation is a Gift* (Albany: State University of New York Press, 2009).

4. It is interesting that all these saturated phenomena are either human creations or describe human activities, thus are all clearly connected to the human in some way (including the excessive phenomenon of revelation which is a *human* experience of the divine). Although Marion claims that they come to us from the "unseen" (as discussed further in this chapter), they are made "visible" by the human recipient.

5. I am here evoking the progression in which ethical consideration tends to be extended: first to "higher" animals (especially those most like us), then other animals, then plants or all living things (biocentrism), finally entire ecosystems (ecosystem and deep ecology). H. Peter Steeves makes a rare attempt to extend ethical consideration even further in "Mars Attacked! Interplanetary Environmental Ethics and the Science of Life," chapter 7 of his *The Things Themselves: Phenomenology and the Return to the Everyday* (Albany: State University of New York Press, 2006), 126–45.

6. Marion is generally rather negative about hermeneutics and has been severely criticized for his neglect or even explicit exclusion of hermeneutics from his project. Two early critiques include Jean Grondin, "La tension de la donation ultime et de la pensée herméneutique de l'application chez Jean-Luc Marion," *Dialogue* 38 (1999) and Jean Greisch, "L'herméneutique dans la 'phénoménologie comme telle': Trois questions à propos de *Réduction et donation*," *Revue de métaphysique et de morale* 96, no. 1 (1991). More recently, the lack of hermeneutics in Marion's account is the main criticism in Shane Mackinlay, *Jean-Luc Marion, Saturated Phenomena, and Hermeneutics* (New York: Fordham University Press, 2010).

7. It first appeared with a collection of other papers by Henry, Chrétien, and Ricoeur responding to Janicaud's claim, translated by Thomas Carlson in Dominique Janicaud et al., *Phenomenology and the "Theological Turn": The French Debate* (New York: Fordham University Press, 2000), 176–216. It is retranslated and reprinted as chapter 2 in Jean-Luc Marion, *The Visible and the Revealed*, trans. Christina Gschwandtner (New York: Fordham University Press, 2002), 18–48.

8. He reiterates this in his summary in "The Banality of Saturation": "Now, the entire question of the saturated phenomenon concerns solely and specifically the possibility that certain phenomena do not manifest themselves in the mode of objects and yet still do manifest themselves. The difficulty is to describe what would manifest itself without our being able to constitute (or synthesize) it as an object (by a concept or an intentionality adequate to its intuition)" (Marion, *The Visible and Revealed*, 122).

9. Marion, *Being Given*, 159.

10. Saturated phenomena are therefore events that escape causality: "[P]henomena as such, namely as given, not only do not satisfy this demand [being understood as effects of a cause], but far from paying for their refusal with their unintelligibility, appear and let themselves be understood all the better as they slip from the sway of cause and the status of effect. The less they let themselves be inscribed in causality, the more they show themselves and render themselves intelligible as such" (ibid., 162).

11. Marion explicates the notion of "counter-experience" in more detail in section 6 of "The Banality of Saturation" (*The Visible and Revealed*, 136–139). Counter-experience turns intentionality back on itself and thus "measures the range of my disappointed vision." It can be detected in the

way in which it "alters" the aim of intentionality (ibid., 137). It is also "marked by the saturation of every concept by intuition" and thus a kind of disappointment of the original aim of intentionality (ibid., 138). Finally, it is an experience of excess that cannot be grasped but whose effect can be felt. Thus, "counter-experience is an issue of the obstinate resistance of what refuses itself to knowledge that is transparent without remainder, of what withdraws into its obscure origin" (ibid., 138). This is taken up again in the conclusion of *Certitudes négatives* (Paris: Grasset, 2010). Counter-experience attests to the saturation of the phenomenon (314).

12. Jean-Luc Marion, *In Excess: Studies of Saturated Phenomena*, trans. Robyn Horner and Vincent Berraud (New York: Fordham University Press, 2002), 113; translation slightly modified.

13. This is first outlined in *Being Given* (225–247) and then examined in more detail in the following study *In Excess* which devotes a chapter to each of the saturated phenomena.

14. Marion repeatedly insists that these give everything to visibility at once on their bedazzling surface that there is no "back" to them. Thus it would be much harder to show how sculptures or music might serve as examples for this type of phenomenon.

15. For this important distinction, see *Being Given*, 367n90. Marion tries to mark the difference between the mere phenomenological possibility of revelation and the actuality of Revelation accessed only by theology by capitalizing only the latter. Several commentators (beginning with Thomas Carlson and most recently Mackinlay) have argued that this distinction cannot be maintained as cleanly as Marion suggests.

16. Marion provides an analysis of Christ's transfiguration as an illustration of saturation in all four respects. See *Being Given*, 234–245.

17. This is a claim Marion later seems to qualify without explicitly admitting this. See his essay on "The Banality of Saturation" discussed later in this chapter.

18. Marion draws heavily on his previous analyses of the saturated phenomenon in his work on *The Erotic Phenomenon*, trans. Stephen E. Lewis (Chicago: The University of Chicago Press, 2007), although he does not identify it explicitly as a saturated phenomenon.

19. For the most recent such essay which establishes an explicit relationship between the Eucharist and the phenomenology of givenness, see "The Phenomenality of the Sacrament—Being and Givenness," in *Words of Life*, 89–102.

20. See chapters 3 and 4 of Marion's recent work *Certitudes négatives*.

21. Yet, Marion does briefly reiterate the four types of saturated phenomena in the conclusion (*Certitudes négatives*, 312–313), thus suggesting that despite his analyses of other saturated phenomena within the book, such as those of the gift and of sacrifice, he has not rescinded the earlier account in terms of the four types of saturation. The phenomenon of revelation is not mentioned in this context.

22. To some small extent, this may even be true of all phenomena to some degree. (See second footnote and later in this chapter.)

23. It is interesting that despite all of Marion's reluctance about hermeneutics, he actually has quite a bit to say about language.

24. Marion, *In Excess*, 33 (for the event) and 123–127 (for the icon).

25. Marion, *In Excess*, 33.

26. Ibid., 126. He also argues that "the face of the other person compels me to believe in my own eternity, like a need of reason or, what comes back to the same thing, as the condition of its infinite hermeneutic" (ibid., 127).

27. This is one of the main arguments of Mackinlay's *Interpreting Excess*. He first outlines his criticism in chapter 2 and then discusses the event specifically in chapter 4.

28. This is the final line of Part V (on the devoted recipient of the saturated phenomenon) of *Being Given* (319). It is striking that Mackinlay, who accuses Marion of collapsing this tension in favor of the phenomenon and making the recipient entirely passive, never cites this phrase which clearly maintains the paradoxical tension. While Mackinlay is right to criticize the lack of hermeneutics in Marion, I think he is wrong in his claim that Marion never considers a more active role for the recipient.

29. Or "is given"—in the French *se donne* can indicate both the reflexive and the passive.

30. Marion uses this and similar imagery in his first depiction of the *adonné* (*Being Given*, 264–265, 283–290). It is this imagery in particular which leads Mackinlay to criticize Marion for assigning too much activity to the phenomenon.

31. *In Excess* describes this role of the recipient in terms of resistance (49–53). This is also the terminology used in "The Banality of Saturation."

32. In fact, in one case Marion depicts this as the opening of a world:

> The event thus attests its nonconstitutability by constituting me, myself, its effect. Whence the third surprise: the event that comes forward with this pleasure is not summed up in it. At issue is the remembrance of the narrator's entire past, not only a reactivation of his memory (secondary retention), but the return of the living present in and through present retention of a past that, at the moment it was lived, was not even remembered or perceived. The event announced by this pleasure provokes, immeasurably beyond, the arising of a world, the world. The event prompts not only the memory of an individual (the narrator), nor just the work in which this past would again become a living present . . . but precisely the total world of history. (*Being Given*, 170)

He contends that events open horizons (ibid., 172).

33. "The Banality of Saturation" is chapter 7 of *The Visible and the Revealed*. Marion quotes Benoist's question "What will you say to me if I say to you that where you see God, I see nothing?" and then responds: "Indeed,

what should I say? Yet the force of the argument can be turned against the one who uses it, for the fact of not comprehending and seeing nothing should not always or even most often disqualify what it is a question of comprehending or seeing, but rather the one who understands nothing and sees only a ruse." *The Visible and Revealed*, 124.

34. Marion, *The Visible and Revealed*, 126; emphasis original.

35. Ibid., 126; emphasis added.

36. Ibid., 131.

37. Ibid., 133. This is the aforementioned "counter-experience." Ibid., 134.

38. Ibid., 136.

39. Ibid., 143; emphasis original.

40. Ibid., 144.

41. "'Christian Philosophy': Hermeneutic or Heuristic?" originally appeared in the collection *The Question of Christian Philosophy Today*, ed. Francis J. Ambrosio, and is retranslated and reprinted as chapter 4 of *The Visible and Revealed*.

42. In fact, Marion presents the idea that the artist makes visible what was previously unseen fairly consistently in his writings on art. He also speaks of the recipient more generally as receiving the "unseen" (via resistance to it) throughout *In Excess* to the point that this becomes the primary function of the *adonné* in this book.

43. *Certitudes négatives*, 313; emphasis original.

44. I have chosen to draw my examples mostly from other pieces in this volume, since the point of the essay is to show primarily that natural phenomena can be interpreted as saturated in Marion's terms and is not an analysis of ecological issues per se. All references to other chapters in the volume will be given in parentheses within the text.

45. It is also interesting that Marion designates Kant's notion of the sublime as a precedent for his analysis of the saturated phenomenon, focusing solely on its sense of being overwhelming. Yet most of Kant's examples for the sublime (at least in the Third Critique) are actually natural phenomena: imposing mountains, great thunderstorms, and so forth. (What is also somewhat ironic is that Kant's analysis seems to imply that we actually perceive ourselves as greater than the phenomenon in the experience of the sublime, while Marion clearly implies the opposite.)

46. See Forrest Clingerman, "Memory, Imagination, and the Hermeneutics of Place," in this volume.

47. See Martin Drenthen, "New Nature Narratives: Landscape Hermeneutics and Environmental Ethics," in this volume.

48. See Janet Donohoe, "The Betweenness of Monuments," in this volume.

49. Bernd Heinrich, *A Year in the Maine Woods, The Trees in my Forest, Mind of the Raven, One Man's Owl*, and various other texts.

50. In fact, Leopold's *Land Ethic* is convincing precisely because of the detailed descriptions with which the book has moved us. Some commentators

have argued that the final chapter of the *Land Ethic* (the only one that usually gets read in environmental ethics classes) only makes sense within and because of those prior descriptions of the book.

51. Annie Dillard, *Pilgrim at Tinker Creek* (New York: HarperCollins, 1974), 35–36. (If this does not qualify as an experience of a saturated phenomenon, I'm not sure what would.)

52. See Brian Treanor, "Narrative and Nature: Appreciating and Understanding the Nonhuman World," in this volume.

53. Marion, *The Visible and Revealed*, 124.

54. Idem, *In Excess*, 51, 52.

55. The closest English translation would be "believing is seeing," but the French contains no reference to "being"—a concept of which Marion is rather critical throughout his work.

56. McGrath stresses the importance of such context in his consideration of Gadamer's account of historically effected consciousness.

57. See Brian Treanor, "Narrative and Nature: Appreciating and Understanding the Nonhuman World," in this volume.

58. See Forrest Clingerman, "Memory, Imagination, and the Hermeneutics of Place," in this volume.

59. Marion, *Being Given*, 167.

60. Ibid., 168.

61. See Mick Smith, "Morrow's Ants: E. O. Wilson and Gadamer's Critique of (Natural) Historicism," in this volume.

62. The "spirit of abstraction" Treanor mentions seems to fit Marion's analysis of the "poor" or "technical" object rather well.

63. This also highlights again that nature is not a technical object (despite Marion's conflation of the two in the example of the tree) and thus may justifiably be regarded as a saturated phenomenon.

64. "It could even be said that the world is covered with an invasive and highly visible layer of poor phenomena (namely, the technical objects produced and reproduced without end), which ends up eclipsing what it covers over." Earlier in the same paragraph he says, "if these phenomena with no or poor intuition assume the status of technically produced objects (which is most frequently the case), their mode of production demands no other intuition than that which gives us their material (a material that itself becomes at once perfectly appropriate to each 'concept' and available in an in principle limitless quantity)." Marion, *The Visible and Revealed*, 125. At times he links this to the inherent disposability of the technical object. The distinction between the "design" of technical objects and the witness to the saturated phenomenon is found on page 143.

65. Idem, *Being Given*, 129.

66. Marion is also very much influenced by Heidegger on this point. See, for example, his analysis in *Being Given*, 126–130, which relies heavily on the "ready-to-hand" or *The Visible and Revealed*, 150.

67. Marion, *The Visible and Revealed*, 151.

68. See Nathan Bell, "Environmental Hermeneutics With and For Others: Ricoeur's Ethics and the Ecological Self," in this volume.

69. See Brian Treanor, "Narrative and Nature: Appreciating and Understanding the Nonhuman World," in this volume.

70. See Forrest Clingerman, "Memory, Imagination, and the Hermeneutics of Place," in this volume.

71. Derrida has done some interesting work on this question in *The Animal That Therefore I am* (New York: Fordham University Press, 2008) and his final seminar at the École des hautes études en sciences sociales published as *The Beast and the Sovereign*, trans. Geoffrey Bennington (Chicago: University of Chicago Press, 2009).

72. See Mick Smith, "Morrow's Ants: E. O. Wilson and Gadamer's Critique of (Natural) Historicism," in this volume.

73. See Brian Treanor, "Narrative and Nature: Appreciating and Understanding the Nonhuman World," in this volume.

74. Marion, *Being Given*, 159; emphasis original.

75. Steeves, *The Things Themselves*, 62.

76. See Brian Treanor, "Narrative and Nature: Appreciating and Understanding the Nonhuman World," in this volume.

77. See Nathan Bell, "Environmental Hermeneutics With and For Others: Ricoeur's Ethics and the Ecological Self," in this volume.

78. In fact, Marion often speaks of the call proceeding from the other (or more generally, the saturated phenomenon) precisely as an "injunction." Both Marion and Levinas would, of course, be far more hesitant than Ricoeur or Bell to establish "similitude" between self and other.

79. See Forrest Clingerman, "Memory, Imagination, and the Hermeneutics of Place," in this volume.

80. See David Utsler, "Environmental Hermeneutics and Environmental/ Eco-Psychology: Explorations in Environmental Identity," in this volume.

81. See Nathan Bell, "Environmental Hermeneutics With and For Others: Ricoeur's Ethics and the Ecological Self," in this volume.

82. See Martin Drenthen, "New Nature Narratives: Landscape Hermeneutics and Environmental Ethics," in this volume.

5. Must Environmental Philosophy Relinquish the Concept of Nature? A Hermeneutic Reply to Steven Vogel
W. S. K. Cameron

1. Steven Vogel first introduces his perspective in an early article, "Marx and Alienation from Nature," *Social Theory and Practice* 14 (1988): 367–387, and develops it at greater length in *Against Nature: The Concept of Nature in Critical Theory* (Albany: State University of New York Press, 1996). I will focus, however, on the fruit Vogel harvests for environmental

philosophy in three later articles: "Nature as Origin and Difference: On Environmental Philosophy and Continental Thought," *Philosophy Today* 42 Supplement (1998): 169–181; "Environmental Philosophy After the End of Nature," *Environmental Ethics* 24, no. 1 (Spring 2002): 23–39; and "The Nature of Artifacts," *Environmental Ethics* 25, no. 2 (Summer 2003): 149–168.

2. For example, as the invention of agriculture, the development of other-worldly religions, or the advent of modern science, industry, or techno-science.

3. Vogel, "Nature as Origin and Difference," 170.

4. Bill McKibben, *The End of Nature* (New York: Random House, [1989] 2006).

5. In a gestalt image, we have a duck/rabbit problem: two individually reasonable interpretations of the data that nevertheless cannot be combined.

6. Lynn White Jr., *Science*, 155:3767 (Mar. 10, 1967): 1205.

7. Aldo Leopold, *A Sand County Almanac* (New York: Ballantine Books, [1949] 1970), 240.

8. Vogel, "The Nature of Artifacts," 152–153.

9. Ibid., 153.

10. Ibid.

11. Ibid., 170.

12. For this and many other examples of extensive cultivation by Native Americans that the new immigrants saw but did not understand, see Charles C. Mann, *1491: New Revelations of the Americas before Columbus* (New York: Alfred A. Knopf, 2005).

13. Vogel, "Environmental Philosophy after the End of Nature," 31.

14. Vogel, "Nature as Origin and Difference," 173.

15. See, for example, William Cronon, "The Trouble with Wilderness; or, Getting Back to the Wrong Nature," in *Uncommon Ground: Rethinking the Human Place in Nature*, ed. William Cronon, 69–90 (New York: W. W. Norton & Co., 1995).

16. Vogel, "Nature as Origin and Difference," 175.

17. Ibid., 176.

18. Ibid., 176–177.

19. Ibid., 177.

20. And perhaps especially when our reservations are so esoteric. I intend thereby no profoundly elitist claim, but I expect that the number of activists (let alone nonactivists) sufficiently motivated to follow these philosophical arguments will always be small. Surely many could master them—just as I might master biology—but the demands of our own fields typically allow us little time to master others.

21. For a description of the episodes, see http://www.history.com/shows/life-after-people.

22. The modal verb "can" is clearly critical: Improvement does occur, but must often overcome resistance, may give way to backsliding, and improvements in one area regularly cause new problems in another. One

need not be as confident as Hegel in the progress of history, for example, to celebrate the fruit of nineteenth- and twentieth-century liberation movements. Yet they all faced resistance, some—such as the end of slavery in the United States—produced powerful backlashes, and finally some apparent improvements—like the reproductive freedom provided by the Pill—came at a real cost: here the emotional and sexual extortion that women faced at the hands of men newly freed from the fear of long-term entanglements. Yet overall, the practice of evaluation is robust and many determinations are justly uncontroversial.

23. It owes its influence to Habermas, whose highly influential early review of Gadamer was sometimes perceptive and elsewhere understandably but critically wrong. See Jürgen Habermas, "A Review of Gadamer's *Truth and Method*," in *Understanding and Social Inquiry*, trans. and ed. F. R. Dallmayr and T. A. McCarthy (Notre Dame: University of Notre Dame Press, 1977). For a brief interpretation of the ensuing Gadamer-Habermas debate, see my "On Communicative Actors Talking Past One Another: The Gadamer-Habermas Debate," *Philosophy Today* 40 (1996): 160–168.

24. W. V. O. Quine, *Word and Object* (New York: MIT Press, 1960).

25. Peter Winch, *The Idea of a Social Science and Its Relation to Philosophy* (Atlantic Highlands, N.J.: Humanities Press International, 1958).

26. Thomas Kuhn, *The Structure of Scientific Revolutions, 2nd Ed.* (Chicago: The University of Chicago Press, 1970).

27. Let me immediately concede: The "three basic moves" I elaborate are distillates of several far longer lines of argument. I hope they clarify essential details, but they are neither explicit in the text as "three moves" nor do they fully capture, much less exhaust, the richness of Gadamer's discussion.

28. G. W. F. Hegel, "Introduction," *The Phenomenology of Spirit*, trans. A. V. Miller (Oxford: Oxford University Press, 1977), 74.

29. Hans-Georg Gadamer, *Truth and Method, 2nd Rev. Ed.*, trans. W. Glen-Doepel, ed. J. Cumming and G. Barden, rev. Joel Weinsheimer and Donald G. Marshal (New York: Crossroads Press, 1991), 277–281. Note that the very word "authority" has, under modern influence, assumed dubious connotations. On Gadamer's view, however, this is a hasty assumption. We often legitimately recognize the authority of experts: They have longer experience or a broader perspective than we, and over time we find their judgments better justified than our own. In the end, for example, I am responsible for the medical treatments I accept. But barring unusual circumstances, I reasonably defer to the superior insight—the legitimate authority—of my doctor.

30. Hegel, *The Phenomenology of Spirit*, 76; Gadamer, *Truth and Method*, 341–357.

31. Gadamer, *Truth and Method*, 443–445.

32. Along with most philosophers in the Western tradition, Gadamer assumes we lack clear cases of animals using language. If his assumption

should turn out to be empirically false, naturally, this claim would need appropriate modification.

33. Actually, sensation does not even determinately capture our immediate environment. See Hegel, "Sense Certainty," chapter 1 of *The Phenomenology of Spirit*.

34. If, as Hegel argues in a reduction of this assumption, our concepts are conceived as tools or a medium, then they don't give us the *real* world but something else—the world as conceptualized, not the world in itself (Hegel, *The Phenomenology of Spirit*, 73–74).

35. Gadamer, *Truth and Method*, 443.

36. Ibid., 353.

37. Ibid., 355.

38. Ibid., 356–357.

39. Although Gadamer highlights this dynamic of application, it remains true—as the traditional view thought of words generally—that concepts can be hypostatized: used only in fixed settings with a narrow, determinate range of meanings or a clear set of necessary and sufficient conditions. Yet this represents an unusual case and is difficult for historical beings to maintain. The translators of the King James Bible rightly followed Koine Greek in referring to God as "thou" and "thee," the then-common familiar English forms of "you." Yet the paradoxical effect of the translation's continuing liturgical use long after conversational English dropped the informal second person had the misleading effect that prayers addressed to a God whom Jesus called "Daddy" took on an unintended formal and artificial tone. One calcifies concepts only at a cost. Living language is dynamic and adaptive— unsurprisingly, for it is our mode of life.

40. Gadamer, *Truth and Method*, 458.

41. Ibid., 447.

42. Ibid.

43. If my dancing looks just like another person's shaking off Northern Ontario blackflies, we can only determinately distinguish the two practices (let alone evaluate either as a model of its type) through language.

44. Steven Vogel, "The Silence of Nature," *Environmental Values* 15 (2006): 145–171.

45. As on Habermas's whom Vogel and I here follow. Habermas explicitly grounds moral obligation on the pragmatic presuppositions of discourse, but he acknowledges that

[t]o the extent that creatures participate in our social interactions, we encounter them in the role of an alter ego as an other in need of protection; this grounds the expectation that we will assume a fiduciary responsibility for their claims. There exists a quasi-moral responsibility toward animals that we encounter in the role (if not *completely* filled) of a second person, one whom we look upon as if it were an alter ego.

Later he continues, "I do not wish to preclude a priori that some vegetarians exhibit a moral sensibility that may prove to be the correct moral intuition under more auspicious social circumstances." See Jürgen Habermas, *Justification and Application*, trans. Ciaran P. Cronin (Cambridge, Mass.: MIT Press, 1993), 109–111.

46. In his *Hitchhiker's Guide*, Adams introduces a fish one puts in one's ear and that feeds on brainwaves. The fish translates any speaker's thoughts into an energy matrix that can be reconstructed (by another fish) as thoughts in the listener's mind. See Douglas Adams, *A Hitchhiker's Guide to the Galaxy* (New York: Crown Books, [1979] 2004).

6. Environmental Hermeneutics and Environmental/Eco-Psychology: Explorations in Environmental Identity
David Utsler

1. Richard Palmer, *Hermeneutics* (Evanston, Ill.: Northwestern University Press, 1969), 8.

2. Ibid., 9.

3. It is important not to conflate environmental psychology and ecopsychology. The differences between the two will be discussed later in this essay.

4. See David Utsler, "Paul Ricoeur's Hermeneutics as a Model for Environmental Philosophy," *Philosophy Today* 53 (2009): 173–178; also, "Who Am I, Who Are These People and What is this Place?: A Hermeneutic Account of the Self, Others, and Environments," in *Placing Nature on the Borders of Religion, Philosophy, and Ethics*, ed. Forrest Clingerman and Mark H. Dixon (Burlington, Vt.: Ashgate, 2011), 139–152.

5. Mitchell Thomashow, *Ecological Identity: Becoming a Reflective Environmentalist*, (Cambridge, Mass.: MIT Press, 1996). See also *The Perceived Self: Ecological and Interpersonal Sources of Self Knowledge (Emory Symposia in Cognition)*, ed. Ulrich Neisser (Cambridge: Cambridge University Press, 1993 [1996]).

6. The term "ecological identity" as it is typically used would seem to refer more narrowly to relationships to natural entities and systems whereas "environmental identity" would seem to encompass broader notions of environment and could include or encompass an ecological component of identity. There is, of course, social ecology, going back to Murray Bookchin, which links environmental problems to social and political problems.

7. Thomashow, *Ecological Identity*, 3.

8. Robert Melchior Figueroa, "Debating the Paradigms of Justice: The Bivalence of Environmental Justice" (PhD diss., University of Colorado Boulder, 1999), 147. This is the unpublished PhD dissertation provided to me by the author.

9. Robert Melchior Figueroa, "Evaluating Environmental Justice Claims," in *Forging Environmentalism: Justice, Livelihood and Contested Environments*, ed. Joanne Bauer (New York: M. E. Sharpe, 2006), 371.

10. See Forrest Clingerman, "Beyond the Flowers and the Stones: 'Emplacement' and the Modeling of Nature." *Philosophy in the Contemporary World* 11 (2004): 17–24.

11. Figueroa, "Evaluating Environmental Justice Claims," 372; emphasis original.

12. For more on narrative and environmental hermeneutics, see Brian Treanor's chapter in this volume, "Narrative and Nature: Appreciating and Understanding the Nonhuman World." See also Treanor, "Narrative Environmental Virtue Ethics: *Phronesis* without a *Phronimos*, *Environmental Ethics* 30 (2008): 361–379.

13. Janet Donohoe's chapter in this volume, "The Betweenness of Monuments" takes up the issue of memory and history in a fashion complementary to Figueroa's idea of environmental heritage. What Donohoe has to say about monuments can bring much to bear on thinking of ways in which the environmental heritages of peoples that have been destroyed by environmental injustices can be preserved in memory.

14. See note 4.

15. See Paul Ricoeur, *Oneself as Another*, trans. Kathleen Blamey (Chicago: The University of Chicago Press, 1992).

16. Ibid., 1.

17. Ibid., 16.

18. Ibid., 1.

19. I am purposely, for the sake of space, leaving out a fuller discussion on narrative and memory that has much to do with this discussion. I refer the reader to the essays in this volume by Forrest Clingerman, Janet Donohoe, Martin Drenthen, and Brian Treanor.

20. Lyric from the song "Natural Science" by Rush. Lyrics by Neil Peart. Music by Geddy Lee and Alex Lifeson. From the album *Permanent Waves*, Mercury Records, 1980.

21. Paul Ricoeur, *Oneself as Another* (Chicago: The University of Chicago Press, 1995), 116.

22. Ibid.

23. Ibid., 2.

24. Ibid., 3.

25. I refer again to the two works cited in note 4.

26. Enrique Salmon, "Sharing Breath: Some Links between Land, Plants, and People," in *Colors of Nature: Culture, Identity and the Natural World*, ed. Alison H. Deming and Lauret E. Savoy (Minneapolis: Milkweed Editions, 2011), 196–197.

27. Ibid., 196. By linking at-risk cultures with an at-risk Earth, Salmon demonstrates the link between traditional environmental concerns and issues

with environmental justice. That this link is demonstrated by indigenous cultures is indicative of the need to listen to the wisdom of others embedded in cultural narratives.

28. Hans-Georg Gadamer, *Truth and Method*, 2nd rev. ed., trans. Joel Weinsheimer and Donald G. Marshall (New York: Continuum Press, 2004), 470; emphasis original.

29. See Forrest Clingerman, "Reading the Book of Nature: A Hermeneutical Account of Nature for Philosophical Theology," *Worldviews: Global Religions, Culture, and Ecology* 13 (2009): 72–91.

30. Gadamer, *Truth and Method*, 469.

31. One could follow an intersection here between hermeneutics and psychology in that human action and its meanings are likewise a concern of hermeneutical inquiry.

32. Robert Gifford, *Environmental Psychology: Principles and Practice*, 4th edition (Colville: Optimal Books, 2007).

33. Susan Clayton and Susan Opotow, eds., *Identity and the Natural Environment: The Psychological Significance of Nature* (Cambridge, Mass.: MIT Press, 2003).

34. Theodore Rozak, *The Voice of the Earth: An Exploration in Ecopsychology*, 2nd edition (Grand Rapids: Phanes Press, Inc., 2001 [1992]), 14.

35. Joseph P. Reser, "Whither Environmental Psychology?: The Transpersonal Ecopsychology Crossroads," *Journal of Environmental Psychology* 15 (1995): 235–257.

36. Joseph P. Reser, "Joseph Reser: The Ecopsychology Interview," *Ecopsychology* 1 (2009): 57–63.

37. Steven J. Holmes, "Some Lives and Some Theories," in *Identity and the Natural Environment*, 34. In-text references omitted. See also Elizabeth Ann Bragg, "Towards Ecological Self: Deep Ecology Meets Constructionist Self-Theory," *Journal of Environmental Psychology* 16 (1996): 93–108.

38. Psychology's focus on the self and narrative theory in hermeneutics finds a meeting point here. As Paul Ricoeur, Richard Kearney, and others have observed, the self is discovered and constructed through narrative and narrative has ethical implications. One way in which environmental psychology can realize its inherent normative elements is to employ the ethical power of narrative.

39. Clayton and Opotow, *Identity and the Natural Environment*. See note 32.

40. Clayton and Opotow, "Introduction: Identity and the Natural Environment," in *Identity and the Natural Environment*, 8.

41. Susan Clayton, "Environmental Identity: A Conceptual and Operational Definition," in *Identity and the Natural Environment*, 45–46.

42. Robert Sommer, "Trees and Human Identity" and Maureen E. Austin and Rachel Kaplan, "Identity, Involvement, and Expertise in the Inner City: Some Benefits of Tree-Planting Projects," in *Identity and the Natural Environment*, 179–204 and 205–225.

43. James Hillman, "A Psyche the Size of the Earth: A Psychological Foreword," in *Ecopsychology: Restoring the Earth, Healing the Mind*, ed. Theodore Rozak, Mary E. Gomes, and Allen D. Kanner (San Francisco: Sierra Club Books, 1995), xvii; emphasis original.

44. Ibid.

45. Rozak et al., *Ecopsychology*.

46. J. Baird Callicott, "Environmental Wellness," in *Beyond the Land Ethic: More Essays in Environmental Philosophy* (Albany: State University of New York Press, 1999), 283–299.

47. Ibid., 285.

48. See http://sites.google.com/site/thegnomeprojectorg/home.

49. Ricoeur, *Oneself as Another*, 172.

50. See Paul Ricoeur, *Freud and Philosophy: An Essay on Interpretation*, trans. Denis Savage (New Haven, Conn.: Yale University Press, 1970). See also, Ricoeur, *The Conflict of Interpretations* (Evanston, Ill.: Northwestern University Press, 1974), 99–208.

51. See Paul van Tongeren and Paulien Snellen, "How Hermeneutics Might Save the Life of (Environmental) Ethics," in this volume.

52. Gadamer, *Truth and Method*, 251.

53. Paul Ricoeur, *From Text to Action: Essays in Hermeneutics II*, trans. Kathleen Blamey and John B. Thompson (Evanston, Ill.: Northwestern University Press, 1991), 119.

54. Gadamer, *Truth and Method*, 295.

55. Hans-Georg Gadamer, *Philosophical Hermeneutics*, trans. and ed. David E. Linge (Los Angeles: University of California Press, 1976), 31.

7. Environmental Hermeneutics with and for Others:
Ricoeur's Ethics and the Ecological Self
Nathan M. Bell

I would like to extend thanks to my fellow contributors in this volume, as well as the editors, for critical feedback on this chapter. Additionally, I would like to thank David M. Kaplan and Robert Figueroa of the University of North Texas, as well as Carl Sachs of the University of Alabama at Birmingham, for comments on previous drafts of this chapter.

1. Ricoeur, *Oneself as Another*, 172; emphasis original.

2. Ibid.

3. Ibid., 179.

4. Ibid.

5. Ibid., 179–180.

6. Roderick Frazier Nash, *Wilderness and the American Mind*, 4th ed. (New Haven, Conn.: Yale University Press, 2001), 25–43.

7. Ibid., 84–86.

8. Other works by Ricoeur which give his general theory of interpretation include, most notably, *Interpretation Theory: Discourse and the Surplus of*

Meaning (Fort Worth: Texas Christian University Press, 1976); *The Rule of Metaphor* (London: Routledge, 2009); and *The Conflict of Interpretations: Essays in Hermeneutics*, ed. Don Ihde (Evanston, Ill.: Northwestern University Press, 1974).

9. Paul Ricoeur, *Hermeneutics and the Human Sciences*, ed. and trans. John B. Thompson (New York: Cambridge University Press, 2007), 175–177.

10. Ibid., 177–178.

11. Ibid., 178.

12. Paul Ricoeur, *From Text to Action: Essays in Hermeneutics II*, trans. Kathleen Blamey and John B. Thompson (Evanston, Ill.: Northwestern University Press, 2007), 150–156.

13. Ricoeur, *Oneself as Another*, 181.

14. Ibid., 189–192.

15. Ibid., 192.

16. Ibid., 193.

17. Ibid.

18. Ibid., 193–194.

19. Ibid., 189–192.

20. Val Plumwood, *Environmental Culture: The Ecological Crisis of Reason* (London: Routledge, 2002), 181.

21. Aldo Leopold, *A Sand County Almanac: And Sketches Here and There* (London: Oxford University Press, 1968), 129–130.

22. I am here thinking of People for the Ethical Treatment of Animals (PETA), which "focuses its attention on the four areas in which the largest numbers of animals suffer the most intensely for the longest periods of time." While I do not wish to align my argument with PETA's overall ideology, I do find it interesting that a group boasting "more than 2 million members and supporters" is focused so explicitly on suffering. People for the Ethical Treatment of Animals, "About PETA," PETA.org, http://www.peta.org/about/default.aspx (accessed June 15, 2011).

23. Ricoeur, *Oneself as Another*, 190.

24. Ibid., 194.

25. Ibid., 196.

26. Ibid., 194.

27. Robert Figueroa and Claudia Mills, "Environmental Justice," in *A Companion to Environmental Philosophy*, ed. Dale Jamieson (Malden, Mass.: Blackwell Publishing, 2003), 427.

28. David Schlosberg, *Defining Environmental Justice: Theories, Movements, and Nature* (Oxford: Oxford University Press, 2009), 56.

29. Ricoeur, *Oneself as Another*, 199.

30. Ibid., 200.

31. Figueroa and Mills, "Environmental Justice," 427.

32. Schlosberg, *Defining Environmental Justice*, 65.

33. Ibid., 59.

34. Schlosberg, *Defining Environmental Justice*, 58–60; Robert Melchior Figueroa, "Bivalent Environmental Justice and the Culture of Poverty," *Rutgers University Journal of Law and Urban Policy* 1 (2003): 36–39.

35. Ricoeur, *Oneself as Another*, 200.

36. Ibid.

37. Schlosberg, *Defining Environmental Justice*, 21–23.

38. Ricoeur, *Oneself as Another*, 200.

39. Ibid.

40. This gives us something like Robert Figueroa's concept of "bivalent environmental justice." See Figueroa, "Bivalent Environmental Justice and the Culture of Poverty."

41. Ricoeur, *Oneself as Another*, 196.

42. Ibid.

43. Figueroa and Mills, "Environmental Justice," 434.

44. David M. Kaplan, "Paul Ricoeur and Development Ethics," in *A Passion for the Possible: Thinking with Paul Ricoeur*, ed. Brian Treanor and Henry Isaac Venema (New York: Fordham University Press, 2010), 119.

45. It should be noted that this line of thinking is not without precedent in environmental justice literature itself. I am thinking mainly of Robert Figueroa's concept of "environmental heritage." Figueroa states that

> environmental identity is closely related to environmental heritage, where the meaning and symbols of the past frame values, practices, and places we wish to preserve for ourselves *as members of a community*. In other words, our environmental heritage is our environmental identity in relation to the community viewed over time.

Robert Melchior Figueroa, "Evaluating Environmental Justice Claims," in *Forging Environmentalism: Justice, Livelihood, and Contested Environment*, ed. Joanne Bauer (Armonk, NY: M. E. Sharpe, 2006), 372; emphasis original.

46. It should be noted that, because of the scope of this chapter, I am currently not addressing the issue of ecological justice, which refers to the application of justice to nonhumans. While I believe this could be addressed using Ricoeur's ethics, the full working out of it, including the tensions between animal recognition and justice, would require more space than it can be given here.

8. Bodily Moods and Unhomely Environments:
The Hermeneutics of Agoraphobia and the Spirit of Place
Dylan Trigg

I acknowledge the support of the "European Platform for Life Sciences, Mind Sciences, and the Humanities" grant by the Volkswagen Stiftung for the project *Narcissus & Echo: Self-Consciousness and the Inter-Subjective*

Body. Thanks also to the editors of this collection and to Dorothée Legrand for their comments on this chapter.

1. Blaise Pascal, *Pensées*, trans. John Warrington (London: Dent and Sons, 1973), 70.

2. Anthony Vidler, *Warped Space: Art, Architecture, and Anxiety in Modern Culture* (Cambridge, Mass.: MIT Press, 2001), 20.

3. Charles Baudelaire, *The Flowers of Evil*, trans. James McGowan (Oxford: Oxford University Press, 1998), 345.

4. Cf. Isaac Marks, *Fears, Phobias, and Rituals: Panic, Anxiety, and their Disorders* (Oxford: Oxford University Press, 1987); David Trotter, "The Invention of Agoraphobia," *Victorian Literature and Culture* 32 (2004): 463–474.

5. Sigmund Freud, *The Standard Edition of the Complete Psychological Works of Sigmund Freud*, Volume III, trans. James Strachey (London: Vintage, 2001), 81.

6. Dianne Chambless and Alan Goldstein, *Agoraphobia: Multiple Perspectives on Theory and Treatment* (Chichester: Wiley and Sons, 1982), 3.

7. Stewart Sadowsky, "Agoraphobia, Erwin Straus and Phenomenological Psychopathology," *The Humanistic Psychologist* 25 (1997): 33.

8. Kirsten Jacobson, "Agoraphobia and Hypochondria as Disorders of Dwelling," *International Studies in Philosophy* 36 (2004): 34.

9. Pascal, *Pensées*, 45.

10. Martin Heidegger, *Being and Time*, trans. Joan Stambaugh (Albany: State University of New York Press, 1996).

11. Ibid., 126.

12. Ibid.

13. Ibid., 129.

14. Ibid., 128–129.

15. John Russon, *Human Experience: Philosophy, Neurosis, and the Elements of Everyday Life* (Albany: State University of New York Press, 2003), 45.

16. Allen Shawn, *Wish I Could be There: Notes from a Phobic Life* (New York: Viking Press, 2007), 117.

17. Ibid.

18. Maurice Merleau-Ponty, *The Phenomenology of Perception*, trans. Colin Smith (New York: Routledge, 2006), 94.

19. Shawn, *Wish I Could be There*, 117.

20. Ibid., 118–119.

21. Merleau-Ponty, *The Phenomenology of Perception*, 107.

22. Shawn, *Wish I Could be There*, 119.

23. Forrest Clingerman, "Memory, Imagination, and the Hermeneutics of Place," in this volume.

24. Ibid., 374.

25. Merleau-Ponty, *The Phenomenology of Perception*, 330.

26. Ibid., 328.

27. Clingerman, "Memory, Imagination, and the Hermeneutics of Place," in this volume.

28. David Utsler, "Paul Ricoeur's Hermeneutics as a Model for Environmental Philosophy," *Philosophy Today* 53 (2009): 174.

29. Merleau-Ponty, *The Phenomenology of Perception*, 330.

30. Shawn, *Wish I Could be There*, 119.

31. Ibid., 133.

32. Joyce Davidson, "Putting on a Face: Sartre, Goffman, and Agoraphobic Anxiety in Social Space," *Environmental and Planning D: Society and Space* 21 (2003): 113.

33. Terry Knapp, *Westphal's "Die Agoraphobie" with Commentary: The Beginnings of Agoraphobia*, trans. Michael T. Schumacher (Lanham, Md.: University Press of America, 1988), 70.

34. Davidson, "Putting on a Face," 119.

35. Sigmund Freud, *The Uncanny*, trans. David Mclintock (Harmondsworth: Penguin, 2003), 125.

36. Jacobson, "Agoraphobia and Hypochondria as Disorders of Dwelling," 37.

37. Shawn, *Wish I Could Be There*, 133.

38. Sadowsky, "Agoraphobia, Erwin Straus and Phenomenological Psychopathology," 33.

39. Cf. Anthony Steinbock, *Home and Beyond: Generative Phenomenology after Husserl* (Evanston, Ill.: Northwestern University Press, 1995).

9. Narrative and Nature: Appreciating and Understanding the Nonhuman World
Brian Treanor

1. Contributing in some small way, perhaps, to the theory of imagination—imagination and narrative being so closely connected—called for by Sean McGrath. See "The Question Concerning Nature" in this volume.

2. Jack Turner, *The Abstract Wild* (Tucson: The University of Arizona Press, 1996), 36. People flocked to the Sierras and other natural sites in the wake of Muir's writing, but they did not "hear the trees speak," that is to say they did not experience the same reverence for the natural world that Muir did. Moreover, in popularizing Yosemite Muir arguably merely altered the method of its degradation (i.e., from conscious exploitation by logging to unconscious exploitation and degradation from overuse).

3. Turner, *The Abstract Wild*, 33.

4. Forrest Clingerman has alerted me to the existence of the Neal Smith National Wildlife Refuge in Iowa, which seeks to restore the tall grass prairie and oak savanna that were indigenous to the region. However, despite this welcome bit of news—and despite complex social, economic, and physical

reasons why one area of land is domesticated while others are left wild—I think the point about the sorts of places we seek to preserve and honor remains.

5. Turner, *The Abstract Wild*, 25.

6. Turner seems to be in good company here, insofar as there are commonsense reasons to suppose that those without personal experience of a phenomenon cannot really understand or appreciate it. Hence the arguments that only survivors, if anyone, can really speak of the Holocaust. See some of the positions considered in Richard Kearney, "Narrative and the Ethics of Remembrance" in *Questioning Ethics: Contemporary Debates in Philosophy*, ed. Richard Kearney and Mark Dooley (London: Routledge, 1999), 18–32.

7. Henry David Thoreau, *Walden* (Princeton, N.J.: Princeton University Press, 1971), 3.

8. Ibid., 300–301. Ultimately, such experience is thoroughly individual. Thus, when Thoreau meets John Field and seeks to help him with his (Thoreau's) experience, the help is ineffectual (451–452).

9. Here we might refer to Kant's categories, Marion's treatment of anamorphosis, and other similar accounts of the way in which our perception and perspective "filter" the world at the most fundamental level.

10. Rosalind Hursthouse, *Ethics, Humans and Other Animals* (London: Routledge, 2000), 165.

11. Ibid., 165.

12. Turner, *The Abstract Wild*, 89.

13. Ibid., 66.

14. Richard Kearney, *On Stories* (London: Routledge, 2002), 139. Also see Marcia Eaton "Fact and Fiction in the Aesthetic Appreciation of Nature," *The Journal of Aesthetics and Art Criticism* vol. 56, no. 2 (Spring 1998): "Vivid imagination may be necessary to enable humans to expand the scales to which they respond aesthetically [and, I would add, ethically]" (154).

15. Paul Ricoeur, *Time and Narrative*, vol. 1, trans. Kathleen McLaughlin and David Pellauer (Chicago: The University of Chicago Press, 1984), 65.

16. Ibid., 64–68.

17. Thoreau, *Walden*, 107.

18. Ricoeur, *Time and Narrative*, vol. 1, 59.

19. Ibid. Strictly speaking, Ricoeur says this of action; but his definition of narrative is based in mimesis as an imitation of action.

20. Richard Kearney, *Paul Ricoeur: The Owl of Minerva* (London: Ashgate, 2004), 114. Achilles, Socrates, and St. Francis are Kearney's examples. Thoreau and Leopold are offered as exemplars of their respective virtues by Phil Cafaro in "Thoreau, Leopold, and Carson: Toward an Environmental Virtue Ethics," in *Environmental Virtue Ethics*, ed. Ronald Sandler and Philip Cafaro (New York: Rowman and Littlefield, 2005), 31–44. I offer Muir as a further example. See my "*Phronesis* without a *Phronimos*: Narrative Environmental Virtue Ethics," *Environmental Ethics* 30, no. 4 (2008): 361–379.

21. Ibid. See also Paul Ricoeur, "Life in Quest of Narrative," in *On Paul Ricoeur: Narrative and Interpretation*, ed. David Wood (London: Routledge, 1994). Virtue ethics recommends itself, in part, because of its sensitivity to particular contexts, while deontology and utilitarianism both, to some extent, remain caught up in the attempt to generate and clarify universal rules. However, insofar as this is a strength of virtue ethics, it is a strength that is utterly dependent on narrative, for it is precisely through narratives—the thin narratives of examples or the thick narratives of literature—that virtue ethics does connect with the context of concrete and particular reality.

22. Kearney, *On Stories*, 25. He is referring here to the use of narrative as part of the "talking cure" of psychoanalysis.

23. Dermot Healy, *The Bend for Home* (London: The Harvill Press, 1996), 57. Cited in Kearney, *On Stories*, 25–26.

24. Jeffrey Meyers, *Hemingway: A Biography* (DeCapo Press, 1999), 138.

25. Sometimes the refiguration of a life is dramatic, as when a life is changed by the reading of a book; however, more often the refigurative changes are subtler in character. It's hard to say what exactly gives a narrative the power to fundamentally reshape a life—it took three readings over fifteen years before *Walden* became one of the books that changed my life. However, while a precise formula for narrative power is not possible, we can say that the impact of a narrative is the result of the interaction between (1) the text itself (its structure, its content, etc.) and (2) the reader of the text (the manner in which she emplots the narrative, the timing of her encounter with the narrative, her existing library of narratives, her prefigurative *ipse-identity*, etc.).

26. Emily Brady, "Imagination and Aesthetic Appreciation of Nature," in *The Journal of Aesthetics and Art Criticism* 56, no. 2 (Spring 1998): 143 ff.

27. Kearney, *On Stories*, 19.

28. See, for example, "In Science We Trust," an article by the editors of *Scientific American*, October 2010: http://www.scientificamerican.com/article.cfm?id=in-science-we-trust-poll (accessed September 26, 2010).

29. Holmes Rolston III, "Does Aesthetic Appreciation of Landscapes Need to be Science-Based?," in *Environmental Ethics: Concepts, Policy, Theory*, ed. Joseph DesJardins (London: Mayfield, 1999), 164 and 166.

30. Allen Carlson, "Appreciation and the Natural Environment," *The Journal of Aesthetics and Art Criticism* 37 (1979): 267–276.

31. See Marcia Eaton, "Fact and Fiction in the Aesthetic Appreciation of Nature," *The Journal of Aesthetics and Art Criticism* 56, no. 2 (Spring 1998): 149–156; and Emily Brady, "Imagination and the Aesthetic Appreciation of Nature," *The Journal of Aesthetics and Art Criticism* 56, no. 2 (Spring 1998): 139–147.

32. Rolston, "Does Aesthetic Appreciation of Landscapes Need to be Science-Based?," 166.

33. Eaton, "Fact and Fiction in the Aesthetic Appreciation of Nature," 150.

34. Ibid.
35. Ibid., 153.
36. Ibid., 150.
37. Ibid., 153.
38. Ibid., 154; emphasis added.
39. Thomas S. Kuhn, *The Structure of Scientific Revolutions* (Chicago: The University of Chicago Press, 1996).
40. Ibid., 170 passim.
41. Ibid., 4.
42. Ibid., 89–90; emphasis added. In some sense, a paradigm shift results in scientists seeing a "different world" (111–112).
43. Ibid., 151; emphasis added.
44. Ibid. Though arguments are, nevertheless, potentially effective.
45. Ibid., 155 and 156; emphasis added.
46. Ibid., 126 and 192.
47. Ibid., 147.
48. Eaton, "Fact and Fiction in the Aesthetic Appreciation of Nature," 151.
49. Gabriel Marcel, *Man against Mass Society*, trans. G. S. Fraser (Chicago: Henry Regnery Company, 1967), 155 ff.
50. Boyd Blundell, "Creative Fidelity: Gabriel Marcel's Influence on Paul Ricoeur," in *Between Suspicion and Sympathy: Paul Ricoeur's Unstable Equilibrium*, ed. Andrzej Wiercinski (Toronto: Hermeneutic Press, 2003), 89–102, and 92. The two passages Blundell cites here are from *Marcel's Man against Mass Society*, 155 and 156, respectively.
51. Michael Pollan, *In Defense of Food: An Eater's Manifesto* (New York: Penguin, 2008).
52. Ibid., 28; emphasis added.
53. "How Facts Backfire" by Joe Keohane in *The Boston Globe* http://www.boston.com/bostonglobe/ideas/articles/2010/07/11/how_facts_backfire/ (accessed on October 7, 2010). Regarding some of the original research about which this article reports, see Brendan Nyhan and Jason Reifler, "When Corrections Fail: The Persistence of Political Misperceptions," *Political Behavior* 32 (2010): 303–330.
54. Keohane, "How Facts Backfire."
55. William James, *Pragmatism* (New York: Dover, 1995), 64.
56. Keohane, "How Facts Backfire."
57. Kearney, *On Stories*, 139.
58. Turner, *The Abstract Wild*, 78; emphasis added.
59. Ibid.
60. This is not to suggest the two examples are identical. In the case of Holocaust testimony, the narrative is told by a human survivor of the Holocaust. However, in the case of narratives about nature or wild animals, the narrative is told by another human rather than by nature or the animal itself. Nevertheless, there is no reason to assume that narratives about animals are any less capable of eliciting empathy or love or respect than actual

experience with animals, for in the case of actual experience the animals also remain mute and foreign. Obviously we can "relate" to some narratives more easily than others; however, our understanding is not limited to those narratives to which we can easily relate. Also, on whether and to what extent nature "speaks," see Scott Cameron "Socrates outside Athens: Plato, the Phaedrus, and the Possibility of 'Dialogue' with Nature," in *Phenomenology 2010, vol. 5: Selected Essays from North America. Part 2: Phenomenology beyond Philosophy*, ed. Lester Embree, Michael Barber, and Thomas J. Nenon (Bucharest: Zeta Books/Paris: Arghos-Diffusion, 2010), 43–68; Forrest Clingerman, "Reading the Book of Nature: A Hermeneutical Account of Nature for Philosophical Theology," *Worldviews* 13 (2009): 72–91; and Steven Vogel, "The Silence of Nature," *Environmental Values* 15 (2006): 145–171.

61. Turner, *The Abstract Wild*, 79. Turner concludes his reflections on the song of the White Pelicans thus: "When I see white pelicans riding mountain thermals, I feel their exaltation, their love of open sky and big clouds. Their fear of lightning is my fear, and I extend to them the sadness of descent. I believe the reasons they are soaring over the Grand Teton are not so different from the reasons we climb mountains, sail gliders into great storms, and stand in rivers with tiny pieces of feathers from a French duck's butt attached to a barbless hook at the end of sixty feet of a sixty-dollar string thrown by a thousand-dollar wand. Indeed, in love and ecstasy we are closest to the Other, for passion is at the root of all life and shared by all life. In passion, all beings are at their wildest; in passion, we—like pelicans—make strange noises that defy scientific explanation" (Turner, *The Abstract Wild*, 79–80). See also "Human and nonhuman beings may share some virtues because we are in some respects similar" (Phil Cafaro, "Thoreau, Leopold, and Carson: Toward an Environmental Virtue Ethics," in *Environmental Virtue Ethics*, ed. Ronald Sandler and Phil Cafaro [New York: Rowman and Littlefield, 2005], 34).

62. As I've argued at length in other settings, the claim that we can come to know something of an other need not bleed into the claim that we can comprehend the other, know it objectively or exhaustively. See my *Aspects of Alterity: Levinas, Marcel and the Contemporary Debate* (New York: Fordham University Press, 2006); "Constellations: Gabriel Marcel's Philosophy of Relative Otherness," *American Catholic Philosophical Quarterly* 80, no. 3 (2006); and "Judging the Other: Beyond Toleration" in *Interpretando la experiencia de la tolerancia* [Interpreting the Experience of Tolerance], ed. Rosemary Rizo-Patrón de Lerner (Lima, Peru: Pontificia Universidad Católica del Perú/Fondo Editorial, 2006). See also Alasdair MacIntyre, *Dependent Rational Animals* (Chicago: Open Court, 1999), 14–15:

Knowledge of others . . . is a matter of responsive sympathy and empathy elicited through action and interaction and without these we could not, as we often do, impute to those others the kind of reasons for their actions that, by making their actions intelligible to us, enable us to respond to them in ways that they too can find intelligible. . . . In the case of the relationship of human being to human being none of this is, or should be, controversial. But I want to suggest, there is no significant difference in the case of the relationship of human beings to members of certain other animal species.

The "responsive activity" between humans and certain nonhuman animals gives us a certain "interpretive knowledge" of their thoughts and feelings (Ibid., 17). While it is true that such interpretation can overreach and become a kind of misleading ascription, the practical experience of many people suggest that this need not always be the case, as MacIntyre suggests in his analysis of dogs and dolphins. Turner agrees, "In our effort to go beyond anthropocentric defenses of nature, to emphasize its intrinsic value and right to exist independently of us, we forget the reciprocity between the wild in nature and the wild in us, between knowledge of the wild and knowledge of the self that was central to all primitive cultures" (Turner, *The Abstract Wild*, 26).

63. "Science is no guarantee that one will [accurately, truthfully] see what is there either." Rolston, "Does Aesthetic Appreciation of Landscapes Need to be Science-Based?," 166.

64. And here I emphasize the *some* of "at least some" cases or experiences, despite Turner's explicit objections to the contrary and his numerous examples, including, most powerfully, encountering pictographs in The Maze, in which the abstract and disinterested stance characteristic of science plays the villain to the experience of gross contact.

65. Mick Smith, "Reading 'Natural History': Hermeneutics, Ethics, and Morrow's Ants" in this volume.

66. Martin Drenthen, "New Nature Narratives" in this volume. Forrest Clingerman, "Environmental Place and Time: Memory, Imagination, and the Hermeneutics of Place" in this volume.

67. Christina M. Gschwandtner, "Might Nature Be Interpreted as a 'Saturated Phenomenon?'" in this volume.

68. It is true that certain logical truths, tautologies, isolated facts, and so forth are generally not narrative. However, the more we begin to think about such facts in relationship to other bits of information, or certainly to lived experience and life, the more narrative comes into play. Once we begin to think about value and meaning, narrative really comes to the fore.

10. The Question Concerning Nature
Sean McGrath

1. "Ecology without Nature" was the title of a lecture Žižek gave in Athens in 2007, which is viewable on YouTube, http://www.youtube.com/watch?v=3h4HHT1bt_A (accessed June 10, 2011). The phrase comes from

Timothy Morton's, *Ecology without Nature* (Cambridge, Mass.: Harvard University Press, 2007). See also idem, *The Ecological Thought* (Cambridge, Mass.: Harvard University Press, 2010); "Thinking Ecology: The Mesh, the Strange Stranger, and the Beautiful Soul," *Collapse* 6 (2010): 265–293; Slavoj Žižek, *In Defence of Lost Causes* (London: Verso, 2008), especially the last chapter, "Unbehagen in der Natur: Ecology Against Nature," available online, http://www.bedeutung.co.uk/magazine/issues/1-nature-culture/ Zizek-unbehagen-natur-ecology-nature/ (accessed July 8, 2011); idem, *Living in the End Times* (London: Verso, 2010). EWN is not a clearly delineated philosophical position. Morton's recent turn to object-oriented ontology stands in uneasy tension with Žižek's transcendental-materialist ontology. What these two have in common is the critique of nature as a subject-sustaining construct. Where Morton seems to be confident that there is matter without subjectivity, Žižek, as a good Lacanian, denies that anything could be meaningfully said to exist in itself.

2. Some environmental hermeneuticians may object to the inclusion of Gadamer in this a-cosmic trajectory. I will not say that Gadamerian hermeneutics is *essentially* a-cosmic and structuralist. I will only note here that (1) hermeneutics as Gadamer conceives it has as its *only* object the Dasein-relative world of culturally generated meanings which Dilthey called "history" (by distinction from material nature), even if the dynamics of interpretive understanding are also held to be at work in the natural sciences— this, in my view, justifies its inclusion in the transcendental-structuralist lineage; (2) Gadamer has next to nothing to say about nature and little connection with the later Heidegger's ecologically crucial turn from Dasein-relative meaning to *physis* in itself ("the fourfold"); (3) the structuralist parallel in Gadamer is the presupposition of Hermann Lang's fusion of Gadamerian hermeneutics and Lacanian psychoanalysis, which is quite famous in Germany (See Hermann Lang, *Language and the Unconscious: Jacques Lacan's Hermeneutics of Psychoanalysis*, trans. Thomas Brockelman [Atlantic Highlands, N.J.: Humanities Press, 1997]). If "being that can be understood is language" (Hans-Georg Gadamer, *Truth and Method*, 2nd edition, trans. J. Weinsheimer and D. G. Marshall [London: Sheed and Ward, 1989], 474), how far are we from Lacan for whom the structure of language exhausts the intelligible and permeates the unconscious itself? Others have read Gadamer as a realist, but this requires several moves which Gadamer, at least in *Truth and Method*, does not make, above all a critique of Kantian representationalism. See, for example, Brice Wachterhauser, *Beyond Being: Gadamer's Post-Platonic Hermeneutic Ontology* (Evanston, Ill.: Northwestern University Press, 1999), 98: "Far from separating the intelligibility of the world from us, or substituting its own intelligibility, the thesis that '[b]eing that can be understood is language' roots language in the world and points instead to its integral connection with the things themselves."

3. Žižek, "Unbehagen in der Natur."

4. Morton, *Ecology without Nature*, 140–205.

5. Žižek, "Unbehagen in der Natur."

6. Ibid.

7. Schellingian nature-philosophy is making a mild comeback. See Iain Hamilton Grant, *Philosophies of Nature after Schelling* (New York: Continuum, 2006); Bruce Matthews, *Schelling's Organic Form of Philosophy: Life as the Schema of Freedom* (Albany: State University of New York Press, 2011).

8. F. W. J. Schelling, *Philosophical Inquiries into the Essence of Human Freedom*, trans. Jeff Love and Johannes Schmidt (Albany: State University of New York Press, 2006), 26. Translation altered.

9. In light of Heidegger's 1929 *Kant and the Problem of Metaphysics*, it is surely not controversial to read *Being and Time* as a version of transcendental philosophy. How else to make sense of Heidegger's statements concerning the dissolution of the problem of the external world or the dependence of the truth of beings on Dasein? See Martin Heidegger, *Being and Time*, trans. John Macquarrie and John Robinson (London: Blackwell, 1962), 43–44.

10. See Sigmund Freud, *Beyond the Pleasure Principle*, trans. James Starchey, in *The Standard Edition of the Complete Psychological Works of Sigmund Freud* (London: Hogarth Press and the Institute of Psycho-Analysis, 1953–1974), chapter 6.

11. See Bruce Fink, *The Lacanian Subject* (Princeton, N.J.: Princeton University Press, 1995); Jacques Lacan, *The Four Fundamental Concepts of Psychoanalysis*, ed. Jacques-Alain Miller, trans. Alan Sheridan (London: Vintage, 1998).

12. Slavoj Žižek, *The Indivisible Remainder: On Schelling and Other Matters* (London: Verso, 1996), 36.

13. Lacan, *Four Fundamental Concepts*, 211.

14. Lacan cited in Marcus Pound, *Žižek: A (Very) Critical Introduction* (Grand Rapids, Mich.: Eerdmans, 2008), 87.

15. Morton, *Ecology without Nature*, 201.

16. See Isabelle Stengers, *Cosmopolotics I*, trans. Robert Bononno (Minneapolis: University of Minnesota Press, 2010); Bruno Latour, *The Politics of Nature: How to Bring the Sciences into Democracy*, trans. Catherine Porter (Cambridge, Mass.: Harvard University Press, 2004); Michael Serres, *The Natural Contract*, trans. Felicia McCarren (Ann Arbor: University of Michigan Press, 1995).

17. Carolyn Merchant, *The Death of Nature: Women, Ecology, and the Scientific Revolution* (New York: Harper, 1980), 278.

18. This is the "Yates thesis" about the esoteric origins of modern science. See Francis Yates, *Giordano Bruno and the Hermetic Tradition* (Chicago: The University of Chicago Press, 1991), and the more nuanced, Ioan Couliano, *Eros and Magic in the Renaissance* (Chicago: The University of Chicago Press, 1987).

19. Schelling and Hegel took the idea from Jakob Boehme. See Cyril O'Regan, *The Heterodox Hegel* (Albany: State University of New York Press, 1994); Glenn Alexander Magee, *Hegel and the Hermetic Tradition* (Ithaca, N.Y.: Cornell University Press, 2001); S. J. McGrath, *The Dark Ground of Spirit: Schelling and the Unconscious* (New York: Routledge, 2012), chapter 2.

20. See Michael Foucault, *The Order of Things: An Archaeology of the Human Sciences* (New York: Random House, 1994), 17–46.

21. Ibid., 17.

22. Lyndy Abraham, *A Dictionary of Alchemical Imagery* (Cambridge: Cambridge University Press, 1998), 70. The "Emerald Table" is at least as old as the eighth century. Throughout the middle ages, it was attributed to Hermes Trismegistus.

23. Paracelsus, *Alchemical Catechism*, http://www.sacred-texts.com/alc/tschoudy.htm (accessed July 8, 2011).

24. See S. J. McGrath, "Hermeneutics and the Unconscious: The Mercurial Play of Interpretation in Phenomenology, Psychoanalysis, and Alchemy," in *The Task of Interpretation: Hermeneutics, Psychoanalysis, and Literary Studies*, ed. Dariusz Skórczewski, Andrzej Wiercinski, and Edward Fiawa (Lublin: Catholic University of Lublin), 45–72.

25. Gilles Deleuze and Felix Guattari, *Anti-Oedipus: Capitalism and Schizophrenia*, trans. Robert Hurley, Mark Seem, and Helen R. Lane (Minneapolis: University of Minnesota Press, 1983), 4.

26. Foucault, *Order of Things*, 27.

27. See Couliano, *Eros and Magic in the Renaissance*, 5:

All is reduced to a question of communication: body and soul speak two languages, which are not only different, even inconsistent, but also *inaudible* to each other. The inner sense alone is able to hear and comprehend them both, also having the role of translating one into the other. But considering the words of the soul's language are phantasms, everything that reaches it from the body—including distinct utterances— will have to be transposed into a phantasmic sequence. Besides— must it be emphasized?—the soul has absolute primacy over the body. It follows that *the phantasm has absolute primacy over the word*, that it precedes both utterance and understanding, of every linguistic message. Whence two separate and distinct grammars, the first no less important than the second: a grammar of the spoken language and a grammar of phantasmic language.

28. See Antoine Faivre, *Access to Western Esotericism* (Albany: State University of New York Press, 1994), 13: "The eye of fire pierces the bark of appearances to call forth significations, 'rapports' to render the invisible visible, the '*mundus imaginalis*' to which the eye of the flesh alone cannot provide access, and to retrieve there a treasure contributing to an enlargement of our prosaic vision."

29. Foucault, *Order of Things*, 30.

30. Ibid.

31. Martin Heidegger, "Memorial Address," in *Discourse on Thinking*, trans. John M. Anderson and E. Hans Freund (New York: Harper, 1969), 43–57.

32. Morton, "Thinking Ecology," 210.

33. It is worth noting that Heidegger's critique of technology belongs to a tradition of twentieth-century response to the myth of progress. Josef Pieper, Erazim Kohak, E. F. Schumacher, and George Grant come to mind as independent thinkers (respectively, a neo-Thomist, a Husserlian, an economist, and a political theorist) who arrived at similar conclusions as Heidegger. See Josef Pieper, *Leisure, the Basis of Culture*, trans. Alexander Dru (San Francisco: Ignatius Press, 2009); E. F. Schumacher, *A Guide for the Perplexed* (New York: Harper, 1978); George Grant, *Technology and Justice* (Toronto: Anansi, 1991); Erazim Kohak, *The Embers and the Stars: An Inquiry into the Moral Sense of Nature* (Chicago: The University of Chicago Press, 1984).

11. New Nature Narratives:
Landscape Hermeneutics and Environmental Ethics
Martin Drenthen

1. John O'Neill, Alan Holland, and Andrew Light, *Environmental Values* (New York: Routledge 2008), 162–164.

2. The term "new nature" may seem odd to an outsider, but it is the most often-used word for ecological restoration projects in the Netherlands. The term expresses the idea that nature is "built" in places where it had been obliterated in the past, much in the same way as the rest of the land was built by humans. In other words, the terminology reveals the deep Dutch conviction that is also expressed in the famous Dutch saying "God created the world, but the Dutch created the Netherlands."

3. See Alison Coleman and Toby Aykroyd, eds., *Conference Proceedings: Wild Europe and Large Natural Habitat Areas* (Prague 2009). http://www.wildheartofeurope.eu/gallery/0/135-proceedings_prague2009.pdf (accessed January 7, 2013).

4. See Franz Höchtl, Susanne Lehringer, and Werner Konold, "'Wilderness': What it Means When it Becomes a Reality—A Case Study from the Southwestern Alps," *Landscape and Urban Planning* 70 (2005): 85–95; Marcel Hunziker, "The Spontaneous Reafforestation in Abandoned Agricultural Lands: Perception and Aesthetic Assessment by Locals and Tourists," *Landscape and Urban Planning* 31 (1995): 399–410.

5. Gerard Müskens and Sim Broekhuizen, *De steenmarter (Martes foina) in Borgharen: aantal, overlast en schade* (Wageningen: Alterra Report no. 1259, 2005).

6. One of the undesirable effects of this development will be that the fear for wild nature will most probably resurface as well. Most Europeans

only know of wolves from fairytales such as Little Red Riding Hood, and are very much prone to unrealistic fears about predation. Already, conservation groups in Germany and the Netherlands are preparing the general public for the arrival of wolves, mainly with education programs and explaining that humans in general do not have anything to fear. But they also started to closely monitor the occasional killing of sheep by domesticated dogs, so that—by the time the wolf finally arrives—wolves cannot be held responsible for all sheep kills.

7. Bas Pedroli, Thomas van Elsen, and Jan Diek van Mansvelt, "Values of Rural Landscapes in Europe: Inspiration or Byproduct?," *NJAS Wageningen Journal of Life Sciences* 54, no. 4 (2007): 431.

8. Edward Relph, *Place and Placelessness* (London: Pion, 1976).

9. Marc Augé, *Non-Places: Introduction to an Anthropology of Supermodernity*, trans. J. Howe (London: Verso 1995).

10. Mahyar Arefi, "Non-Place and Placelessness as Narratives of Loss: Rethinking the Notion of Place," *Journal of Urban Design* 4, no. 2 (1999): 179–194.

11. David Lowenthal, "Authenticities Past and Present," *CRM Journal: The Journal of Heritage Stewardship* 5, no. 1 (2008): 6–17.

12. Kimberly Dovey, "The Quest for Authenticity and the Replication of Environmental Meaning," in *Dwelling, Place & Environment: Towards a Phenomenology of Person and World*, ed. David Seamon and Robert Mugerauer (New York: Columbia University Press, 1995), 33–49.

13. Arjen Buijs, "Public Support for River Restoration: A Mixed-Method Study into Local Residents' Support for and Framing of River Management and Ecological Restoration in the Dutch Floodplains," *Journal of Environmental Management* 90, no. 8 (2009): 2680–2689.

14. Hans Renes, "Landschap in het nieuws: De Hedwigepolder binnenkort ontpolderd?," *Historisch Geografisch Tijdschrift* 27, no. 1 (2009): 25–28.

15. "Ontpoldering raakt de Zeeuwse ziel," *De Volkskrant*, December 17, 2007.

16. The question that one could ask today is whether all claims about regional identity and landscape heritage are credible and should be taken seriously in their current form. Some have noted that some of the claims regarding the Zeeland identity suffer from memory defect, and only refer to the situation after 1953. The fight against the sea may have been important in the history of this landscape, but it is only a part of it. The sea has also provided the Zeelanders with fisheries and fertile soil throughout history. Properly understood, the history of humans and the sea is one of give and take. The *Drowned Land of Saeftinghe* could be seen as a reminder that one cannot base one's identity on the idea that the sea can be subjected. The sea will always be a major presence here: Humans can choose to aspire to dominate the sea, but such a choice would be self-deceit. Zeelanders may cherish the idea of being fighters against the sea, but, in fact, their

history is better understood as a negotiation process between the landscape's inhabitants and the forces of nature. The current Zeeland identity claim could be seen as a symptom of modern Zeelanders not being aware enough of the role that nature has had in their very own history, and of the many ways that the border of land and sea has provided a place for humans to make a living.

17. The idea that the flooding of land with *salt* water goes against the very nature of Zeeland is somehow put in perspective by considering the fact that the salinity of the water is deemed irrelevant when the land is consciously flooded with saline water by farmers to grow saline-loving crops such as *salicornia*, again confirming that the real worry of many Zeelanders is the perceived loss of human control over nature.

18. Since then, many new developments took place. In early 2012, after intensive negotiations with the European Commission, the Dutch government finally agreed to partially flood the Hedwige Polder. Deputy Agriculture Minister Henk Bleker informed the European Commission and parliament of the decision that a third of the land (roughly one hundred hectares) will be flooded, and another two hundred hectares of reclaimed land on alternative sites will also be "sacrificed" as compensation measure. Neither Flanders nor the European Commission accepted the revised plan; European Commissioner for the Environment Janez Potočnik even launched an infringement procedure against the Netherlands regarding the delayed flooding or partial flooding of the Hedwige Polder "despite clear legal obligations." In June 2012 the Dutch Cabinet fell again, after having been in power for only one and a half year. In December 2012 the newly elected government decided that the Hedwige Polder will be flooded in 2016.

19. Emma Marris, "Conservation Biology: Reflecting the Past," *Nature* 462, no. 7269 (2009): 30–32.

20. Nico Roymans, Fokke Gerritsen, Cor Van der Heijden, Koos Bosma, and Jan Kolen, "Landscape Biography as Research Strategy: The Case of the South Netherlands Project," *Landscape Research* 34, no. 3 (2009): 337–359.

21. Willem van Toorn, *Leesbaar landschap* (Amsterdam: Querido, 1998), 66.

22. Max Oelschlaeger, *The Idea of Wilderness: From Prehistory to the Age of Ecology* (New Haven, Conn.: Yale University Press, 1991); J. Baird Callicott and Michael P. Nelson, eds., *The Great New Wilderness Debate* (Athens: University of Georgia Press, 1998).

23. See Ned Hettinger, "The Problem of Finding a Positive Role for Humans in the Natural World," *Ethics and the Environment* 7, no. 1 (2002): 109–123; see also William Cronon, "The Riddle of the Apostle Islands; How Do You Manage a Wilderness Full of Human Stories?," in *The Wilderness Debate Rages on: Continuing the Great New Wilderness Debate*, ed. Michael P. Nelson and J. Baird Callicott (Athens: University of Georgia Press, 2008), 632–644.

24. Robert Elliott, *Faking Nature: The Ethics of Environmental Restoration* (New York: Routledge, 1997).

25. Eric Katz, "The Big Lie: The Human Restoration of Nature," *Research in Philosophy and Technology* 12 (1992): 231–241.

26. For example, Eric Higgs, *Nature by Design: Human Agency, Natural Process and Ecological Restoration* (Cambridge, Mass.: MIT Press, 2003); William Jordan III, *The Sunflower Forest: Ecological Restoration and the New Communion with Nature* (Berkeley: University of California Press, 2003); Andrew Light, "Ecological Restoration and the Culture of Nature: A Pragmatic Perspective," in *Environmental Ethics: An Anthology*, ed. Andrew Light and Holmes Rolston III (London: Blackwell 2003), 398–411.

27. O'Neill, Holland, and Light, *Environmental Values*; Martin Drenthen, *Grenzen aan wildheid; wildernisverlangen en de betekenis van Nietzsches moraalkritiek voor de actuele milieu-ethiek* (Budel: Damon, 2003); Martin Drenthen, "Wildness as Critical Border Concept: Nietzsche and the Debate on Wilderness Restoration," *Environmental Values* 14, no. 3 (2005): 317–337.

28. Ed Casey, *Getting Back into Place: Toward a Renewed Understanding of the Place-World* (Bloomington: Indiana University Press, 1993); Robert Mugerauer, *Interpreting Environments: Tradition, Deconstruction, Hermeneutics* (Austin: University of Texas Press, 1995); Anne Whiston Spirn, *The Language of Landscape* (New Haven, Conn.: Yale University Press, 1998); Mick Smith, *An Ethics of Place: Radical Ecology, Postmodernity and Social Theory* (Albany: State University of New York Press, 2001).

29. Martin Drenthen, "Ecological Restoration and Place Attachment: Emplacing Non-places?," *Environmental Values* 18, no. 3 (2009): 285–312; Jim Cheney, "Postmodern Environmental Ethics: Ethics of Bioregional Narrative," *Environmental Ethics* 11, no. 2 (1989): 117–134.

30. For example, Van Toorn, *Leesbaar landschap*.

31. Karina Hendriks and Henk Kloen, *IVN Handleiding leesbaar landschap* (Culemborg: CLM, 2007); see also Martin Drenthen, "Reading Ourselves through the Land."

32. Roymans et al., "Landscape Biography as Research Strategy: The Case of the South Netherlands Project"; see also Tim Ingold, "The Temporality of Landscape," *World Archaeology* 25, no. 2 (1993): 152–174.

33. For a reflection on the notion of layeredness in the landscape, see also Robert Mugerauer's chapter in this volume.

34. "[M]any types of ecological restoration [. . .] can be considered a response stemming from an ethics of memory, when memories challenge the state of the present" (Forrest Clingerman, "Environmental Amnesia or the Memory of Place? The Need for Local Ethics of Memory in a Philosophical Theology of Place," in *Religion and Ecology in the Public Sphere*, ed. Celia Deane-Drummond and Heinrich Bedford-Strohm [New York: T. & T. Clark, 2011], 141–159).

35. O'Neill, Holland, and Light, *Environmental Values*, 162–164.

36. Ibid.

37. Drenthen, "Ecological Restoration and Place Attachment."

38. Wouter Helmer and Willem Overmars, "Genius of Place," *Aarde & Mens* 2, no. 2 (1998): 3–10.

39. See Higgs, *Nature by Design*.

40. Glenn Deliège, "Restoring or Restorying Nature?," in *Nature, Space and the Sacred*, ed. S. Bergmann, P. Scott, M. Jansdotter Samuelsson, and H. Bedford-Strohm (Farnham: Ashgate, 2009).

41. See Higgs, *Nature by Design*; and Jordan, *The Sunflower Forest*.

42. It may appear strange to speak of "facts" in a defense of a hermeneutical ethic. Facts are often understood as the "objective" substratum underneath all interpretations of the world—those features of reality that exist independent of any specific interpretation. Arguing that interpretations should acknowledge the relevant ecological facts would imply the introduction of a thoroughly *un*-hermeneutical element. One can solve this paradox by first acknowledging that facts are not as "objective" as one may think—facts are only considered *relevant* from a certain specific interpretational perspective that renders them such. Facts in this view, then, refer not so much to the objective substrate that remains when the world is stripped from all interpretation, but rather those features of the world that each interpretation has to somehow incorporate because they have a kind of "unruliness" that cannot be ignored without undermining the interpretation whole.

43. http://www.waalweelde.nl.

44. See Drenthen, "Ecological Restoration and Place Attachment."

45. There is some irony in writing about the hermeneutics of the Dutch River landscape in English language in an international academic volume. The interpretational character of our relation with nature may be universal, but the moral meanings involved in this relation are essentially particularistic. See Martin Drenthen, "NIMBY and the Ethics of the Particular," *Ethics, Place & Environment* 14, no. 3 (2010): 321–323.

46. Bruce Janz, "Thinking Like a Mountain: Ethics and Place as Travelling Concepts," in *New Visions of Nature: Complexity and Authenticity, ed.* Martin Drenthen, Jozef Keulartz, and Jim Proctor (Dordrecht: Springer, 2009), 181–195. See also Cheney, "Postmodern Environmental Ethics."

47. Hans-Georg Gadamer, *Wahrheit und Methode. Grundzüge einer philosophischen Hermeneutik* (Tübingen: Mohr, 1975).

48. See Martin Drenthen, "The Paradox of Environmental Ethics: Nietzsche's View of Nature and the Wild," *Environmental Ethics* 21, no. 2 (1999): 163–175.

49. The stories that we tell to ourselves—about who we are and what our lives are about—are built on the contexts that we always already find ourselves in. The most well-known is the way that our life stories get shaped by the stories and narratives we hear around us: Our culture surround us with a body of narratives—our holy texts, our dearest works of literature and art, and so on—that provide us with words and storylines with which we can tell ourselves who we are and what our life is about. Paul Ricoeur elaborates

on this idea in terms of "emplotment." In an interesting paper, Forrest Clingerman has extended on this idea by focusing on the role of "emplace-ment." He argues that we do not just understand ourselves from the context of the stories surrounding us but also from the meaningful places we find ourselves in. Whereas texts help us to find the plot of our lives, meaningful places also provide context from which we understand ourselves. See Forest Clingerman, "Beyond the Flowers and the Stones: 'Emplacement' and the Modeling of Nature," *Philosophy in the Contemporary World* 11, no. 2 (2004): 17–24. See also Martin Drenthen, "Reading Ourselves through the Land: Landscape Hermeneutics and Ethics of Place," in *Placing Nature on the Borders of Religion, Philosophy, and Ethics*, ed. Forrest Clingerman and Mark Dixon (Farnham: Ashgate Publishing Ltd, 2011). See also David Utsler, "Paul Ricoeur's Hermeneutics as a Model for Environmental Philosophy," *Philosophy Today* 53, no. 2 (2009): 173–178.

12. Memory, Imagination, and the Hermeneutics of Place
Forrest Clingerman

1. Edward Casey, *Getting Back into Place* (Bloomington: Indiana University Press, 1993), 4.

2. Paul Van Tongeren and Paulien Snellen, "How Hermeneutics Might Save the Life of (Environmental) Ethics," in this volume.

3. For more information, see Forrest Clingerman, "Modeling the Book of Nature: Seeking the Depth of Nature in a Scientific World," in *Religion, Science, and Public Concern: Discourses on Ethics, Ecology, and Genomics*, ed. Willem Drees (Leiden: University of Leiden Press, 2009); and "Wilderness as the Place between Philosophy and Theology: Questioning Martin Drenthen on the Otherness of Nature," *Environmental Values* 19, no. 2 (May 2010): 211–232.

4. Brian Treanor, "Narrative and Nature," in this volume.

5. Tim Cresswell, *Place: A Short Introduction* (Malden, Mass.: Blackwell, 2004), 7.

6. Arto Haapala, "Aesthetics, Ethics, and the Meaning of Place," *Filozofski Vestnik* 20 (1999): 260.

7. Martin Heidegger, "Building Dwelling Thinking," in *Basic Writings*, ed. David Farrell, 2nd edition (San Francisco: Harper San Francisco, 1993), 343–363. For discussions on whether only humans dwell, see Julian Young, *Heidegger's Later Philosophy* (Cambridge: Cambridge University Press, 2002), 63–104; Forrest Clingerman, "The Intimate Distance of Herons: Theological Travels through Nature, Place, and Migration," *Ethics, Place & Environment* 11 (2008): 313–325.

8. Cf. Paul Ricoeur, *Time and Narrative, Volume 1* (Chicago: The University of Chicago Press, 1984); "The Text as Dynamic Identity," in *Identity of the Literary Text*, ed. Mario J. Valdés and Owen Miller (Toronto: University of Toronto Press, 1985).

9. For a more detailed discussion of place along these lines, see Forrest Clingerman, "Beyond the Flowers and the Stones: 'Emplacement' and the Modeling of Nature," *Philosophy in the Contemporary World* 11 (2004): 17–24; "Reading the Book of Nature: A Hermeneutical Account of Nature for Philosophical Theology," *Worldviews: Global Religions, Culture, and Ecology* 13 (2009): 73–76.

10. Jozef Keulartz, "Using Metaphors in Restoring Nature," *Nature and Culture* 2, no. 1 (Spring 2007): 27–48. Quotation is taken from p. 28. He writes that metaphors have a cognitive function, ". . . operating as mechanisms for the translation of something abstract into something concrete and shedding light on new and unknown phenomena through familiar ones. In short, metaphors are heuristic devices crucial for creating and conceptualizing novel ideas and new knowledge" (27). Metaphors are ". . . also discursive tools that enable communication and negotiation with others through the world. Metaphors then are also diplomatic devices that facilitate interaction between different disciplines and discourses" (27). Third, they have a normative function, meaning they are not simply about thinking and talking, but also acting.

11. Anton Marko Žižeks, *Readers of the Book of Life* (New York: Oxford University Press, 2002).

12. Treanor, "Narrative and Nature," in this volume; "Turn Around and Step Forward: Ideology and Utopia in the Environmental Movement," *Environmental Philosophy* 7 (2009): 27–46.

13. Martin Drenthen, "New Nature Narratives," in this volume; "Reading Ourselves through the Land: Landscape Hermeneutics and Ethics of Place," in *Placing Nature on the Borders of Religion, Philosophy and Ethics*, ed. Forrest Clingerman and Mark H. Dixon (Aldershot, UK: Ashgate, 2011), 123–138.

14. Clingerman, "Reading the Book of Nature," 78.

15. Paul Ricoeur, *Oneself as Another*, see esp. 1–3.

16. Dieter Teichert, "Narrative, Identity and the Self," *Journal of Consciousness Studies* 11 (2004): 185.

17. Paul Ricoeur, "Narrative Identity," in *On Paul Ricoeur: Narrative and Interpretation*, ed. David Wood (London: Routledge, 1991), 195.

18. Ibid., 188.

19. Ibid.

20. Ibid., 198.

21. S. H. Clark, "Narrative Identity in Ricoeur's *Oneself as Another*," in *Ethics and the Subject*, ed. Karl Simms (Atlanta: Rodopi, 1997), 92.

22. David Utsler, "Paul Ricoeur's Hermeneutics as a Model for Environmental Philosophy," *Philosophy Today* 53 (2009): 174.

23. Augustine, *The Confessions*, trans. Chadwick, 235.

24. This section draws on some of the ideas I discuss in "Environmental Amnesia or the Memory of Place? The Need for Local Ethics of Memory in a Philosophical Theology of Place," in *Religion and Ecology in the Public Sphere* (New York: T. & T. Clark, 2011), 141–159. That work was concerned

with the theology and ethical dimensions of memory that are complementary to the overall discussion of the present work.

25. While memory and environmental philosophy is not a common topic, there is a great deal of work on memory and place. Issues such as memorials, emotional connections to place, and so forth all deal with the need for individual and collective memory of environments. See Janet Donohoe's chapter in this volume.

26. Janet Donohoe, "Where Were You When . . .? On the Relationship between Individual and Collective Memory," *Philosophy in the Contemporary World* 16 (Spring 2009): 105–113.

27. Edward Casey, "Levinas on Memory and the Trace," in *The Collegium Phaenomenologicum*, ed. John Sallis, Giuseppina Moneta, and Jacques Taminiaux (Dordrecht: Kluwer, 1988), 241. Interestingly, there are parallels between this definition and Paul Tillich's definition of the symbol. The trace, however, seems to contain at least an element of the physical, despite its use as a constitutive part of memory.

28. My description of the trace focuses on natural elements. This discussion is rendered more complex when confronted with the topic of how humans might seek to build memory and explicit traces in environments, as seen through monuments, for instance. Thus there is interesting overlap between this description of the trace and Donohoe's call for a hermeneutics of monuments in "The Betweenness of Monuments," in this volume.

29. Dylan Trigg, "Altered Place: Nostalgia, Topophobia, and the Unreality of Memory," *Existential Analysis* 18 (2007): 157. While I am in agreement with Trigg on this point, my concern here is to show the continuity of past, present, and future, whereas Trigg's is to show the discontinuity.

30. Sara Ebenreck, "Opening Pandora's Box: Imagination's Role in Environmental Ethics," *Environmental Ethics* 18 (1996): 3.

31. Ibid., 12.

32. Paul Ricoeur, "Imagination in Discourse and Action," in *From Text to Action*, trans. Kathleen Blamey and John B. Thompson (Evanston, Ill.: Northwestern University Press, 1991), 174.

33. Cf. George Taylor, "Ricoeur's Philosophy of Imagination," *Journal of French Philosophy* 16 (2006): 93–104.

34. Ibid., 177.

35. Ricoeur, "Imagination in Discourse and Action," 181.

36. Treanor, "Turn Around and Step Forward," 40.

37. Tracey Stark, "Richard Kearney's Hermeneutic Imagination," *Philosophy and Social Criticism* 23, no 2 (1997): 122.

13. The Betweenness of Monuments
Janet Donohoe

My thanks to the participants in the Environmental Hermeneutics seminar of 2010–2011, and in particular to Christina Gschwandtner and Forrest

Clingerman for their exceedingly helpful comments and suggestions on an earlier version of this chapter.

1. Hannah Arendt, *The Human Condition* (Chicago: The University of Chicago Press, 1958).

2. See Arendt, *The Human Condition*, 19, 55, 173.

3. Paul Ricoeur, *Memory, History, Forgetting*, trans. Kathleen Blamey and David Pellauer (Chicago: The University of Chicago Press, 2004), 351.

4. See J. Donohoe, "Where Were you When?: On the Relationship between Individual and Collective Memory," *Philosophy in the Contemporary World* 16 (2009): 105–113.

5. W. James Booth, *Communities of Memory: On Witness, Identity, and Justice* (Ithaca, N.Y.: Cornell University Press, 2006), 73.

6. Booth, *Communities of Memory*, 74.

7. The Amy Biehl Foundation has recently erected a new memorial to Amy that is a larger marble cross elevated on a marble base, so it is higher than the original. The new memorial also includes an inscription that valorizes Amy's life and her commitment to racial equality.

8. Many discuss this duty to remember in terms of the paradox of remembering the immemorial. I side with Richard Kearney in this debate in that I think we must not remain silent even in the face of the paradoxicality of speech. This is part of the reason why the narrativity of monuments to victims has taken on such a prominent role. Monuments in such cases are not for the sake of glorification but are an additional way of attempting to narrate that which interrupts or disrupts more documentary or direct historical narrative.

9. Booth, *Communities of Memory*, 79.

10. Ibid., 103.

11. Ibid., 79.

12. Ibid., 94.

13. Arendt, *The Human Condition*, 173.

14. Ibid., 174. I fully recognize that there are ways in which Arendt's position is problematic and have addressed those elsewhere. See Janet Donohoe, "Dwelling with Monuments," *Philosophy and Geography* 5 (2002): 235–242.

15. Ricoeur, *Memory, History, Forgetting*, 166.

16. Monuments serve as an in-between for nature and the built environment in many ways. Some of those ways include how monuments are frequently in a place that is set apart from the everyday as a place for reflection and possibly for an experience of the sublime. Monuments remind us of our natural return to the earth insofar as they address the death of the human, but they also have the overtones of our conquest over the earth in our pretensions to immortality. Brian Treanor noted in our discussion another interesting point about their status as in-between in that often the natural places that we have set aside for preservation as parks or wilderness areas are called national monuments, for example, White Sands National Monument in New Mexico

or Muir Woods National Monument in California. National monuments differ from national parks in that they most often include some kind of culturally significant element such as Native American artifacts, Aztec ruins, or fossils, but some of the national monuments are simply smaller areas of natural interest such as barrier reefs or grasslands. In so naming these places, we indicate on the one hand our respect for and desire to preserve the natural environment, but also perhaps our desire to control or delimit what is deemed natural, bringing these natural places into the built or human environment and thus attempting to take them from the wild, natural environment. Those places that include cultural aspects are preserved in part as testimony to human effect on the landscape as evidenced by the preservation of William Clark's signature in sandstone in Pompey's Pillar National Monument in Montana. It is important to note that the nomenclature is inconsistent and seems often to have served political purposes more than anything else. How we have so designated these places is a topic for another work, but their very designation helps to see a possible way in which we conceive of monuments as taking up a place between natural and built environments.

17. I have elaborated upon these kinds of questions before in a different way in the following: J. Donohoe, "Rushing to Memorialize," *Philosophy in the Contemporary World* 13 (2006): 6–12; and J. Donohoe, "Dwelling with Monuments."

18. Paul Ricoeur, *From Text to Action: Essays in Hermeneutics II* (Evanston, Ill.: Northwestern University Press, 2007), 249.

19. Ibid., 251.

20. Ibid.

21. Ibid., 252.

22. Ibid., 254.

23. Ricoeur, *Memory, History, Forgetting*, 349.

24. Ibid., 357.

25. Ibid.

26. Ibid. This notion is one that is elaborated upon by Ricoeur in one of his last works, *Living up to Death*, trans. David Pellauer (Chicago: The University of Chicago Press, 2009).

27. Ibid., 358.

28. Ibid., 359.

29. For those who are anxious about the way in which this is egological, I would say that it is indeed egological because it cannot be otherwise. While I can mourn the death of the stranger, I do so in large part because it reflects my own death and fear of death back to me. This is not something for which we should be made to feel guilty. Nor does it mean that we take the death of another lightly or that we might be inclined to be cavalier about it—quite the opposite in fact.

30. Ricoeur, *Memory, History, Forgetting*, 361.

31. Ibid., 363.
32. Ibid., 367.
33. Tzvetan Todorov, "The Abuses of Memory," *Common Knowledge* 5 (1996): 6–26. I would like to thank Forrest Clingerman for directing me to this interesting and helpful article from Todorov.
34. Ibid., 14.
35. Ricoeur, *Memory, History, Forgetting*, 89.
36. For more on this, see Donohoe, "Rushing to Memorialize," 6–12.
37. Ricoeur, *Memory, History, Forgetting*, 347.

14. My Place in the Sun
David Wood

1. See Descartes' *res extensa*.
2. "Having a place means that you know what a place means." This famous quote was from an interview with Gary Snyder in *The Paris Review* in 1996. http://www.theparisreview.org/interviews/1323/the-art-of-poetry-no-74-gary-snyder (accessed January 30, 2013).
3. Ed Casey, *The Fate of Place: A Philosophical History* (Berkeley: University of California Press, 1998); Ed Casey, *Getting Back into Place* (Bloomington: Indiana University Press, 2009).
4. Even German colonialism made use of this sense, however disingenuously: Foreign Minister Bernhard von Bülow used it to launch Germany's policy of *Realpolitike*. In a parliamentary debate, he argued that "[i]n one word: we do not want to place anyone into the shadow, but we also claim our place in the sun" [1897].
5. Emmanuel Levinas, "Ethics as First Philosophy," in *The Levinas Reader*, ed. Sean Hand (Oxford: Blackwell, 1997), 82.
6. John Locke, *Second Treatise on Government* (Cambridge: Cambridge University Press, 1988), sect. 27.
7. Consider Ezra Pound's question: Would you prefer an empty cell, or one with an open sewer running through it? And stories of prisoners befriending the cell mouse. Anything to establish relation.
8. If a home is essentially a shared space, and a basic right, solitary confinement would be a violation of such a right, facilitated perhaps by a confusion between space and place.
9. Things get more problematic with Borges' discussion of exactly rewriting Cervantes' *Don Quixote* as a different book. These issues become more concrete when one thinks of intellectual property rights to popular songs, for example.
10. This criterion could be and was used to dispossess indigenous people of their land in Australia and in North America when their way of owning territory—nomadic, for example—did not meet the "mixing labor" specification.

11. It is rare but not unheard of for parties stranded in remote locations to eat their dead companions. Whether this is legal, it is understandable. Killing those who would otherwise be dead tomorrow for dinner today is equally understandable under the circumstances, even though it is hard to imagine a legal system that would condone it.

12. See, for example, George Santayana, *The Realms of Being* (New York: Charles Scribner's Sons, 1942). I owe this reference to my colleague John Lachs.

13. It is worth noting how irresistible spatial analogies are. Bergson, noted for lamenting our tendency to think of time spatially, writes of memory as an inverted cone, a pyramid.

14. A common book title. See, for instance, Thomas Huxley's [1863] *Man's Place in Nature* (New York: Modern Library, 2001), and Teilhard De Chardin [1966], *Man's Place in Nature* (New York: Harper Collins, 2000). Each advocates an evolutionary approach, with the latter extending evolution into a spiritual dimension, and speculates on the future. Charles F. Hockett, in his *Man's Place in Nature* (New York: McGraw-Hill, 1973), offered an account of the overlap between human and animal linguistic capacity. While many components are variously shared with nonhumans, only humans have the full complement of linguistic capacities.

15. Reuven Firestone, "Jerusalem: Jerusalem in Judaism, Christianity, and Islam," in *Encyclopedia of Religion* Second Edition (2005), 4838. This article is a brilliant account of the various sanctifications of Jerusalem in its service to different religions.

16. These quotes are not cited for their historical accuracy (or otherwise) but for their exemplary status in explaining how Jerusalem can be claimed by all sides.

17. http://www.thefreedictionary.com/Jerusalem.

18. http://www.jewishvirtuallibrary.org/jsource/vie/Jerusalem1.html.

19. http://www.goisrael.com/Tourism_Eng/Tourist+Information/ Discover+Israel/Cities/Jerusalem.htm.

20. http://en.wikipedia.org/wiki/Religious_significance_of_Jerusalem.

15. How Hermeneutics Might Save the Life of (Environmental) Ethics
Paul van Tongeren and Paulien Snellen

The title alludes, of course, to S. Toulmin's famous article "How Medicine saved the Life of Ethics," *Perspectives in Biology and Medicine* 25 (1982) 4: 736–750. Parts of this chapter are shortened and reworked versions of sections from: Paul van Tongeren, "Ethics and the Limits of Philosophy," in *Gadamer's Hermeneutics and the Art of Conversation*, ed. Andrew Wiercinski. International Studies in Hermeneutics and Phenomenology, Volume 2. (Berlin: Lit Verlag), 2011. The first author expresses his gratitude to NIAS, the Netherlands Institute for Advanced Study, for providing him with the opportunity, as Fellow-in-Residence 2010/2011, to complete this chapter.

1. See, for example, Clare Palmer, "An Overview of Environmental Ethics," in *Environmental Ethics: An Anthology*, ed. A. Light and H. Rolston III (Malden, Mass.: Blackwell Publishing, 2003), 15–37; and Martin Drenthen, "Milieu-ethiek," in *Lexicon van de ethiek*, ed. M. Becker et al. (Assen: Van Gorcum, 2007), 218–221.

2. Bernard Williams, *Ethics and the Limits of Philosophy* (London: Fontana, 1985).

3. Ibid., 3.

4. Ibid., 74.

5. Ibid., 174.

6. Plato, "Apology," *Complete Works*, trans. John M. Cooper (Indianapolis: Hackett Publishing Company, 1997), 38a.

7. Williams, *Ethics and the Limits of Philosophy*, 74.

8. Ibid.

9. Ibid., 26.

10. Ibid., 52.

11. Ibid., 63.

12. Ibid., 64.

13. Ibid., 67.

14. Ibid., 70.

15. Ibid., 71.

16. Ibid., 72.

17. Ibid., 74.

18. Ibid., 84.

19. Ibid., 110.

20. Ibid., 80.

21. Ibid., 99.

22. Ibid., 111.

23. Hans-Georg Gadamer, "Hermeneutik als praktische Philosophie," in *Vernunft im Zeitalter der Wissenschaft* (Frankfurt: Suhrkamp, 1976), 78–109.

24. Williams, *Ethics and the Limits of Philosophy*, 93.

25. cf. Paul van Tongeren, "Moral Philosophy as a Hermeneutics of Moral Experience," *International Philosophical Quarterly* 34, no. 2 (1994): 199–214.

26. Williams, *Ethics and the Limits of Philosophy*, 88.

27. Ibid., 110.

28. Ibid., 186.

29. Martin Drenthen, "Ecological Restoration and Place Attachment: Emplacing Non-Places?" *Environmental Values* 18, no. 3 (2009): 285–312. See also Drenthen's contribution to this volume.

30. Williams, *Ethics and the Limits of Philosophy*, 129.

31. Ibid., 118.

32. Ibid., 184.

33. Ben A. Minteer, "Intrinsic Value for Pragmatists?" *Environmental Ethics* 23, no. 1 (2001): 58.

34. Gadamer, *Truth and Method*, 2nd rev. ed., 336; and Gadamer, *Wahrheit und Methode, Grundzüge einer philosophischen Hermeneutik* (Tübingen: Mohr, 1975), 355.

35. Williams, *Ethics and the Limits of Philosophy*, 174.

36. Ibid., 178.

37. Ibid., 182.

38. Ibid., 196.

39. See Paul van Tongeren, "The Relation of Narrativity and Hermeneutics to an Adequate Practical Ethic," *Ethical Perspectives* 1, no. 2 (1994): 57–71.

40. Williams, *Ethics and the Limits of Philosophy*, 112.

41. Gadamer, "Hermeneutik als praktische Philosophie," 108. English translation: "Understanding always harvests a broadened and deepened self-understanding. But that means hermeneutics is philosophy, and as philosophy it is practical philosophy." Hans-Georg Gadamer, "Hermeneutics as Practical Philosophy," in *The Gadamer Reader: A Bouquet of the Later Writings*, ed. Richard E. Palmer (Evanston, Ill.: Northwestern University Press, 2007), 245.

42. Williams, *Ethics and the Limits of Philosophy*, 168.

A Bibliographic Overview of Research in Environmental Hermeneutics

As suggested by both the introduction and the individual chapters of this volume, environmental hermeneutics continues to evolve. The works of environmental hermeneutics presented here are, in fact, based on works that have been discussed and debated in environmental philosophy for years—if not decades. To fully appreciate the emergence of environmental hermeneutics as a self-conscious dialogue, then, it is important to note some of the earlier work that has been done. Obviously, environmental hermeneutics attempts to address the early issues of environmental ethics and philosophy, including the works of seminal figures such as Aldo Leopold and John Muir, as well as more recent philosophers. But there is also a small, growing literature that is explicitly hermeneutical in its orientation.

This bibliography includes only items that are directly relevant to philosophical hermeneutics and environmental thought. It also concentrates on English language works. Certainly a larger bibliography could include works that are influential but not directly addressed to the topic, such as *Sand County Almanac* or *Truth and Method*. Thus, while it makes no claims to being exhaustive, the editors do hope that this initial bibliography presents an opportunity to further explore the emerging field of environmental hermeneutics.

Ablett, Phillip G., and Pamela K. Dyer. "Heritage and Hermeneutics: Towards a Broader Interpretation of Interpretation." *Current Issues in Tourism* 12 (2009): 209–233.
Balabanski, V. S. "Ecological Hermeneutics as a Daughter of Feminism: Reflections on the Earth Bible Project." *Women-Church* 40 (2007): 145–149.

Bell, Nathan, M. *The Green Horizon: An (Environmental) Hermeneutics of Identification with Nature through Literature.* Master's thesis, University of North Texas, 2010.

Beringer, Almut. *The Moral Ideals of Care and Respect: A Hermeneutic Inquiry into Adolescents' Environmental Ethics and Moral Functioning* (Akademische Hochschulschriften Reihe XX, Philosophie). Frankfurt, Germany: Peter Lang Publishing Group, 1994.

Buitendag, Johan. "Nature as Creation from an Eco-Hermeneutical Perspective: From a 'Natural Theology' to a 'Theology of Nature.'" *Herv. teol. stud.* 65 (2009): 1–10.

Cameron, W. S. K (Scott). "Can Cities be both Natural and Successful? Reflections Grounding Two Apparently Oxymoronic Aspirations." In *The Natural City: Re-Envisioning the Built Environment*, edited by Ingrid Leman Stefanovic and Stephen Bede Scharper, 36–49. Toronto: University of Toronto Press, 2011.

———. "Socrates outside Athens: Plato, the Phaedrus, and the Possibility of 'Dialogue' with Nature." In *Phenomenology 2010, Volume 5: Selected Essays from North America, Part 2*, edited by Lester Embree, Michael Barber, and Thomas J. Nenon, 40–65. Bucharest: Zeta Books, 2010.

———. "Tapping Habermas's Discourse Theory for Environmental Ethics." *Environmental Ethics* 31 (Winter 2009): 339–357.

———. "Wilderness in the City: Not Such a Long Drive after All." *Environmental Philosophy* 3/2 (2006): 28–33.

Clingerman, Forrest. "Beyond the Flowers and the Stones: 'Emplacement' and the Modeling of Nature." *Philosophy in the Contemporary World* 11 (2004): 17–24.

———. "Environmental Amnesia or the Memory of Place? The Need for Local Ethics of Memory in a Philosophical Theology of Place." *Religion and Ecology in the Public Sphere*, edited by Celia Deane-Drummond and Heinrich Bedford-Strohm, 141–159. New York: T. & T. Clark, 2011.

———. "From Artwork to Place: Finding the Voices of Moreelse, Bacon and Beuys at the Hermeneutical Intersection of Culture and Nature." *Environmental Philosophy* 8 (2011): 1–24.

———. "Interpreting Heaven and Earth: The Theological Construction of Nature, Place, and the Built Environment." In *Nature, Space and the Sacred*, edited by Sigurd Bergmann, Peter Scott, Heinrich Bedford Strohm, and Maria Jansdotter, 45–54. Burlington, Vt.: Ashgate Publishing, 2009.

———. "The Intimate Distance of Herons: Theological Travels through Nature, Place, and Migration." *Ethics, Place & Environment* 11 (2008): 313–325.

———. "Reading the Book of Nature: A Hermeneutical Account of Nature for Philosophical Theology." *Worldviews: Global Religions, Culture, and Ecology* 13 (2009): 72–91.

———. "Wilderness as the Place between Philosophy and Theology: Questioning Martin Drenthen on the Otherness of Nature." *Environmental Values* 19 (2010): 211–232.

Coolen, Maarten. "Toward a Hermeneutics of Nature: On the Necessity of Enduring Distance." In *Ecology, Technology and Culture*, edited by Wim Zweers and Jan J. Boersema, 118–126. Cambridge, UK: The White Horse Press, 1994.

Coronel, D. A., J. M. Alves da Silva, and A. Leonardi. "Hermeneutical and Philosophical Consideration about Ethics and Sustainable Development." *Global Journal of Human Social Studies* 10/2 (2010): 30–37.

Cronon, William. "A Place for Stories: Nature, History, and Narrative." *Journal of American History* 78 (1992): 1347–1376.

Deliege, Glenn. "Restoring or Restorying Nature?" In *Nature, Space and the Sacred*, edited by Sigurd Bergmann, Peter Scott, Heinrich Bedford Strohm, and Maria Jansdotter, 189–199. Aldershot: Ashgate, 2009.

———. "Toward a Richer Account of Restorative Practices." *Environmental Philosophy* 4 (2007): 135–147.

De Moura Carvalho, Isabel Cristina, Mauro Grün, and Maria Rita Avanzi. Paisagens da compreensão: contribuições da hermenêutica e da fenomenologia para uma epistemologia da educação ambiental [Understanding Landscape: Contributions from Hermeneutic and Phenomenology toward an Epistemology of Environmental Education]. *Cad. CEDES* 29/77 (2009): 99–115.

Donohoe, Janet. "Where Were You When...? On the Relationship between Individual and Collective Memory." *Philosophy in the Contemporary World* 16 (2009): 105–113.

Drenthen, Martin. "Ecological Restoration and Place Attachment: Emplacing Non-Places?" *Environmental Values* 18 (2009): 285–312.

———. "Fatal Attraction: Wildness in Contemporary Film." *Environmental Ethics* 31 (2009): 297–315.

———. *Grenzen aan wildheid: wildernisverlangen en de betekenis van Nietzsches moraalkritiek voor de actuele milieu-ethiek* [Bordering Wildness: The Desire for Wilderness and the Meaning of Nietzsche's Critique of Morality for Environmental Ethics]. Budel: Damon, 2003.

———. "New Wilderness Landscapes as Moral Criticism: A Nietzschean Perspective on Our Contemporary Fascination with Wildness." *Ethical Perspectives* 14 (2007): 371–403.

———. "NIMBY and the Ethics of the Particular." *Ethics, Place & Environment* 14 (2010): 321–323.

———. "The Paradox of Environmental Ethics: Nietzsche's View of Nature and the Wild." *Environmental Ethics* 21 (1999): 163–175.

———. "Reading Ourselves through the Land: Landscape Hermeneutics and Ethics of Place." In *Placing Nature on the Borders of Religion,*

Philosophy, and Ethics, edited by Forrest Clingerman and Mark H. Dixon. Aldershot: Ashgate, 2011.

———. "Wildness as a Critical Border Concept: Nietzsche and the Debate on Wilderness Restoration." *Environmental Values* 14 (2005): 317–337.

Drenthen, Martin, Mirjam de Groot, and Wouter de Groot. "Public Visions of the Human/Nature Relationship and their Implications for Environmental Ethics." *Environmental Ethics* 33 (2011): 25–44.

Dupré, Louis. *Passage to Modernity: An Essay on the Hermeneutics of Nature and Culture*. New Haven, Conn.: Yale University Press, 1993.

Falter, Reinhard, and Jürgen Hasse. "Landschaftsgeografie und Naturhermeneutik—Zur Ästhetik erlebter und dargestellter Natur" [Landscape Geography and Hermeneutic of Nature—On the Aesthetics of Lived and Depicted Nature]. *Erdkunde* 55/2 (2001): 121–137.

Gare, Arran. "MacIntyre, Narratives, and Environmental Ethics." *Environmental Ethics* 20 (1998): 3–21.

Gebauer, Michael, and Ulrich Gebhard, eds. *Naturerfahrung: Wege zu einer Hermeneutik der Natur* [Nature Experience: Towards a Hermeneutics of Nature]. Kusterdingen: Die Graue Edition, 2005.

Grassie, William. *Reinventing Nature: Science Narratives as Myths for an Endangered Planet*. Doctoral dissertation, Temple University, 1994.

Grove-White, Robin, and Bronislaw Szerszynski. "Getting Behind Environmental Ethics." *Environmental Values* 1 (1992): 285–296.

Habel, Norman C., and Peter Trudinger, eds. *Exploring Ecological Hermeneutics*. Atlanta: Society of Biblical Literature, 2008.

Horrell, David G., Cherryl Hunt, and Christopher Southgate. "Appeals to the Bible in Ecotheology and Environmental Ethics: A Typology of Hermeneutical Stances." *Studies in Christian Ethics* 21 (2008): 219–238.

Horrell, David G., Cherryl Hunt, Christopher Southgate, and Francesca Stavrakopoulou, eds. *Ecological Hermeneutics: Biblical, Historical, and Theological Perspectives*. London: T. & T. Clark International, 2010.

Ihde, Don. *Technology and the Lifeworld*. Bloomington: Indiana University Press, 1990.

Irrgang, Bernhard. *Hermeneutische Ethik: Pragmatisch-ethische Orientierung in technologischen Gesellschaften* [Hermeneutical Ethics: Pragmatical-Ethical Orientation in Technological Societies]. Wissenschaftliche Buchgesellschaft: Darmstadt, 2007.

Jardine, David W. *To Dwell With a Boundless Heart: Essays in Curriculum Theory, Hermeneutics, and the Ecological Imagination*. New York: Peter Lang, 1998.

Joldersma, Clarence W. "How Can Science Help Us Care for Nature? Hermeneutics, Fragility, and Responsibility for the Earth." *Educational Theory* 59/4 (2009): 465–483.

Junges, José Roque, and Lucilda Selli. "Bioethics and Environment: A Hermeneutic Approach." *Journal International de Bioéthique* 19 (2008): 105–119, 198–199.

Keller, David R. "Ecological Hermeneutics." Paper presented at Twentieth World Congress of Philosophy, Boston, MA, August 10–15, 1998. Retrieved October 14, 2011, at http://www.bu.edu/wcp/Papers/Envi/ EnviKell.htm.

Kidder, Paul. "Philosophical Hermeneutics and the Ethical Function of Architecture." *Contemporary Aesthetics* 9 (2011). Retrieved September 10, 2012, at http://www.contempaesthetics.org/newvolume/pages/article .php?articleID=618.

Kockelkoren, Petran. "The House in the Cat's Claws: A Framework for a Hermeneutics of Nature." In *Ecology, Technology and Culture*, edited by Wim Zweers and Jan J. Boersema. Cambridge, UK: The White Horse Press, 1994.

Lijmbach, Susanne. "Potter's Bull and Castrated Pigs: Considering the Impossibility of a Hermeneutic Natural Science." In *Ecology, Technology and Culture*, edited by Wim Zweers and Jan J. Boersema. Cambridge, UK: The White Horse Press, 1994.

Markoš, Anton. *Readers of the Book of Life: Contextualizing Developmental Evolutionary Biology*. Oxford: Oxford University Press, 2002.

Markoš, Anton, et al. *Life as Its Own Designer: Darwin's Origin and Western Thought*. Dordrecht: Springer, 2009.

Medeiros, Paul Joseph. *Juxtaposing Aldo Leopold and Martin Heidegger: Interpretation, Time, and the Environment*. Doctoral dissertation, Duke University, 2000.

Michelfelder, Diane P. "Contemporary Continental Philosophy and Environmental Ethics: A Difficult Relationship?" In *Rethinking Nature: Essays in Environmental Philosophy*, edited by Bruce V. Foltz and Robert Frodeman. Bloomington: Indiana University Press, 2004.

Mugerauer, Robert. "Deleuze and Guattari's Return to Science as a Basis for Environmental Philosophy." In *Rethinking Nature: Essays in Environmental Philosophy*, edited by Bruce V. Foltz and Robert Frodeman. Bloomington: Indiana University Press, 2004.

———. *Interpretations on Behalf of Place: Environmental Displacements and Alternative Responses*. Albany: State University of New York Press, 1994.

———. *Interpreting Environments. Tradition, Deconstruction, Hermeneutics*. Austin: University of Texas Press, 1995.

Patterson, Michael E., Alan E. Watson, Daniel R. Williams, and Joseph R. Roggenbuck. "An Hermeneutic Approach to Studying the Nature of Wilderness Experiences." *Journal of Leisure Research* 30 (2008): 423–452.

Pile, S. "Depth Hermeneutics and Critical Human Geography." *Environment and Planning D: Society and Space* 8 (1990): 211–232.

Postma, Dirk. *Why Care for Nature? In Search for an Ethical Framework for Environmental Responsibility and Education*. Dordrecht: Springer, 2006.

Sammel, Ali. "An Invitation to Dialogue: Gadamer, Hermeneutic Phenomenology, and Critical Environmental Education." *Canadian Journal of Environmental Education* 8 (2003): 155–168.

Schönherr, Hans-Martin. *Die Technik und die Schwäche. Ökologie nach Nietzsche, Heidegger und dem 'schwachen' Denken.* [Technology and Weakness: Ecology after Nietzsche, Heidegger and 'Weak' Thought]. Wien: Edition Passagen, 1989.

———. "Ökologie als Hermeneutik—Ein wissenschaftstheoretischer Versuch" [Ecology as Hermeneutics—a Scientific Attempt]. *Philosophia Naturalis* 24 (1987): 311–332.

———. *Von der Schwierigkeit, Natur zu Verstehen. Entwurf einer negativen Ökologie* [The Difficulty of Understanding Nature: A Model for a Negative Ecology]. Frankfurt am Main: Fischer-Verlag Reihe Perspektiven, 1989.

Seamon, David, and Robert Mugerauer, eds. *Dwelling, Place & Environment: Towards a Phenomenology of Person and World.* New York: Columbia University Press, 1985.

Smith, Mick. "Environmental Antinomianism: The Moral World Turned Upside Down?" *Ethics and the Environment* 5 (2000): 125–139.

———. "The Face of Nature: Environmental Ethics and the Boundaries of Contemporary Social Theory." *Current Sociology* 49 (2001): 49–65.

———. "Hermeneutics and the Culture of Birds: The Environmental Allegory of 'Easter Island.'" *Ethics, Place & Environment* 8 (2005): 21–38.

———. "Lost for Words? Gadamer and Benjamin on the Nature of Language and the 'Language' of Nature." *Environmental Values* 10 (2001): 59–75.

———. "On 'Being' Moved by Nature." In *Emotional Geographies*, edited by J. Davidson, L. Bondi, and M. Smith, 219–230. Aldershot: Ashgate, 2005.

Smith, Nicholas H. *Strong Hermeneutics: Contingency and Moral Identity.* London: Routledge, 1997.

Stables, Andrew. "The Landscape and the 'Death of the Author.'" *Canadian Journal of Environmental Education* 2 (1997): 104–113.

Stefanovic, Ingrid Leman. *Safeguarding Our Common Future: Rethinking Sustainable Development.* Albany: State University of New York Press, 2000.

Taghvaei, Hassan. "Tacit Knowledge and Deep Ecology: A Hermeneutic Approach to the Concept of Tacit Environmental Knowledge in Landscape Architecture." *Environmental Sciences* 6 (2008): 111–122.

Treanor, Brian. "Environmentalism and Public Virtue." *Agricultural and Environmental Ethics* 23 (2010): 9–28.

———. "Narrative Environmental Virtue Ethics: *Phronesis* without a *Phronimos.*" *Environmental Ethics* 30 (2008): 361–379.

———. "Turn Around to Step Forward: Ideology and Utopia in the Environmental Movement." *Environmental Philosophy* 7 (2010): 27–46.

Trepl, Ludwig. "Ökologie als konservative Naturwissenschaft: Von der schönen Landschaft zum funktionierenden Ökosystem" [Ecology as a

Conservative Science: From Landscape Beauty to Functional Ecosystems].
Urbs et Regio. Kasseler Schriften zur Geographie und Planung 65 (1997):
467–492.

———. "Stadtnatur - in ökologisch-funktionalistischer und hermeneutischer
Betrachtung" [City Nature from a Functional-ecological and Hermeneutic
Perspective]. In *Rundgespräche der Kommission der Ökologie, Band 3:
Stadtökologie*, edited by Bayerische Akademie der Wissenschaften, 53–58.
München: F. Pfeil, 1992.

Trigg, Dylan. "Altered Place: Nostalgia, Topophobia, and the Unreality of
Memory." *Existential Analysis* 18 (2007): 155–169.

Utsler, David. "Paul Ricoeur's Hermeneutics as a Model for Environmental
Philosophy." *Philosophy Today* 53 (2009): 173–178.

———. "Who am I, Who Are These People, and What is This Place? A
Hermeneutic Account of the Self, Others, and Environments." In *Placing
Nature on the Borders of Religion, Philosophy, and Ethics*, edited by Forrest
Clingerman and Mark H. Dixon, 139–151. Aldershot: Ashgate, 2011.

Van Buren, John. "Critical Environmental Hermeneutics." *Environmental
Ethics* 17 (1995): 259–275.

Contributors

Nathan M. Bell is a doctoral candidate in the Department of Philosophy and Religion Studies at the University of North Texas. His main research interest is the application of hermeneutics to environmental philosophy, with further interests in eco-phenomenology, environmental justice, and literature and the environment. He has presented extensively on various aspects of hermeneutic application to environmental thought and is the author of one of the first graduate theses on environmental hermeneutics (MA from the University of North Texas).

John van Buren is professor of philosophy and director of environmental policy at Fordham University. He is the author of *The Young Heidegger: Rumor of the Hidden King* (Indiana University Press, 1994), editor of *Reading Heidegger from the Start: Essays in His Earliest Thought* (SUNY Press, 1994), Heidegger's 1923 lecture course *Ontology—The Hermeneutics of Facticity* (Indiana University Press, 1999), and the Heidegger anthology *Supplements—From the Earliest Essays to Being and Time and Beyond* (SUNY Press, 2002).

Scott Cameron is a Canadian born on the prairies and raised in Northern Ontario. He left Queen's University to counsel alcoholic youth in Thunder Bay before pursing his PhD at Fordham University among the workaholics of New York City. When that ended, he took a one-year post at Loyola Marymount University in Los Angeles where he continues to teach and do research in his eighteenth year. While his early work tied hermeneutics to critical theory, a sabbatical allowed him to develop his current interest: using hermeneutical tools to resolve long-standing perplexities in environmental philosophy.

Forrest Clingerman is associate professor of philosophy and religion at Ohio Northern University. Along with Mark H. Dixon, he is coeditor of *Placing*

Nature on the Borders of Philosophy, Religion, and Ethics (Ashgate, 2011). In addition, he has published a number of articles on environmental thought. His main research focus is on issues of place in environmental philosophy and theology, but he has also written on topics of pedagogy in higher education.

Janet Donohoe is professor of philosophy at the University of West Georgia. She is the author of *Husserl on Ethics and Intersubjectivity: From Static to Genetic Phenomenology* (Humanity Books, 2004) as well as many articles ranging from feminist phenomenology to Heidegger and Benjamin to interpretations of the place of home. Her current research is grounded in phenomenology, using that method as an approach to issues of collective memory and the built environment with a focus on monuments.

Martin Drenthen is associate professor of philosophy at Radboud University Nijmegen (The Netherlands). Together with Jozef Keulartz and Jim Proctor, he coedited *New Visions of Nature: Complexity and Authenticity* (Springer, 2009). In English and Dutch publications, he has written about the significance of Nietzsche's critique of morality for environmental ethics, the concept of wildness in debates on ecological restoration, and ethics of place. His most recent research focuses on the relationship between landscapes, cultures of place, and moral identity.

Christina M. Gschwandtner is associate professor of philosophy at the University of Scranton. She holds a PhD in French phenomenology from DePaul University (Chicago) and a PhD in ecological theology from the University of Durham (Durham, UK). She has published in the areas of religious phenomenology and hermeneutics. She is author of *Reading Jean-Luc Marion: Exceeding Metaphysics* (Indiana University Press, 2007). Her most recent book is titled *Postmodern Apologetics?* (Fordham University Press, 2012).

Sean J. McGrath researches in the areas of phenomenology, hermeneutics, metaphysics, and the history of ideas. The author of several works on German philosophy, including *The Early Heidegger and Medieval Philosophy: Phenomenology for the Godforsaken* (Catholic University of America, 2006) and *The Dark Ground of Spirit: Schelling and the Unconscious* (Routledge, 2012), he has been teaching environmental philosophy for many years.

Robert Mugerauer is professor and dean emeritus in the Departments of Architecture, Urban Design and Planning, and adjunct in Landscape Architecture at the University of Washington. He first advocated an environmental hermeneutics with "Language and the Emergence of Environment" in *Dwelling, Place, Environment* (1985), then followed with *Interpretations on Behalf of Place* (SUNY Press, 1994) and *Interpreting Environments* (University of Texas Press, 1995). His current research applies continental thought and dynamic complexity to biocultural environments and issues of well-being. Recent publications include *Heidegger and Homecoming* (University of Toronto, 2008) and *Environmental Dilemmas* (with Lynne Manzo; Lexington Books, 2008).

Mick Smith is associate professor and Queen's National Scholar at Queen's University in Kingston, Ontario. His latest book is titled *Against Ecological Sovereignty: Ethics, Politics and Saving the Natural World* (Minneapolis: University of Minnesota Press, 2011). He is a founding editor of *Emotion, Space and Society* and a regular contributor to journals such as *Environmental Ethics*.

Paulien Snellen, MA, graduated cum laude with a specialization in philosophical ethics from the Radboud University Nijmegen (The Netherlands) and is currently a PhD student in philosophy at the Rijksuniversiteit Groningen (RUG) where she is writing a thesis titled "The Moral Relevance of Weakness of Will: A Dispositional Account." Her work focuses on topics such as moral motivation. She also studies philosophical questions of irrational behavior, such as the weakness of will and self-deception. Her interest in these topics is often placed in conversation with environmental issues.

Paul J. M. van Tongeren is professor of moral philosophy at Radboud University Nijmegen (The Netherlands), and special professor of ethics at the University of Louvain (Belgium). He studied theology in Utrecht (The Netherlands) and philosophy in Louvain (Belgium), where he also earned his PhD. His doctoral dissertation was on Nietzsche's critique of morality. He has published widely on Nietzsche and on ethics.

Brian Treanor is associate professor of philosophy and director of environmental studies at Loyola Marymount University. He has published and presented widely in the fields of ethics, environmental philosophy, and philosophy of religion and is the author of *Aspects of Alterity: Levinas, Marcel, and the Contemporary Debate* (Fordham University Press, 2006) and co-editor of *A Passion for the Possible: Thinking with Paul Ricoeur* (Fordham University Press, 2010). He is currently completing a manuscript applying hermeneutic resources to environmental virtue ethics.

Dylan Trigg is a research fellow at The Centre de Recherche en Épistémologie Appliquée, Paris. He earned his PhD at the University of Sussex, where he also taught philosophy for several years. His research concerns the phenomenology of embodiment and place, with interests including memory and materiality, spatial phobias, and environmental aesthetics. In addition to several articles, Trigg is the author of *The Memory of Place: A Phenomenology of the Uncanny* (Ohio University Press, 2012) and *The Aesthetics of Decay: Nothingness, Nostalgia and the Absence of Reason* (Peter Lang, 2006).

David Utsler is a PhD candidate at the University of North Texas in the Department of Philosophy and Religion Studies. His current areas of research focus on hermeneutics, critical theory, and their application to environmental philosophy. He has published essays in environmental hermeneutics and presented several conference papers on environmental hermeneutics and environmental justice. He is currently working on a manuscript (along with Robert Melchior Figueroa of UNT) on the application of hermeneutics to environmental justice studies and activism.

David Wood has taught philosophy in Europe and the United States for over thirty years and has published sixteen books, among them: *Time after Time* (Indiana University Press, 2007); *The Step Back: Ethics and Politics After Deconstruction* (SUNY Press, 2005); *Thinking After Heidegger* (Polity Press, 2002); and *Philosophy at the Limit* (Routledge, 1990). In addition to teaching at Vanderbilt University, where he is centennial professor of philosophy, and Joe B. Wyatt distinguished university professor, he also codirects (with Beth Conklin) a research program in ecology and spirituality for the Centre for the Study of Religion and Culture. Dr. Wood's teaching and research interests lie in the possibilities of reading and thinking opened up by contemporary continental philosophy and by nineteenth-century German thought. He is also an environmental artist.

Index

Abbey, Edward 92
abstraction 9, 182, 187, 192–93, 198,
 205, 271–72, 306
activism 9, 11, 151, 267, 311
adonné 87, 94
aesthetic 11, 18–20, 23, 30, 54, 109,
 111, 116, 164, 189–192, 203–210,
 216, 223, 263
agoraphobia 160–177
agriculture 4, 36, 129, 226–229, 231,
 234
alchemy 216–223
aletheia 67
alienation 8, 30, 109, 174, 232
alter-modern 202, 216
American 17, 18, 33, 104–6, 145, 182,
 233, 281, 285, 288
anarchism 24, 40, 41, 62
animals (non-human) 82, 98, 100,
 110–12, 119, 131, 141–42, 147–52,
 183–84, 196–97, 219, 264, 286
Anselm, Saint 68
anthropocentrism 6–9, 24, 99, 101–6,
 149, 171, 196–97, 226, 233–35,
 236–37, 254, 311
anthropogenic 226, 235
anthropology 20–22, 42

aporia 6, 13, 47, 99–100
application 7–9, 18–21, 33, 91, 94,
 140–41, 158, 208, 223, 254
architecture 3, 6, 20, 65, 216
Arctic National Wildlife Refuge
 (ANWR) 183
Arendt, Hannah 59, 80, 264, 265, 270,
 271, 274
Aristotle 18, 32, 143, 185, 213, 217,
 299–300, 307
art/the artistic 1–2, 19, 29, 35, 53–54,
 78–87, 90, 132, 146, 182–84, 189,
 219–24, 249–50, 310
artist(s) 3, 25, 54, 76–79, 90, 94, 98–99,
 190–91, 270
as-if 189
atmosphere 30, 95, 261
attention 43, 49, 116, 138, 186, 220, 237,
 286
Auerbach, Frank 78–80
Augustine, Saint 2, 216, 256, 257, 260
authenticity 123, 208, 210, 227, 232
authority 28, 40–41, 59, 89, 109, 111,
 148–53, 207, 215, 271, 307
autopoiesis 69, 73
awareness 43, 48, 89, 111, 227, 257, 258,
 261, 294

Bachelard, Gaston 75, 76
Badiou, Alain 201
Baudelaire, Charles 161
beauty 304, 306
being-in-the-world 18, 21, 111, 118, 145, 163–68, 174–75, 241
being-toward-death 274–75
beings 8–10, 24, 28, 43, 51–60, 69, 72, 104–6, 132, 176, 185, 196, 205, 209, 222–24, 239, 241, 248–50, 253, 275, 289, 306, 312
belief 11, 75, 98, 116, 136–37, 155, 175, 186–91, 194–96, 200, 204, 225, 231–32, 239, 292–93, 301–4
Bell, Nathan M. 66, 97, 100–1
Biehl, Amy 267–69, 275
biocentrism 8, 22, 24, 99–100
biodiversity 226, 229–32, 236–37, 262, 303
biology/the biotic 19–23, 26, 40, 46, 69, 71, 91, 129, 205, 238, 274
bioregionalism 19, 24, 28, 29–30, 35, 238
Booth, James 266–69
borders (and boundaries) 47, 161–63, 182, 227–30, 286
Brady, Emily 188–89
Brower, David 183, 199
Bruntland Commission 23
buddhism 205, 210
van Buren, John 12
business (as a practice) 20, 25–27, 34

calculation 104, 201–2, 216–17, 222–24, 257, 281
Callicott, J. Baird 9, 137–40
capitalism 36, 45, 50, 64, 205, 213, 223
care/care ethics 41, 147, 198, 225, 238, 239, 274, 306
Carlson, Allen 189, 190
Cartesianism 52, 55, 87, 90, 99, 102, 106, 211, 217–19, 223

causation 51, 95–96, 145
certainty 53, 90, 96–98, 181, 270
Chauvin, Rémy 38
chemistry 54, 70–72, 104, 189, 193, 205, 220
children 33, 37, 93, 138, 190–91, 264, 272
choice 31, 37–41, 211, 302
Christianity 2, 24, 42, 75, 77, 90, 105, 216, 219–23, 293, 294
citizenship 32, 75, 105, 184, 239, 268
Clayton, Susan 135, 136
climate 23, 30, 95–97, 104, 129, 181, 196, 234, 262
climate change 95–97, 129, 196, 262
Clingerman, Forrest 66, 91, 95, 98, 101, 126, 127, 169, 170
commonsense 62, 187
communication (as a practice) 8, 13, 33, 132
communism 36, 45, 48, 289
community 8–10, 14, 19, 22–27, 33, 37, 63, 119, 126–27, 137, 152–56, 191, 207, 211, 215, 239, 266–69
conformity 32–34, 40, 58, 231
consciousness 4, 43–55, 58–60, 66, 73, 83–84, 91, 100, 170, 209–11
conservation 20, 63, 192, 225–227, 230–234, 237, 277
constructivism 42, 102, 108, 111, 169, 249, 251
cosmology 202, 210, 213–17
creativity 11, 29–30, 33–36, 59, 61, 171, 211, 261–62, 273
culture 6–10, 30, 42, 44–45, 50, 99, 102, 114, 117, 119, 131, 143, 155–56, 164, 203–17, 226, 231–33, 236–40, 252, 260, 263, 269, 271–73, 311
cross-cultural 33, 111
cybernetics 37–38, 45–47, 60–62

danger 5, 30, 46–47, 162, 173, 203
Darwin, Charles 42, 45, 106, 295
Dasein 6, 10, 55, 111, 118, 163, 165, 274
death 37, 86, 92, 103, 115, 161–63, 174,
    183, 209, 215, 265–69, 272–76, 283,
    286, 293–95
deconstruction 201, 205
deep ecology 21, 42, 134, 202, 205
Deleuze, Gilles 222
democracy 32, 34, 45, 117, 276
deontology 299–302, 308
Derrida, Jacques 41, 56, 80, 101, 289, 291
Descartes, René 52, 208
development 3, 23–28, 31, 36, 45, 52,
    63, 73, 75, 125, 134–35, 144, 155–57,
    202–3, 216, 219, 222, 232, 238,
    285, 310
dialogical 7–10, 13, 117, 124–28,
    168, 249
dialogue 2, 6–8, 11, 13–14, 26–27, 30–34,
    58–59, 81, 124–28, 133–35, 140, 158,
    166–68, 252, 274
dichotomy 8, 106, 124, 137–39, 208–9
Dillard, Annie 3, 92–93
Dilthey, Wilhelm 2, 19, 43–60, 95,
    208, 282
displacement 30, 75, 156, 175
domination 53, 108, 292
Donahoe, Janet 92, 258
Drenthen, Martin 66, 98, 101, 199,
    251, 259
dualism 6–9, 102–6, 111, 119–20, 170
duties 262, 269, 277–79, 300, 308
dwelling 18, 22–26, 35, 77, 136, 140,
    215, 231, 248–50, 255–59, 282, 292

earth 18–19, 67–68, 74, 77, 100, 105–7,
    125, 131, 137, 149, 203, 207, 215–21,
    270, 283, 292, 296, 310
Eaton, Marcia 189–200
Eberneck, Sara 260–61

ecofeminism 24, 42, 202, 215
ecohermeneutics 4
ecological citizenship 239
ecological identity 125–27
ecological knowledge 99, 190, 192, 238
ecological restoration 226–27, 231–34,
    237–39
ecological self 66, 141–59
ecology 20–21, 35, 42, 60–66, 134,
    189–92, 201–7, 214, 216, 239
ecology without nature 203–7
ecophenomenology 9, 202–5, 215–16
ecosystem 4, 19, 23–24, 30, 82, 91, 95,
    98, 101, 105, 109, 125, 129–32,
    182–83, 190–92, 199, 229, 231,
    235–37, 248, 303
Elliot, Robert 105, 106, 233
embodiment 10, 66, 162, 166, 202,
    245–46, 250, 254, 256, 259
emergence (in nature) 43, 69, 227
empathy 20, 184
emplaced 10, 126, 202, 241, 250,
    255–57, 263
emplotment 185–88, 192, 250, 278
energy 36, 70, 79, 114, 170, 207, 285–86
Enlightenment, the 52, 111, 116, 215
environmental ethics 5, 18, 21–24, 28,
    103, 139–41, 177, 225, 239–41, 259–60,
    297, 298–312
environmental heritage 126
environmental hermeneutics 3–4, 7,
    10–13, 18, 34, 81, 123–24, 140, 159,
    254, 297, 310, 312
environmental identity 124–26, 138–42,
    252–54
environmental imagination 257, 263
environmental justice 126, 153, 158
environmental philosophy 1–14, 17–18,
    65, 102–3, 115, 120, 123, 137–43, 147,
    152, 158, 177, 202, 232–33, 240,
    248–51, 258, 260, 263, 303

environmental psychology 124–28, 133–35, 138–40
environmental sciences 20, 30
environmentalism 17, 19, 39, 62–64, 93, 103, 104–8, 145, 201–8, 213–15, 264, 304
epistemology 8, 17–18, 21, 27–28, 34, 40, 44, 49, 52, 163, 181, 190, 215, 223
eschatology 38, 41–46
ethical experience 298, 303–8, 312
ethical intentions 141–43, 148, 155–58
ethical obligations 153, 261
evolution 19, 20, 37–43, 49, 55, 75, 105, 119, 204–7, 217, 295
existence 2, 6, 43, 44, 54–57, 61, 98, 126, 139, 166, 192, 207–12, 231, 249–50, 273, 282–84
existential, the 22, 55, 167–69, 172, 176, 254, 276, 279, 282
extinction (of a species) 177, 231, 264, 303

facts/facticity 1, 6, 10, 28, 33, 62, 102, 148, 186, 194–96, 238, 269, 273–74, 290–94, 301–3
falsity/falsehood 169, 189–90, 232, 270
Fichte, Johann Gottlieb 201
Figueroa, Robert M. 125–27, 153–56
food 93, 151, 193
forestry 17–18, 21–35, 92, 208
Foucault, Michel 52, 61, 76, 217, 223
Francis of Assisi, Saint 186
freedom 36–39, 58–61, 110, 115, 201, 285, 300, 310
Freud, Sigmund 161, 174, 203, 209
Fukuyama, Francis 50, 203

Gadamer, Hans-Georg 2–9, 12–17, 36, 46–61, 76, 103, 111–20, 124–27, 131–32, 139–40, 201, 209, 240, 249, 303, 306, 310

Galileo Galilei 215
Gelassenheit 203, 216, 224
genetics 31, 38–41, 44, 54, 62–63, 204
geography (as a discipline) 18, 20–22, 65, 282
geology (as a discipline) 20
Gestell 224
Gifford, Robert 133
God 24, 36, 48, 52, 87, 114, 214–17, 219, 221, 295
Grondin, Jean 46
gross contact 182–200
Guattari, Felix 222
guilt 268, 283, 304

Habermas, Jürgen 2, 18, 32
harm 151–52, 157, 171, 263
health 10, 24, 109, 137–38, 283
Healy, Dermot 187, 188
Hedwidge Polder 228, 230, 231, 237
Hegel, Georg Wilhelm Friedrich 36, 42, 48–52, 56, 108, 111–13, 116, 206, 211, 290
Heidegger, Martin 2, 6, 12, 17, 35, 55–56, 65–67, 76, 87, 111–12, 118, 163–66, 202–16, 223–24, 249, 265, 274–76, 296
heritage landscapes 101, 225, 231–36
hermeneutic circle 89, 143–45, 198
hermeneutic ontology 46, 52, 55–61, 96
hermeneutical ethics 66, 248, 297, 303–4, 308–11
Higgs, Eric 237
Hillman, James 137
historical consciousness 43–52
horizon 7, 11, 13, 50, 55–59, 85, 88–90, 94, 120, 123, 140, 146, 249, 266, 290
human other 97, 101, 141–42, 148–52, 158–59
human-nature relationship 103–6, 141, 236–39

humanism 44–48, 52, 56–57, 216
hunting 21, 22, 150
Husserl, Edmund 83
Hyams, Edward 36–41, 61–64

idealism 49, 56, 108, 110, 118, 216,
    221, 291
idem 129–30, 185, 253
identity 12, 19, 34, 52, 65–66, 78, 87, 101,
    123–47, 152, 155–58, 167, 170–71,
    185, 196, 207–12, 221, 225–38,
    251–57, 269–70, 291–92
Ihde, Don 12
imagination 66, 94, 98, 101, 156,
    160, 184, 188–92, 215, 223, 237,
    245–63, 290
indigenous (peoples, cultures, practices)
    23, 33, 131, 155, 156. *See* native
    peoples
industrialization 143, 155, 234
intelligibility 46–47, 167, 201–5,
    208–10, 222
intentionality 18–21, 87, 150, 165,
    168, 170
interdependence 202–6, 215–17
interdisciplinarity 4, 20, 26
interpretative completeness 50–60
interpretative materialism 46, 96
intersubjectivity 5–8, 107, 164–65,
    258, 291
intimacy 94, 131–32, 175, 186
intuition 28, 83, 87–89, 111, 116, 187,
    291, 302, 307
ipse 130, 185, 253
Islam 77, 294

James, William 31, 136, 195
Jordan, William 237, 293
judgment (as a hermeneutic practice) 27,
    33, 66, 113, 147–51, 186, 277
Jung, Carl 75–76

justice 10, 11, 86–88, 97, 117, 125–26,
    142–43, 152–58, 163, 270, 277–78

Kant, Immanuel 31, 84–86, 116, 201,
    204, 208, 212, 299–300, 307–9
Kaplan, David 136, 156
Katz, Eric 105, 233
Kearney, Richard 184–87, 196–97, 262
Kierkegaard, Søren 80, 116
kosmos 201, 216–19, 221
Kuhn, Thomas 111, 191

Lacan, Jacques 42, 201–14
landscapes 3, 4, 19, 65–66, 77–78, 91, 98,
    101, 110, 170–72, 199, 225–41,
    247–51, 256–59, 271, 289, 305
language 8, 10, 18, 21, 28, 44–47, 52–69,
    78, 85, 96, 103, 107, 111–20, 131–32,
    155, 164, 184, 199, 204, 208–11, 257,
    261, 276, 285, 296
Latour, Bruno 42, 107, 206, 214–15
law 11, 20–22, 37, 41, 43, 277, 287
learning (as a hermeneutic practice) 42,
    116–19, 129, 209, 277
Lefebvre, Henri 75
Leopold, Aldo 3, 11, 92, 131, 150, 186
Levinas, Emmanuel 84–87, 100–1, 148,
    275, 283
liberty 24, 36–37, 268
lifeworld 20–21, 66, 75, 80–81
literature 20, 64, 75–76, 125, 133, 146,
    184, 189, 254
locality 75, 162, 226–32, 238–39, 250,
    255, 284–86
logic 4, 156, 160, 171, 174, 261, 293
logos 32, 56, 211
love 37, 94, 174, 182–90, 196–97

MacIntyre, Alasdair 185
magus 217–22
mapping 25, 281, 286, 289

Marcel, Gabriel 192
Marion, Jean-Luc 82–101
Marx, Karl 36, 41, 50, 108, 118, 289
materialism 40–42, 46–50, 58, 60, 62, 96
mathematics 75, 96, 211
McGrath, Sean J. 91, 94
McKibben, Bill 104, 115
meaning-construction 291–92
measurement 281
mediated experience 5, 182–84
mediation 27, 35, 161, 249–57
memory 12, 66, 69, 73, 78, 86, 101, 127,
    168, 245–49, 251–72, 275–79, 289–92
Merchant, Carolyn 215
metanarrative 35, 45, 203
metaphysics 56–57, 106, 109, 160, 205,
    214, 223
mimesis 185–86, 277
misunderstanding (as a hermeneutic
    concept) 115, 193, 310
modernity 202, 209–10, 216
mortality 55, 155, 274, 275
Morton, Timothy 45, 201–17, 224
Mugerauer, Robert 12
Muir, John 3, 11, 92–93, 182–86, 197–99
mythology 75, 95, 184, 207, 293

narrative 12, 17–26, 32–35, 43–48, 66,
    93–101, 127, 141, 148, 181–200,
    225–96
native peoples 17, 25–26, 30, 33, 91,
    237, 283–88. *See also* indigenous
    (peoples, cultures, practices)
naturalism 209, 290–91
negative certainty 90, 95–96
networks 75
Nietzsche, Friedrich 1
nihilism 91, 214–16
nominalism 208, 222
non-humanity 19, 22–24, 30, 33, 38, 82,
    97, 100–1, 104, 119, 131–32, 136,

    141–42, 147–52, 158–59, 181, 188,
    197, 214, 226, 249, 256, 283, 286,
    297–99, 311
nothingness 211, 221
nuclear energy 36, 106
Nussbaum, Martha 185

object-oriented ontology 201
objectivism 33, 34, 251, 291–92
objectivity (as a hermeneutic concept)
    39, 50, 55–59, 90, 170, 265, 281
oceans 67, 68, 73, 95, 227, 285, 296
onto-theology 223
ontology 4, 11, 46, 52, 55–61, 68–69, 96,
    132, 164, 172–76, 201–9, 216,
    222–24, 253, 263, 289, 291
organicism 206
otherness 107, 128–32, 136, 170, 249,
    252–55, 278

painting 29, 54, 68–69, 78–79, 82–84,
    94, 100
palimpsest 199, 233–34, 236–37
Palmer, Richard 123–24, 334
paradigm shifts 112, 191
Pascal, Blaise 160–66
perception (as a practice) 17, 33–34, 73,
    99, 101, 125, 146, 150, 170, 176, 190,
    208, 232–33, 258, 261
phenomenology 33, 76, 83, 89–90, 100,
    163, 282, 304
physis 213, 220, 223
place 1–10, 13–14, 18–24, 29, 37–39,
    43–44, 52, 58–59, 65–70, 77, 80, 86,
    98, 101–7, 112, 118, 126–29, 139–40,
    144, 149, 160–93, 202–10, 215–17,
    225–96, 303, 312
Plato 112, 216–17, 223
poetry 213, 224, 285
poiesis 182, 188
Pollan, Michael 193

pollution 21, 129, 153, 154, 177
positivism 28–29, 33, 40–45, 49, 52–53
poststructuralism 102–9, 112, 115, 117
pragmatism 18, 307
psychoanalysis 42, 75, 139, 167, 201, 209

racism 154
Rancière, Jacques 276
Rawls, John 155, 301–2, 307–8
reality 5, 28–30, 33, 45–49, 76, 98, 124,
    140, 160, 166–74, 187–92, 196–99,
    209–10, 222, 239, 246, 258–61, 289,
    300, 304–5, 311
recreation 19, 22–26
Regan, Tom 184
reinterpretation 10, 115, 145, 270, 273
relativism 28, 33–34, 53, 108–10,
    159, 177
religion 4, 18, 23, 30, 40–42, 63, 67,
    83–85, 105, 116, 128, 144, 197, 207,
    214, 293, 294
renaturing 226–28, 231
responsibility 5, 99, 109, 148–51, 217,
    268–79, 284, 304, 306
restoration 20, 225–27, 231–39, 246, 262
rewilding 234
Ricoeur, Paul 2, 6, 7, 12, 17–20, 35, 66,
    124, 127–59, 185–86, 199, 249–54,
    261–66, 270, 271–79
ritual 246, 273, 292
rivers 66, 68, 92, 98, 197, 229–30, 238–39,
    282
Rolston, Holmes 19, 99, 189–91,
    198–200
romanticism 206, 224

sacredness 20–22, 288, 292–94
Sadowsky, Stewart 162, 176
Schleiermacher, Friedrich 2, 240
science 4, 18–22, 39, 40–49, 53–63, 90, 99,
    107, 134, 176, 181–82, 188–93,
    198–204, 209, 211, 213, 215–19,
    222–23, 232, 251, 303
secularization 41, 77, 106, 246, 295
self-consciousness 43, 46, 51–52, 55,
    57, 100
self-identity 145, 205, 210, 251–55,
    274
self-understanding 19, 49, 55–60, 101,
    124–27, 138–39, 141–47, 171, 186,
    237, 239, 248, 254, 310–12
selfhood 12, 101, 125–32, 136, 141, 144,
    148, 151, 253–56
semiotics 201, 208, 251
sentiment 40, 231, 304, 309
Serres, Michel 214
Shakespeare, William 45
Shawn, Allen 166–72, 175
similitude 148–51
Singer, Peter 184
skepticism 120, 299, 301–3, 309
Smith, Mick 95–98, 199
Snellen, Paulien 13, 248
sociobiology 38–40, 43, 49
Socrates 28, 37, 186, 223, 298–99,
    303, 306
speculative idealism 47–49, 52
Stengers, Isabelle 214
structuralism 206–10
subjectivism 35, 53, 56, 251
superorganism 38, 44, 52, 61–62
sustainability 23, 28, 129, 256, 262

techne 31, 220
technique 2, 31, 309
technology 45, 90–91, 96–97, 107, 203,
    213–14, 219, 223–24
Teilhard de Chardin, Pierre 295
text 2–3, 7, 19, 38, 47, 51–55, 58, 66,
    101, 123, 145–47, 186, 221, 226,
    234–35, 240, 249, 250–51, 255–56,
    259, 263, 291, 305–7

theology 4, 11, 20–22, 85, 90, 216, 249, 263
van Tongeren, Paul 13, 248
totalitarianism 61
tradition 2, 10–13, 18, 28, 31, 48, 51–52, 64, 68, 104, 108, 111–14, 131, 194, 198, 241, 251, 260–62, 265–66, 271, 279, 281, 294, 304, 308, 311
translation 94, 111, 119, 132, 216
trauma 277, 290
Treanor, Brian 66, 93–94, 98–100, 249, 251, 262, 277
Trigg, Dylan 260
truth 6, 8, 28–41, 46–67, 96, 110, 116–18, 182, 185–95, 199–201, 214–15, 262, 269–71, 286, 292, 294
Turner, Jack 182–86, 196–200
typology 21, 25

urbanization 234
utilitarianism 24, 217, 220, 301–8
Utsler, David 101, 171

validity 53, 143, 159, 164, 186
values 22–27, 32–35, 39–40, 52–53, 59, 99, 113, 125–26, 155, 164–67,

170, 189, 207, 231, 232, 237, 261–63, 282
victimhood 187, 199, 211, 269, 272, 277–79
violence 130, 214, 275, 285, 287, 288
virtues 103, 184, 262
voice 26, 88, 131, 153, 154–55, 176, 260, 266, 270

war 62, 271–76, 288–89
well-being 24, 31, 62, 137–38, 172–73
wilderness 5, 19, 33, 106, 107, 144–45, 182–83, 202, 205–6, 226, 232–33, 246
wildness 95, 182–87, 231, 248
Williams, Bernard 297–312
Wittgenstein, Ludwig 106, 111–12
world-soul 217, 219
worldview 115, 129, 162, 196, 260

Yellowstone 182
Yosemite 182–83, 197–98

Zeeland 228–30
Zeitgeist 203
Žižek, Slavoj 201–16

gROUNDWORKS |
ECOLOGICAL ISSUES IN PHILOSOPHY AND THEOLOGY

Forrest Clingerman and Brian Treanor, *Series editors*

Interpreting Nature: The Emerging Field of Environmental Hermeneutics
Forrest Clingerman, Brian Treanor, Martin Drenthen,
and David Utsler, eds.

*The Noetics of Nature: Environmental Philosophy and the Holy Beauty
of the Visible*
Bruce V. Foltz

Environmental Aesthetics: Crossing Divides and Breaking Ground
Martin Drenthen and Jozef Keulartz, eds.

The Logos of the Living World: Merleau-Ponty, Animals, and Language
Louise Westling